# Advances in Carbon Capture, Utilization and Storage Technologies (CCUS)

# Advances in Carbon Capture, Utilization and Storage Technologies (CCUS)

Editors

**Weifeng Lv**
**Tiyao Zhou**
**Yongbin Wu**
**Xiaoqing Lu**

Basel • Beijing • Wuhan • Barcelona • Belgrade • Novi Sad • Cluj • Manchester

*Editors*

Weifeng Lv
Institute of Porous Flow and
Fluid Mechanics, University
of Chinese Academy of
Sciences
Langfang
China

Tiyao Zhou
Research Institute of
Petroleum Exploration and
Development, PetroChina
Beijing
China

Yongbin Wu
Research Institute of
Petroleum Exploration and
Development, PetroChina
Beijing
China

Xiaoqing Lu
School of Materials Science
and Engineering, China
University of Petroleum (East
China)
Qingdao
China

*Editorial Office*
MDPI
St. Alban-Anlage 66
4052 Basel, Switzerland

This is a reprint of articles from the Special Issue published online in the open access journal *Energies* (ISSN 1996-1073) (available at: https://www.mdpi.com/journal/energies/special_issues/ GALJTV640F).

For citation purposes, cite each article independently as indicated on the article page online and as indicated below:

Lastname, A.A.; Lastname, B.B. Article Title. *Journal Name* **Year**, *Volume Number*, Page Range.

**ISBN 978-3-7258-1125-0 (Hbk)**
**ISBN 978-3-7258-1126-7 (PDF)**
**doi.org/10.3390/books978-3-7258-1126-7**

© 2024 by the authors. Articles in this book are Open Access and distributed under the Creative Commons Attribution (CC BY) license. The book as a whole is distributed by MDPI under the terms and conditions of the Creative Commons Attribution-NonCommercial-NoDerivs (CC BY-NC-ND) license.

# Contents

About the Editors . . . . . . . . . . . . . . . . . . . . . . . . . . . . . . . . . . . . . . . vii

Preface . . . . . . . . . . . . . . . . . . . . . . . . . . . . . . . . . . . . . . . . . . . . ix

**Xiuxiu Pan, Linghui Sun, Xu Huo, Chun Feng and Zhirong Zhang**
Research Progress on $CO_2$ Capture, Utilization, and Storage (CCUS) Based on Micro-Nano Fluidics Technology
Reprinted from: *Energies* **2023**, *16*, 7846, doi:10.3390/en16237846 . . . . . . . . . . . . . . . . . . . . 1

**Shumin Ni, Weifeng Lv, Zemin Ji and Kai Wang**
$CO_2$ Mineralized Sequestration and Assistance by Microorganisms in Reservoirs: Development and Outlook
Reprinted from: *Energies* **2023**, *16*, 7571, doi:10.3390/en16227571 . . . . . . . . . . . . . . . . . . . . 19

**Jia Liu, Dengzun Yao, Kai Chen, Chao Wang, Chong Sun, Huailiang Pan, Fanpeng Meng, et al.**
Effect of $H_2O$ Content on the Corrosion Behavior of X52 Steel in Supercritical $CO_2$ Streams Containing $O_2$, $H_2S$, $SO_2$ and $NO_2$ Impurities
Reprinted from: *Energies* **2023**, *16*, 6119, doi:10.3390/en16176119 . . . . . . . . . . . . . . . . . . . . 43

**Jinhong Cao, Ming Gao, Zhaoxia Liu, Hongwei Yu, Wanlu Liu and Hengfei Yin**
Research and Application of Carbon Capture, Utilization, and Storage–Enhanced Oil Recovery Reservoir Screening Criteria and Method for Continental Reservoirs in China
Reprinted from: *Energies* **2024**, *17*, 1143, doi:10.3390/en17051143 . . . . . . . . . . . . . . . . . . . . 56

**Xu Huo, Linghui Sun, Zhengming Yang, Junqian Li, Chun Feng, Zhirong Zhang, Xiuxiu Pan, et al.**
Mechanism and Quantitative Characterization of Wettability on Shale Surfaces: An Experimental Study Based on Atomic Force Microscopy (AFM)
Reprinted from: *Energies* **2023**, *16*, 7527, doi:10.3390/en16227527 . . . . . . . . . . . . . . . . . . . . 79

**Ming Gao, Zhaoxia Liu, Shihao Qian, Wanlu Liu, Weirong Li, Hengfei Yin and Jinhong Cao**
Machine-Learning-Based Approach to Optimize $CO_2$-WAG Flooding in Low Permeability Oil Reservoirs
Reprinted from: *Energies* **2023**, *16*, 6149, doi:10.3390/en16176149 . . . . . . . . . . . . . . . . . . . . 102

**Xiang Qi, Tiyao Zhou, Weifeng Lyu, Dongbo He, Yingying Sun, Meng Du, Mingyuan Wang, et al.**
Front Movement and Sweeping Rules of $CO_2$ Flooding under Different Oil Displacement Patterns
Reprinted from: *Energies* **2024**, *17*, 15, doi:10.3390/en17010015 . . . . . . . . . . . . . . . . . . . . 123

**Zangyuan Wu, Qihong Feng, Yongliang Tang, Daiyu Zhou and Liming Lian**
Experimental Study on Carbon Dioxide Flooding Technology in the Lunnan Oilfield, Tarim Basin
Reprinted from: *Energies* **2024**, *17*, 386, doi:10.3390/en17020386 . . . . . . . . . . . . . . . . . . . . 148

**Mohammad Hossein Golestan and Carl Fredrik Berg**
Simulations of $CO_2$ Dissolution in Porous Media Using the Volume-of-Fluid Method
Reprinted from: *Energies* **2024**, *17*, 629, doi:10.3390/en17030629 . . . . . . . . . . . . . . . . . . . . 161

**Reyhaneh Ghorbani Heidarabad and Kyuchul Shin**
Carbon Capture and Storage in Depleted Oil and Gas Reservoirs: The Viewpoint of Wellbore Injectivity
Reprinted from: *Energies* **2024**, *17*, 1201, doi:10.3390/en17051201 . . . . . . . . . . . . . . . . . . . **182**

# About the Editors

**Weifeng Lv**

Weifeng Lv graduated from the Department of Chemistry at Tsinghua University in July 2005. The following year, he served as a visiting scholar at the French Institute of Petroleum, and in 2020, he obtained a doctoral degree from Nanjing University. Since 2005, he has been serving as the Director of the Enhanced Oil Recovery Research Center at the Research Institute of Petroleum Exploration and Development. Since 2023, he has been appointed as the Vice President of the Research Institute of Petroleum Exploration and Development. He has extensive experience in teaching and researching topics such as green and efficient development and carbon transformation of oil and gas fields, enhanced oil recovery, digital core, CCUS, and more. In the past five years, he has published twenty-one relevant papers, obtained over ten invention patents, authored five software copyrights, contributed to one industry standard, and authored two monographs. He has led more than ten national major scientific and technological projects, national key research and development programs, youth science fund projects, and major scientific and technological projects of PetroChina. He has served as an editor for journals such as Special Topics and Reviews, Porous Media, and Energies. He has received more than ten awards at the provincial and ministerial levels, including the Sun Yueqi Youth Science and Technology Award. In particular, his long-term research on three-phase relative permeability has made significant breakthroughs. In 2014, he won the first prize at the provincial and ministerial level for the three-phase relative permeability experimental platform and testing technology.

**Tiyao Zhou**

Tiyao Zhou, In 2010, Tiyao Zhou graduated from the China University of Petroleum (Beijing), where he also obtained a PhD degree. And since then, he has been working at the Research Institute of Petroleum Exploration and Development (RIPED). In 2022, he was appointed as a technical expert in improving reservoir recovery through gas flooding. His general interests include the following: CCUS, oil recovery enhancement, gas flooding, miscible/immiscible flooding, GAGD, tight oil, and reservoir engineering. He has recently focused on key technology research on the synergy between $CO_2$ flooding for enhanced oil recovery and geological storage. At present, he is undertaking innovative research in gas flooding technology in two projects of CNPC, including the CCUS project and the enhanced oil recovery project.

**Yongbin Wu**

Yongbin Wu, In 2005, Yongbin Wu graduated from the China University of Petroleum (Beijing) with a bachelor's degree in petroleum engineering. In 2007, he graduated from the Research Institute of Petroleum Exploration & Development, with a degree in development of gas & oilfields, where he also obtained a PhD degree. Since 2007, he has been affiliated with the department of thermal recovery, RIPED of PetroChina. His general interests include the following: heavy oil CCUS, foamy oil, EOR, electrical heating, and mechanical engineering of oil recovery process. He has received 12 awards, published 23 papers, and been granted 35 patents of invention.

**Xiaoqing Lu**

Xiaoqing Lu received his Ph.D. degree from City University of Hong Kong in 2011. Then, he joined the faculty at China University of Petroleum (UPC), where he is now a professor. His research interests are focused on the design and screening of materials in energy and environment, including the capture, separation, and conversion of carbon dioxide in nanoporous adsorbent materials, and the mechanisms of carbon dioxide flooding and carbon dioxide geological storage.

# Preface

Carbon capture, utilization, and storage (CCUS) is a vital technology for the large-scale industrial reduction of carbon dioxide emissions and is considered a foundational technology for achieving carbon neutrality in the next 30 to 50 years, gaining widespread consensus globally. CCUS technology primarily encompasses four areas: carbon capture, carbon transportation, carbon utilization, and carbon storage. From the current technological development trends in carbon capture, sources such as steel, cement, and coal-fired power plants with low-concentration carbon dioxide emissions are increasingly receiving attention. Moreover, direct carbon capture from the air is an area of focus. Regarding carbon transportation, pipeline delivery is the most economical method, whereas the long-distance safe transportation of supercritical carbon dioxide with impurity gases remains a challenge. In addition, with respect to carbon utilization, geological methods that enhance oil recovery hold the most potential for scalable applications; yet, researching the mechanisms of carbon dioxide enhanced oil recovery under porous media conditions and efficient recovery methods is a priority for the upcoming period. For carbon storage, increasing storage capacity and accelerating safer mineralization storage through microbial facilitation are significant research directions in the future.

From the papers collected in this CCUS Special Issue, researchers are utilizing more advanced technologies, such as atomic force microscopy and molecular dynamics simulations, to conduct finer mechanistic studies in micro- and nano-spaces, pore dead-ends, and shale surfaces. They are exploring more economical, efficient, and safe CCUS technologies. We have every reason to believe that, with the collective effort of the vast scientific elite, the green low-carbon Earth that humanity hopes for will arrive sooner rather than later.

We sincerely thank everyone who has helped.

**Weifeng Lv, Tiyao Zhou, Yongbin Wu, and Xiaoqing Lu**
*Editors*

*Review*

# Research Progress on $CO_2$ Capture, Utilization, and Storage (CCUS) Based on Micro-Nano Fluidics Technology

Xiuxiu Pan [1,2], Linghui Sun [2,3,*], Xu Huo [1,2], Chun Feng [3] and Zhirong Zhang [1,2]

1. University of Chinese Academy of Sciences, Beijing 100049, China; panxiuxiu22@mails.ucas.ac.cn (X.P.); huoxu21@mails.ucas.ac.cn (X.H.); zhangzhirong20@mails.ucas.ac.cn (Z.Z.)
2. Institute of Porous Flow and Fluid Mechanics, Chinese Academy of Sciences, Langfang 065007, China
3. Research Institute of Petroleum Exploration & Development, Beijing 100083, China; fengchun123@petrochina.com.cn
* Correspondence: sunlinghui@petrochina.com.cn

**Abstract:** The research and application of $CO_2$ storage and enhanced oil recovery (EOR) have gradually emerged in China. However, the vast unconventional oil and gas resources are stored in reservoir pores ranging from several nanometers to several hundred micrometers in size. Additionally, $CO_2$ geological sequestration involves the migration of fluids in tight caprock and target layers, which directly alters the transport and phase behavior of reservoir fluids at different scales. Micro- and nanoscale fluidics technology, with their advantages of in situ visualization, high temperature and pressure resistance, and rapid response, have become a new technical approach to investigate gas–liquid interactions in confined domains and an effective supplement to traditional core displacement experiments. The research progress of micro–nano fluidics visualization technology in various aspects, such as $CO_2$ capture, utilization, and storage, is summarized in this paper, and the future development trends and research directions of micro–nano fluidics technology in the field of carbon capture, utilization, and storage (CCUS) are predicted.

**Keywords:** microfluidic and nanofluidic chip; phase behavior; confined fluids; CCUS

## 1. Introduction

Current global conventional oil and gas resources are tending to depletion, and unconventional oil and gas is gradually becoming a strategic replacement energy source. Its contribution and position in global oil and gas production are increasingly prominent, and in the future, it will shoulder the three major missions of global energy security, stable oil and gas production, and green and low-carbon development [1]. Developing unconventional oil and gas resources can not only accelerate the achievement of the "dual carbon" goal but also to some extent solve the energy supply and demand contradiction in our country. Compared to other enhanced oil recovery (EOR) methods that increase energy supply and promote carbon neutrality by sequestering greenhouse gases, $CO_2$ flooding for improving oil recovery in unconventional reservoirs has both economic and social benefits. However, although $CO_2$ flooding for unconventional reservoirs has been ongoing for several years, the understanding of the multi-phase flow and behavior mechanism of multiscale reservoirs under displacement is still lacking. The reason for this is that unconventional reservoirs exhibit the characteristics of coexisting micro–nano pores across scales [2], with pore size distributions ranging from several nanometers to several micrometers, and even millimeter-scale fractures.

The average free path of nanoscale confined fluids is comparable to the pore size, and this confinement in a small space restricts molecular thermal motion (as shown in Figure 1), leading to phenomena different from the bulk phase, such as surface adsorption, density differences, increased capillary forces, and the contraction of phase envelopment lines. These phenomena not only affect oil production and gas injection process selection [3] but also impact the efficiency and cost of oil and gas extraction [4,5]. Therefore, gaining a comprehensive understanding of the transition of wall wetting in micro–nanochannels, the mechanical characteristics of fluids, and their impact on phase behavior is of significant engineering importance for applications such as $CO_2$ geological storage and petroleum extraction [6,7]. $CO_2$ flooding has become one of the most important extraction methods since the early 1980s due to its excellent extraction and dissolution capabilities. However, the development of $CO_2$-enhanced oil and gas recovery in China still lags behind foreign countries, as shown in Figure 2, and currently, $CO_2$-enhanced oil and gas recovery and geological sequestration face the challenge of the aforementioned multiscale pore size distribution. As a complex multiphase dissolution and migration method, $CO_2$ flooding and geological storage are influenced by various factors such as reservoir and fluid physical properties, flow parameters, etc. The key to its technical application lies in understanding the mechanisms of gas–liquid interactions across micro- and nanoscales. In cracks and large pores (>50 nm [8]), fluid flow exhibits bulk phase characteristics, while in nanopores (<50 nm), the flow behavior deviates significantly from the bulk phase [9,10]. Existing theoretical models for studying the multiscale mechanisms of multiphase flow mainly rely on molecular simulations, digital rock models, and equations of state (EOSs). These methods require certain assumptions and lack sufficient experimental data support. In addition, the results of traditional theoretical simulations are inconsistent with existing microfluidic experimental results [11,12] and are depicted in Figure 3a. Most experimental methods simulate the limiting effects of pure component fluid static and dynamic behaviors based on a single nanopore size. Therefore, further understanding the multiscale pore-scale oil and gas seepage and migration mechanisms is the basis for developing effective enhanced oil recovery techniques and large-scale $CO_2$ sequestration, as well as a prerequisite for multiscale reservoir flow simulation and parameter optimization [13].

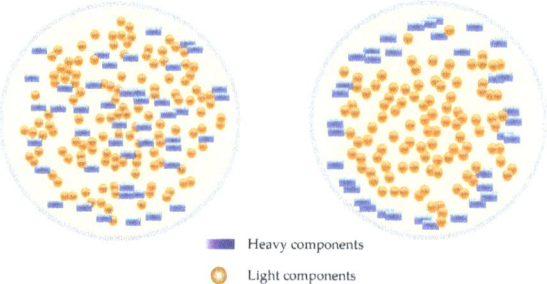

Figure 1. Comparison between bulk-phase fluids and nanoscale reservoir matrix fluids.

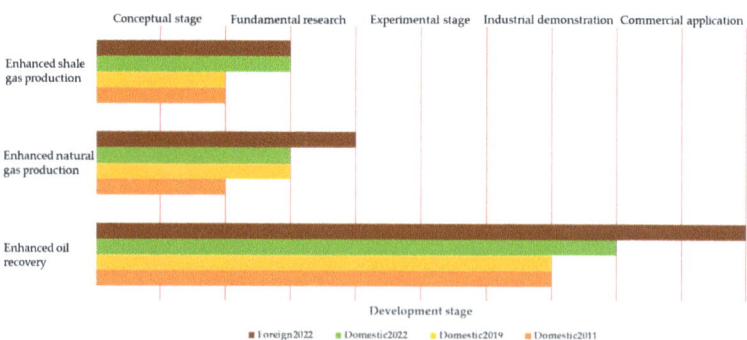

Figure 2. CCUS at home and abroad to strengthen oil and gas technology development level.

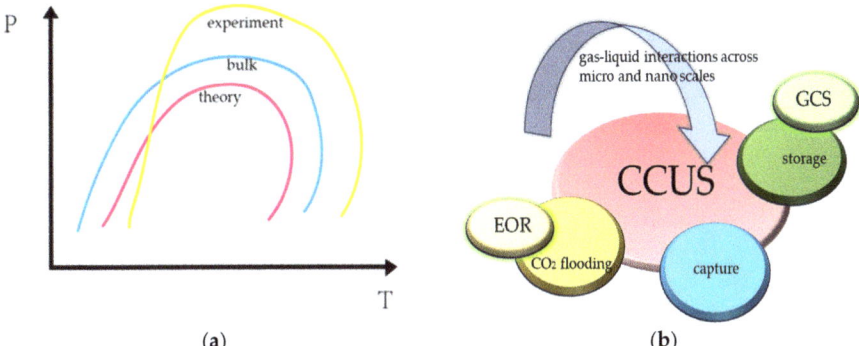

**Figure 3.** (a) Bulk, the theory and experiment of the phase envelope; (b) $CO_2$ capture, utilization, and storage.

Thanks to the advances in chip processing technology, micro–nanofluidics technology has the advantages of the fine description of nanoscale pores and in situ visualization detection, providing a new experimental perspective for the study of unconventional oil and gas microscale seepage and phase characteristics [14]. It has become one of the most promising tools for visual analysis of oil and gas. Currently, it has been successfully applied in the petroleum and natural gas industry for physicochemical analysis [15], heavy oil recovery [16,17], hydraulic fracturing [18], chemical flooding [19–21], gas flooding [22], and other studies on multiphase flow mechanisms. It serves as an effective complement to traditional oil displacement mechanism evaluation techniques such as theoretical simulation and experimental analysis.

This paper summarizes the current research status in the field of micro–nano fluidics technology applied to $CO_2$ capture, storage, and utilization, as shown in Figure 3b, and identifies the shortcomings in current research.

## 2. Micro–Nanofluidic Devices and Chip Manufacturing

Extensive theoretical research has indicated that when the pore size reaches the subnanometer scale [12,23–27], the interactions and wettability characteristics between the liquid and solid substrates intensify the flow–solid interaction in nanoscale pore throats [28], and the phase characteristics and seepage behavior of the fluid are significantly altered. The properties of reservoir fluids and wettability [29,30] further complicate the confinement effects, necessitating experimental tools to address this highly coupled phenomenon. In recent years, with the development of unconventional oil and gas resources, micro–nanofluidic experimental platforms [31–34], as shown in Figure 4, have been introduced into the petroleum and natural gas industry. Different imaging modes of optical microscopy (bright field, dark field, and fluorescence) enable the in situ visualization monitoring of micro–nanochannel samples. In addition, spectroscopic methods such as infrared Raman can be used for the fine characterization of micro–nanofluidic systems [35]. The materials used for the chips include silicon, glass, ceramics, and other polymers, such as polydimethylsiloxane (PDMS) [36]. However, the high-temperature and high-pressure experimental conditions in the oil and gas industry and the research requirements for organic nonpolar fluids make silicon the most ideal manufacturing material. As depicted in Figure 5, the micro–nanofluidic chip transfers the characteristics of reservoir porous media micro–nanochannels onto a silicon substrate using techniques from the semiconductor industry [37], and the silicon substrate is bonded to a glass plate to form enclosed channels [38]. The chip channels can be modified on the wall surface [39,40] or filled with modified nanoparticles [41] to simulate the real conditions of reservoir porous media and address the fundamental issues of reservoir dynamics and fluid flow [42]. The so-called nanochannels have at least one dimension (width and depth) within the range of 1 to

100 nm. During processing, care must be taken to avoid channel collapse caused by capillary or van der Waals forces. Due to the high aspect ratio and challenging fabrication of high-throughput nanochannels, current research focuses mainly on low-aspect-ratio channels with nanoscale depth and micrometer-scale width [38].

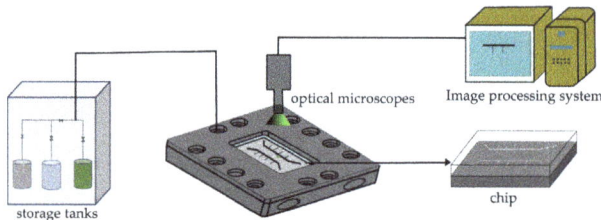

**Figure 4.** Schematic diagram of microfluidic and nanofluidic devices.

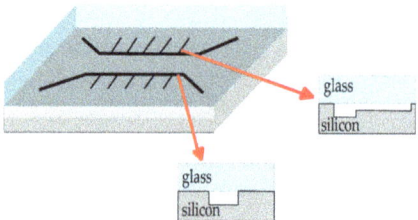

**Figure 5.** Chip model.

## 3. Research on CO$_2$ Capture

Carbon capture is the process of collecting greenhouse gases, such as those emitted from fossil fuels, through fixed emission channels, and then compressing and transporting them for utilization or storage. As the first step in the CCUS process, carbon capture technology is crucial for the long-term feasibility of CCUS technology. Existing research has demonstrated the potential of micro–nanofluidic technology in CO$_2$ capture. Currently, CO$_2$ capture technologies mainly include absorption, adsorption [43–45], and membrane separation. Absorption methods involve physical and chemical absorption, which utilize organic solvents or alkaline solutions with high solubility, selectivity, and stability to separate CO$_2$. Adsorption methods mainly rely on carbon-based materials, zeolites, nanomaterials, and other chemical adsorbents to capture CO$_2$. The mechanisms of solvent absorption, porous medium absorption, and separation dissolution related to CO$_2$ capture using organic solvents and alkaline solutions, as well as nanomaterials, can be further studied using micro–nanofluidic control technology, as demonstrated in previous research [46]. In particular, the carbon capture applications of deep eutectic solvents (DESs) and ionic liquid analogues [47] hold significant engineering significance.

## 4. Research on CO$_2$ Enhanced Oil and Gas Recovery

Phase behavior is crucial in gas displacement processes as it determines gas–liquid equilibrium, swelling effects, gas solubility, and minimum miscibility pressure [48,49]. The complexity of phase behavior in unconventional reservoir fluids is attributed to the diversity of oil and gas components and the influence of porous media and reservoir rock properties [50]. In unconventional reservoirs, the ratio of pore surface area to volume increases significantly, leading to phenomena such as bubble point depression, dew point trajectory deviation, and variations in critical parameters, which directly result in a rapid decline in initial production rates (approximately ten times faster than conventional reservoirs) [14]. Although theoretical simulations [27,51,52] and instrumental analyses [53,54] have provided reliable means for studying restricted fluid phase behavior, as shown in

Table 1, experimental studies to validate the effectiveness of correction calculations are still in their early stages. Micro–nanofluidic technology allows for high-resolution imaging of fluids in nanochannels under high-temperature and high-pressure conditions, providing visual support for understanding the phase transition characteristics and flow behavior of fluids in confined spaces. It serves as a new experimental approach to investigate and validate predictions of nanoscale properties and the phase behavior of fluids in multi-scale media.

**Table 1.** The research methods of the phase behavior.

| | Research Methods | Principles | Advantages | Disadvantages |
|---|---|---|---|---|
| Experimental methods | Rock core displacement experiment [55] | Analyze recovery rate and other data based on experimental parameters | Realistic rock core displacement with high fidelity | Long cycle, not visually observable, and poor repeatability |
| | Adsorption–desorption method [56,57] | Isotherm adsorption line | Direct approach to observe the phase transition point and critical behavior of hydrocarbons in nanoporous carbon | Most are non-realistic rock cores |
| | Differential scanning calorimetry (DSC) [58] | Determining material thermal properties by measuring the rate of heat release or absorption during temperature increase | Simulating the bubble point temperature of confined hydrocarbons | Difficulty in controlling and measuring phase transition rates |
| Theoretical simulation | Molecular simulation — Molecular dynamics (MD) [59] | Numerically solving the classical equations of motion allows for the determination of the phase trajectory of a molecular system and the characterization of its macroscopic thermodynamic properties | High accuracy for simple components | High computational cost, difficulty in calculating near the critical point |
| | Molecular simulation — Monte Carlo (MC) [12] | Repeatedly sample different configurations of a molecular system and calculate the total energy of each configuration | | |
| | The equation of state [27,51] | Phase equilibrium and force calculation | Consider capillary forces and critical displacement | Depend on the equation of state |
| | Density functional theory (DFT) [52,60] | Fluid molecular density as the fundamental variable to describe the thermodynamics of a system | Exhibits minimal discrepancies with molecular simulation results | Limited to simple molecular density statistics |
| Instrumental analysis | Nuclear magnetic resonance online scanning (NMR) [61] | $^1H$ and $^{13}C$ nuclei resonate in the magnetic field | Full-scale pore observation in the range of nanometers to millimeters | In situ imaging of oil or water requires phase shielding, resulting in low imaging accuracy and high costs |
| | CT scanning | Three-dimensional image scanning and reconstruction | Digital rock | High sample requirements and high costs |

## 4.1. Phase Behavior

In oil and gas extraction, bubble nucleation and vaporization are more favorable in larger pores, while they are constrained in nanoscale porous matrices, significantly affecting the fluid critical parameters and occurrence states. For instance, with the enhancement of confinement effects, the bubble point pressure is significantly reduced, while the upper dew point pressure increases and the lower dew point pressure decreases. Such phase behavior differences may have a significant impact on oil and gas recovery or data fitting.

Currently, methods for predicting phase behavior in nanoscale confined spaces include density functional theory [62] (DFT), molecular simulations [63–65], modified equation of state considering various parameter influences (such as the Peng–Robinson equation) [66,67], or a combination of the aforementioned methods [24]. However, the predicted results sometimes remain controversial [68], necessitating experimental verification of prediction accuracy. For instance, in the case of the offset phenomenon in confined space bubble points, the most commonly used method is to modify critical properties in vapor–liquid equilibrium calculations and consider the capillary pressure in the modified equation of state. The bubble point pressure lines generated using these two different methods exhibit slight differences, particularly with significant discrepancies near the critical point. Therefore, more experimental data are required for the accurate comparison and calibration of these methods. Experimental methods include differential scanning calorimetry (DSC) [58], the constant volume method [69], or traditional PVT experiments to determine the bubble point through volume–pressure curves. The testing apparatus is generally a plunger or piston-type PVT instrument, which consumes a large amount of oil sample and requires long equilibration time. Micro–nanofluidic chip devices have small thermal mass and rapid response, enabling the continuous measurement of multiple saturation pressures [70]. By confining the fluid sample within a closed cavity at the end using $CO_2$ and oil properties, the precise isolation of the fluid sample within the cavity is achieved, compensating for the existing experimental limitations. Currently, research on the evaporation and condensation of single-component fluids at ambient temperature and pressure based on micro–nanofluidic technology is relatively well-developed (as shown in Table 2).

Table 2. The investigation of phase behavior in single-component systems.

| Experimental Objective | Component | Size | Phenomenon |
| --- | --- | --- | --- |
| Evaporation | n-pentane [71] | 50 μm 145 nm | Nanoporous capillary confinement enhances liquid capture and significantly impedes liquid evaporation, reducing the evaporation rate by approximately 16 fold. |
| | n-pentane [72] | 100 nm 5 μm | The confinement effect leads to the preferential evaporation of the fluid in microchannels over nanochannels. |
| Condensation | propane [11] | 30 nm 50 nm 500 nm | The condensation pressure for the 50 nm chip is close to the prediction of the Kelvin equation, while for the 30 nm chip, the condensation pressure is significantly lower than the predicted value. |
| | propane [73,74] | 70 nm 100 nm | There exist disparities between the two length scales. |
| | n-butane [12] | 2 nm | The deviation of the condensation pressure from the bulk phase reaches as high as 22.9%. |
| Bubble point temperature | n-hexane, n-octane, n-heptane [50] | 4 nm 20 nm 50 nm 100 nm | The bubble point temperatures for 4 nm and 10 nm confinement exhibit significant deviations compared to those for 100 nm and 50 nm confinement. The confinement effect is more pronounced in the 4 nm channel, leading to a noticeable increase in the bubble point temperature. |
| Dew point pressure | n-butane [75] | 4 nm 10 nm 50 nm | At 4 nm, the dew point pressure exhibits a deviation of up to 14% compared to the bulk phase. |

Factors influencing the phase behavior of multi-component systems are more complex compared to single-component systems [72], and research in this area is relatively scarce. Zhang et al. [76] conducted experimental studies on the static phase behavior and fluid flow behavior of pure $CO_2$ and $CO_2$-$C_{10}$ systems in shale dual-scale porous reservoirs. The results showed significant changes in the static behavior of the fluid from the bulk phase to the nanoscale. Jatukaran et al. [77,78] used chip simulations to investigate the evaporation rates of methane, propane, and butane ternary hydrocarbon mixtures in nanoscale pores, revealing a 3000-fold reduction in evaporation rate, which further decreased at low temperatures. Alfi et al. [79] studied the bubble point temperatures of butane, hexane, and heptane mixtures using etched chips with depths of 10 nm, 50 nm, and 100 nm, and observed significant deviations in the bubble point temperatures at the 10 nm scale. Despite the progress made in multi-component phase-behavior studies across different length scales, these studies have not considered the mass transfer diffusion and the impact of flow on evaporation. Therefore, further improvement is needed in this area of research.

### 4.2. Seepage

Capillary forces, as the main driving force of seepage, make the seepage effect in nanopores more significant. Multiple studies have shown that the pore size of reservoir rocks reaches the nanoscale, and capillary pressure increases to several megapascals. The seepage of fluids in microchannels can be described as:

$$q = \frac{P_{ca}}{f} = wh\frac{dl}{dt} \quad (1)$$

where $q$ is the mass flow rate, $P_{ca}$ represents the capillary force, $f$ is the flow resistance, $w$ represents the width of the nanochannel, $h$ represents the depth of the nanochannel, and $l$ represents the seepage distance.

The capillary pressure in a circular channel can be represented as:

$$P_{ca} = \frac{2\gamma \cos\theta}{r} \quad (2)$$

However, microfluidic chip designs often have a high aspect ratio with channel widths in the micrometer range and depths in the nanometer range ($w \gg h$). Therefore, the fluid flow in such channels resembles that of a flat plate flow. Consequently, combining Equations (1) and (2), the capillary pressure $P_{ca}$ in these channels can be expressed as:

$$P_{ca} = \frac{2\gamma \cos\theta}{h} \quad (3)$$

where $\gamma$ represents the surface tension, and $\theta$ is the contact angle.

Flow resistance:

$$f = \frac{12\mu l}{wh^3} \quad (4)$$

Therefore, the seepage distance can be obtained as shown in Equation (5):

$$l = \sqrt{\frac{\gamma h \cos\theta t}{3\mu}} \quad (5)$$

where $\mu$ is the fluid viscosity, $t$ represents the time.

According to the Washburn equation [80], seepage is closely related to surface tension and viscosity, and as the pore size decreases, the surface tension and viscosity of fluids deviate significantly from the bulk phase [81]. However, there are differences in the results of seepage experiments in confined spaces. For example, Li et al. [82] studied the effect of surface tension, viscosity, and contact angle on seepage in polar and non-polar mixtures at the 8 nm scale using microfluidic chips, and the experimental results showed that the Washburn equation was generally applicable in confined spaces. Another study showed that the oil uptake rate of shale in a 34 nm channel was only 40% of the theoretical calculation [83]. The reason for this is that the strong interaction between the nanochannel wall and the fluid increases the resistance, thereby slowing down the self-seepage process.

*4.3. Diffusion Mass Transfer Study*

The effectiveness of $CO_2$ for enhanced oil recovery depends on its potential to alter the properties of the reservoir oil, known as the phase-mixing ability, which in turn depends on the mass transfer capability between $CO_2$ and oil [84,85]. Studying the mass transfer and reaction kinetics in gas–liquid two-phase flow is of great significance for improving oil recovery.

The key parameter in the diffusion process is the molecular diffusion rate. At the nanoscale, the size of substance molecules approaches the limiting size, leading to a deviation between nanoscale diffusion rate and bulk diffusion rate, even though some convection [86] is restricted, resulting in relatively high mass flow rates [87], which affects mass transfer. However, under supercritical conditions, the interfacial effects are significantly weakened, and both convection and diffusion transport are enhanced. Previous studies on oil and gas mass transfer and diffusion laws have mostly relied on traditional PVT experiments or bubble towers to determine the kinetic data in the liquid phase. Under certain conditions, after gas makes contacts with oil, it diffuses into oil under the influence of concentration difference, and a diffusion model is used to fit the pressure drop curve to obtain the diffusion coefficient [88]. However, such methods not only have low heat and mass transfer efficiency but also have difficulties in separating diffusion from convection in large-scale measurement systems. For example, convection caused by changes in liquid-phase density due to gas dissolution is often ignored, leading to uncertainties and inaccuracies in measurements of phase, composition, velocity, and concentration.

Microfluidic technology, as a small-scale and non-invasive method for precise measurement of gas–liquid diffusion coefficients, is necessary for revealing quantitative information about gas–liquid contact and mass transfer diffusion. This method can calibrate the relationship between different concentrations and fluorescence emission intensity and convert fluorescence intensity into pH values using a standard curve [89]. The Fick diffusion equation is used to fit the change in concentration with diffusion distance to obtain the diffusion coefficient. Qiu et al. [90] used a fluorescence-based microfluidic chip method to measure the diffusion coefficient of $NO_2$ in $H_2O_2$ solution. Hu et al. [91] studied liquid flow in microscale pores at various flow rates and heat fluxes, analyzing the relationship between liquid supply capacity in the two-phase region and enhanced heat transfer in porous structures. The above studies indicate that at the nanoscale, the classical Fick diffusion law still has some applicability, but the diffusion coefficient of molecules will be significantly smaller than the macroscopic theoretical prediction.

*4.4. Minimum Miscibility Pressure*

The minimum miscibility pressure (MMP) is a key parameter for evaluating and optimizing $CO_2$ flooding in oil reservoirs, representing the lowest pressure at which gas and oil form a miscible phase. When the pressure (P) is lower than MMP, $CO_2$ and oil do not mix, and the displacement process is dominated by mobility and capillary forces, leading to fingering and premature breakthrough. On the other hand, when P is higher than MMP, $CO_2$ and oil reach a near-miscible or miscible state, the gas–oil interface disappears, the crude oil expands, and hydrocarbons evaporate. In the miscible flooding process, the capillary forces caused by pore size differences are significantly reduced, resulting in a significant increase in oil recovery efficiency [92] (as shown in Figure 6). Conventional reservoirs are associated with convection-dominated transport, while unconventional reservoirs are associated with diffusion-dominated transport [93], indicating that pore size to some extent affects MMP. The determination of MMP is generally based on the solubility theory, which can be expressed as:

$$\delta = \sqrt{T\left(\frac{\partial P}{\partial T}\right)_V - P} \qquad (6)$$

$$P = \frac{RT}{v-b} - \frac{a}{v(v+b)+b(v-b)} \tag{7}$$

where $v$ is the molar volume, $\delta$ is the solubility, $R$ is the universal gas constant, and $a$ and $b$ represent the pressure and volume terms, respectively, in the Peng-Robinson equation of state.

$$a = \frac{0.45724 R^2 T_{cp}^2}{P_{cp}} \left[1 + k\left(1 - \sqrt{\frac{T}{T_{cp}}}\right)\right]^2 \tag{8}$$

$$b = \frac{0.0778 R T_{cp}}{P_{cp}} \tag{9}$$

where $k$ is a function of the eccentric factor, $T_{cp}$ and $P_{cp}$ are the supercritical temperature and pressure in the confined space, and their relationship with the critical temperature and pressure can be expressed as:

$$T_{cp} = T_c - T_c \left(0.22958 \frac{(T_c/P_c)^{\frac{1}{3}}}{r} - 0.014378 \frac{(T_c/P_c)^{\frac{2}{3}}}{r^2}\right) \tag{10}$$

$$P_{cp} = P_c - P_c \left(0.22958 \frac{(T_c/P_c)^{\frac{1}{3}}}{r} - 0.014378 \frac{(T_c/P_c)^{\frac{2}{3}}}{r^2}\right) \tag{11}$$

where $r$ is the pore radius, and $T_c$ and $P_c$ are the critical temperature and pressure.

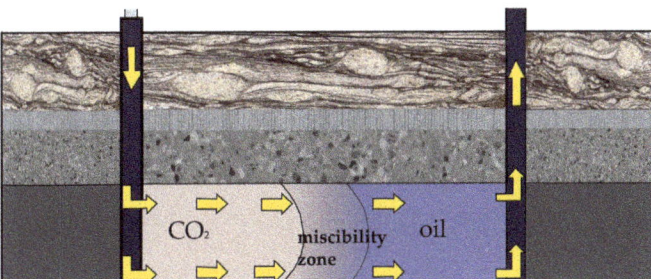

Figure 6. $CO_2$ miscible displacement mechanism.

In conclusion, the solubility $\delta$ is a function of the critical temperature and pressure parameters $T_c$, $P_c$, the molar volume $v$, and the pore radius $r$. This indicates that the transition from bulk phase to nanochannel critical properties dominates the miscibility of fluids in confined spaces [94].

Traditional experimental methods for determining MMP mainly include the slim tube experiment (ST) [95], rising bubble method (RBA) [96], and the disappearance of interfacial tension method (VIT) [97]. However, these experimental methods are time-consuming, subjective, lack quantitative information, and are not suitable for complex mixtures [98]. Theoretical methods simulate the situations of immiscible and miscible phases by coupling the Navier–Stokes equation, interface tracking equation, and convection–diffusion equation, or calculating critical parameters using state equations. For example, Zhang et al. [99] calculated that the critical temperature and pressure decrease as the pore radius decreases based on the van der Waals state equation and the semi-analytical state equation. The Peng–Robinsion EOS (PR EOS) proposes different fitting formulas based on the magnitude of the eccentricity factor, thereby exhibiting better performance near the fluid's critical point. As a result, the PR EOS is currently the most popular equation of state in the petroleum industry [100,101]. However, these calculations are computationally expensive and depend on boundary conditions. In recent years, methods such as nuclear magnetic resonance and CT scanning have emerged for determining MMP, which can measure the minimum miscibility pressure of oil and gas without invasive procedures. However, the high cost of equipment limits their large-scale application [102]. Therefore, the existing methods are unable to meet the requirements for the in situ detection of oil–gas miscibility pressure at the micro- and nanoscale. Microfluidic experimental platforms estimate MMP by observing the appearance of the interface between $CO_2$ and oil [103], or quantitatively observe the mixing process of $CO_2$ and crude oil in microchannels using fluorescence imaging. The design of quasi-static (dead-end) microfluidic chips can also avoid complex experimental operations and potential measurement errors caused by pressure drop in dynamic flow, enabling a large number of tests to be performed in a short period of time. Compared to standard ST tests, this method provides a sufficient number of data points, and existing experiments have demonstrated its effectiveness [104].

*4.5. Displacement Efficiency*

The spreading law refers to the behavior of fluid displacement in different scales, including the flow and transport mechanisms in nanoscale pores and fractures. Gas injection for enhanced oil recovery (EOR) includes $CO_2$ [105–107], natural gas [108], and nitrogen [109], among which $CO_2$ has been widely studied due to its high displacement efficiency and applicability in low-permeability reservoirs [110]. The improvement of oil recovery by $CO_2$ involves the flow of oil and gas in nanoscale pores, gas dissolution, and diffusion [111] in low-permeability matrix, oil swelling [112], wettability alteration [113], the reduction in interfacial tension [114], and interaction between fractures and matrix. Therefore, a comprehensive understanding of the flow and transport mechanisms in porous media and fractures at different scales is crucial for the development of effective techniques to enhance oil recovery. In China, there is an urgent need to scale up the application of $CO_2$ flooding technology for stable and increased oil production. So far, various $CO_2$ injection schemes have been developed by the industry and academia for carbon sequestration and oil and gas extraction. Fluids in nanochannels enter the supercritical state in a sequence of smaller pores before entering larger ones, which exhibits a positive influence on the microscale spreading efficiency and flow path of $CO_2$, while the curvature of porous media may have a negative impact. Therefore, studying the miscible and immiscible displacement patterns across scales is of great engineering significance for understanding the displacement processes in real reservoirs. However, few experimental studies have considered the impact of cross-scale effects on unconventional reservoir displacement spreading.

Conventional core displacement experiments rely on analyzing changes in core mass before and after displacement or using characterization techniques such as nuclear magnetic resonance, which are time-consuming and have low repeatability [115]. Additionally, the spatial distribution and surface morphology of reservoir fractures are complex, and the unique flow patterns are the focus of displacement studies. For example, in actual production, as the reservoir pressure decreases, $CO_2$ in the oil grows and condenses into large gas bubbles that escape from the matrix into the fractures. The oil in the fractures is drawn into the matrix and reoccupies the pore space, significantly affecting the oil recovery in the matrix phase and hindering the geological storage of $CO_2$ in the reservoir. However, existing technologies are unable to directly observe and fully simulate the complex flow characteristics of reservoirs.

Microfluidic experiments based on micro- and nanofluidic chips with different matrices (as shown in Figure 7) can not only investigate the spreading laws of oil displacement in micro- and nanoscales but also directly observe the coupled mechanisms of oil flow and phase behavior under different displacement modes. For example, Lu et al. [116] studied the multiphase transport mechanisms during $CO_2$ imbibition under different wettability conditions, and the experimental results showed that the presence of water phase can slow down the rapid transport and gas channeling of $CO_2$ from the matrix to the fractures, thereby improving oil recovery. Huang et al. [117] used a high-pressure visualization system at the pore scale to study the influence of the presence of aqueous phase on $CO_2$ exsolution under different initial states, pressure drop rates, and wettability conditions [118]. Guo et al. [119] found that the residual saturation of the displacing fluid increases with the increase in the number of capillary tubes. Zhong et al. [120] studied the pressure threshold effect of $N_2$ immiscible displacement and $CO_2$ miscible displacement in confined spaces of 60 nm.

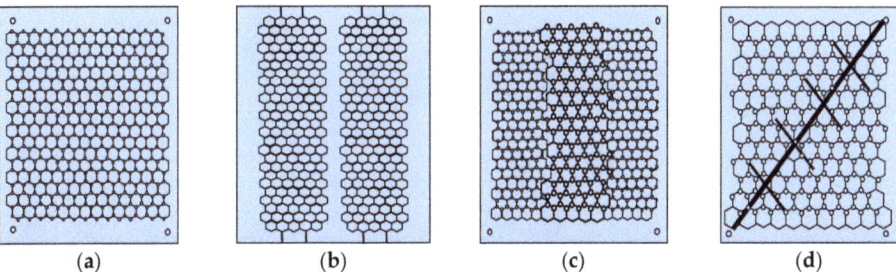

**Figure 7.** $CO_2$ displacement model of the chip (**a**). Homogeneous matrix model (**b**). Heterogeneous matrix model (**c**). Heterogeneous matrix model (**d**). Cracks in the chip model.

## 5. Research on $CO_2$ Geological Sequestration

$CO_2$ geological sequestration primarily involves $CO_2$-enhanced oil (natural gas) recovery and storage, $CO_2$ displacement for coalbed methane storage, and $CO_2$ storage in saline aquifers [121]. Among them, reservoir storage and saline aquifer storage utilize mechanisms such as structural trapping, capillary trapping, dissolution trapping, and mineral trapping (as shown in Figure 8). The sequestration process is controlled by various parameters such as fluid properties, interfacial tension, wetting [122,123], solubility, and pore throat characteristics. Due to the coupling phenomena of fluid flow, geochemistry, and biogeology in porous media during $CO_2$ sequestration, traditional core displacement experiments are insufficient to understand the pore-scale mechanisms of $CO_2$ storage, and there are significant changes in thermophysical properties near the critical point. Therefore, accurate data on the interaction between $CO_2$-saline water-porous media are key to improving storage capacity and the reliability of numerical simulations for geological sequestration [124]. Conventional methods are time-consuming, and there is a greater need for rapid, accurate, and reproducible methods to precisely describe key data. Micro–nanofluidic experiments can be used to optimize numerical and modeling methods from the pore scale to the reservoir scale, making it an ideal experimental approach for studying $CO_2$ sequestration.

*5.1. $CO_2$-Enhanced Oil and Gas Recovery and Sequestration Technology*

During the process of $CO_2$-enhanced oil and gas recovery and storage, the majority of $CO_2$ is stored in underground structural and capillary spaces, a portion dissolves in residual underground fluids, and a small fraction is mineralized in underground rocks. Some $CO_2$ is also produced to the surface along with oil and gas flow. The dissolution of supercritical $CO_2$ into the saline water of the target reservoir, the physicochemical reactions between acidic saline water and solid matrix, the parameters of the aquifer, and the presence of fractures [125], as well as the information on the geometry of the pores introduce uncertainties to large-scale $CO_2$ storage applications. Additionally, capillary trapping, deposition structures, and the permeability of the target layer significantly affect the migration pathways of $CO_2$ [126]. Micro–nano fluidics technology provides a new approach for simulating the migration of $CO_2$ in complex geological structures and porous media.

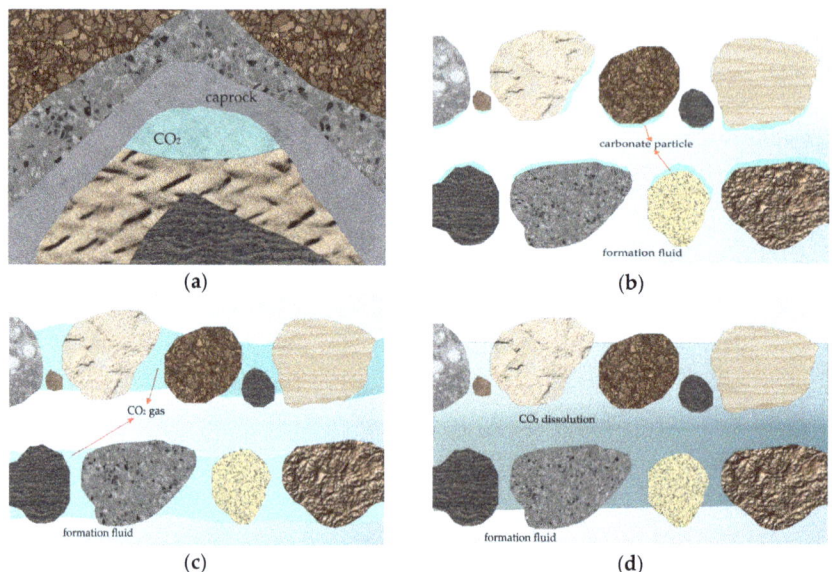

**Figure 8.** Mechanisms of $CO_2$ geological storage. (**a**) Structural trapping. (**b**) Mineralization trapping. (**c**) Confinement space trapping. (**d**) Dissolution trapping.

*5.2. Saline Aquifer $CO_2$ Sequestration Technology*

Saline aquifers have great potential for $CO_2$ storage, but the density difference between $CO_2$ and the surrounding fluids increases the risk of upward migration and leakage. Salt precipitation in porous media can hinder carbon capture and storage in saline aquifers. Therefore, when $CO_2$ is dissolved into the formation brine, saline aquifer storage has a higher level of safety. Injected $CO_2$ diffuses in the porous media, displaces formation water, and is eventually stored underground after undergoing a series of physical and chemical reactions [127,128]. The dissolution of $CO_2$ in water depends on temperature, pressure, and salinity conditions [129]. Therefore, developing more effective methods to represent the interaction between $CO_2$ and water and the behavior of water and gas in porous media is the basis for calculating the amount of carbon storage in saline aquifers [130]. CT scanning is commonly used to observe sediment distribution and changes in pore size and permeability to determine the impact of salt precipitation on geological carbon storage (GCS) [131]. Additionally, scanning electron microscopy (SEM) and X-ray diffraction (XRD) can analyze the products of the interaction between nanoscale porous media and $CO_2$ [132]. Microfluidic technology is an emerging method for studying salt precipitation and has been used to investigate the effects of two configurations of precipitates and pore structure on salt crystal growth rate [133], as well as co-confocal Raman spectroscopy for solubility studies [134].

## 6. Limitations

(1) The dimensions of microfluidic chips currently mainly focus on longitudinal depth differences, and the manufacturing of high aspect ratio chips with higher throughput still faces technological and cost issues. In addition, there are discrepancies between existing microfluidic experimental results and theoretical simulations (such as mixed-phase pressures and phase envelopes), and the reasons for these discrepancies still need further exploration. Furthermore, there is no unified definition for the size boundary of the confinement effect, and extensive experimental research and simulation predictions are still needed.

(2) Most existing studies mainly focus on the phase behavior of small molecule hydrocarbons or binary mixtures at a single-pore-size scale. However, the multi-component nature of

reservoir fluids further complicates the confinement effect, and the mass transfer diffusion between components cannot be ignored. There are still certain deficiencies in the study of multi-component mixture phase behavior across scales based on micro-fluidic technology. In addition, idealized pore geometries and pure fluid systems are not sufficient to represent the inherent complexity of multi-component mixtures in unconventional reservoirs.

(3) Target formations for $CO_2$ geological sequestration often contain brine, and the interaction between $CO_2$ and pore media and formation water under high temperature and pressure conditions still needs further clarification. The rate of salt growth and in situ distribution of precipitation at the pore scale require extensive experimental research.

## 7. Prospects

(1) Pores below 10 nm approach the scale of $CO_2$ and hydrocarbon molecules more closely, and the effects of fluid continuity, solid–liquid interface properties, and critical parameters on phase behavior may have important implications for recovery and storage rate analysis.

(2) Micro–nano fluidics experiments can provide support for future studies on $CO_2$ migration in water and tight caprock, and serve as a reference for the design of $CO_2$ geological storage and simulation of supercritical $CO_2$ dissolution processes.

(3) The development of micro–nano fluidics technology should be combined with innovation from multiple disciplines to achieve more precise fluid control and response. For example, the simulation of reservoir wettability in $CO_2$-enhanced oil recovery and geological storage can refer to surface modification of chips from disciplines such as biomedicine. The study of droplet condensation, adsorption, and separation in $CO_2$ capture processes can also draw inspiration from micro–nano fluidic technologies. Additionally, insights can be gained from filtration and separation applications in other disciplines.

(4) In the future, micro–nano flow control experiments can investigate how the efficiency of $CO_2$ flooding and storage can be improved by studying the effects of additives such as surfactants, foams, and nanoparticles, as well as changes in physical and chemical conditions. They can also explore emerging fields such as microbial enhanced oil recovery or geothermal energy to enhance the understanding of fine-scale mechanisms in pore structures and establish connections with reservoir-scale phenomena.

(5) Supercritical $CO_2$ is in a state that exists between gas and liquid at critical conditions. It has wide-ranging industrial applications. In the future, research on supercritical $CO_2$ based on micro–nanofluidic experimental platforms can refer to low-carbon extraction technology using supercritical $CO_2$ as an organic solvent, the facile control of morphological structure in the synthesis of nanoparticles, supercritical drying technology, carriers for nanoscale drug synthesis, and reaction media for organic synthesis under special conditions.

## 8. Conclusions

The capture, utilization, and storage of $CO_2$ have broad application prospects, and the application of micro–nanofluidics technology in $CO_2$ capture is just beginning to emerge and requires further exploration. Although the application of $CO_2$-enhanced oil and gas recovery has been ongoing for many years, it is currently limited to the study of simple mixtures at a single scale due to manufacturing processes and fluid control limitations. There is a significant lack of research on more realistic scenarios such as crude oil degassing, $CO_2$ extraction, and stripping. Additionally, there is a scarcity of research on caprock leakage in $CO_2$ geological storage, and further clarification is needed on the mechanisms of $CO_2$ interaction with saltwater and rocks. In conclusion, there is a tremendous demand for research on $CO_2$ capture, utilization, and storage based on micro–nanofluidics technology, making it an indispensable and important tool in future experimental developments.

**Author Contributions:** Writing—original draft, writing—review and editing, conceptualization, X.P.; supervision, X.H., C.F. and Z.Z.; conceptualization, L.S. All authors have read and agreed to the published version of the manuscript.

**Funding:** This research was supported by The Major Project of CNPC's "CCUS oil displacement geological body fine description and reservoir engineering key technology research" (No. 2021ZZ01-03).

**Data Availability Statement:** Not appliable.

**Acknowledgments:** The support given by The State Key Laboratory of Enhanced Oil Recovery of Open Fund Funded Project, Major Special Projects of CNPC is acknowledged.

**Conflicts of Interest:** The authors declare no conflict of interest.

## References

1. Zou, C.; Tao, S.; Bai, B.; Yang, Z.; Zhu, R.; Hou, L.; Yuan, X.; Zhang, G.; Wu, S.; Pang, Z.; et al. Differences and Relations between Unconventional and Conventional Oil and Gas. *China Pet. Explor.* **2015**, *20*, 1.
2. Loucks, R.G.; Reed, R.M.; Ruppel, S.C.; Jarvie, D.M. Morphology, Genesis, and Distribution of Nanometer-Scale Pores in Siliceous Mudstones of the Mississippian Barnett Shale. *J. Sediment. Res.* **2009**, *79*, 848–861. [CrossRef]
3. Sheng, J.J. Enhanced oil recovery in shale reservoirs by gas injection. *J. Nat. Gas Sci. Eng.* **2015**, *22*, 252–259. [CrossRef]
4. Bao, B.; Zhao, S.; Xu, J. Progress in studying fluid phase behaviours with micro- and nano-fluidic technology. *CIESC J.* **2018**, *69*, 4530–4541.
5. Tan, S.P.; Qiu, X.; Dejam, M.; Adidharma, H. Critical point of fluid confined in nanopores: Experimental detection and measurement. *J. Phys. Chem. C* **2019**, *123*, 9824–9830. [CrossRef]
6. Wang, Z.; Pereira, J.-M.; Gan, Y. Effect of Wetting Transition during Multiphase Displacement in Porous Media. *Langmuir* **2020**, *36*, 2449–2458. [CrossRef] [PubMed]
7. Wang, F.; Nestler, B. Wetting transition and phase separation on flat substrates and in porous structures. *J. Chem. Phys.* **2021**, *154*, 094704. [CrossRef]
8. Alfi, M.; Nasrabadi, H.; Banerjee, D. Experimental investigation of confinement effect on phase behavior of hexane, heptane and octane using lab-on-a-chip technology. *Fluid Phase Equilibria* **2016**, *423*, 25–33. [CrossRef]
9. Jin, Z.; Firoozabadi, A. Thermodynamic Modeling of Phase Behavior in Shale Media. *SPE J.* **2016**, *21*, 190–207. [CrossRef]
10. Bhatia, S.K.; Bonilla, M.R.; Nicholson, D. Molecular transport in nanopores: A theoretical perspective. *Phys. Chem. Chem. Phys.* **2011**, *13*, 15350–15383. [CrossRef]
11. Parsa, E.; Yin, X.; Ozkan, E. Direct Observation of the Impact of Nanopore Confinement on Petroleum Gas Condensation. In Proceedings of the SPE Annual Technical Conference and Exhibition, Houston, TX, USA, 28–30 September 2015; D031S031R008.
12. Yang, Q.; Bi, R.; Banerjee, D.; Nasrabadi, H. Direct Observation of the Vapor–Liquid Phase Transition and Hysteresis in 2 nm Nanochannels. *Langmuir* **2022**, *38*, 9790–9798. [CrossRef]
13. Baek, S.; Akkutlu, I.Y. Enhanced Recovery of Nanoconfined Oil in Tight Rocks Using Lean Gas ($C_2H_6$ and $CO_2$) Injection. *SPE J.* **2021**, *26*, 2018–2037. [CrossRef]
14. Zhong, J.; Wang, Z.; Sun, Z.; Yao, J.; Yang, Y.; Sun, H.; Zhang, L.; Zhang, K. Research advances in microscale fluid characteristics of shale reservoirs based on nanofluidic technology. *Acta Pet. Sin.* **2023**, *44*, 207.
15. Yarin, A.L. Novel nanofluidic and microfluidic devices and their applications. *Curr. Opin. Chem. Eng.* **2020**, *29*, 17–25. [CrossRef]
16. Wang, H.; Wei, B.; Hou, J.; Liu, Y.; Du, Q. Heavy oil recovery in blind-ends during chemical flooding: Pore scale study using microfluidic experiments. *J. Mol. Liq.* **2022**, *368*, 120724. [CrossRef]
17. Onaka, Y.; Sato, K. Dynamics of pore-throat plugging and snow-ball effect by asphaltene deposition in porous media micromodels. *J. Pet. Sci. Eng.* **2021**, *207*, 109176. [CrossRef]
18. Liang, T.; Xu, K.; Lu, J.; Nguyen, Q.; DiCarlo, D. Evaluating the Performance of Surfactants in Enhancing Flowback and Permeability after Hydraulic Fracturing through a Microfluidic Model. *SPE J.* **2020**, *25*, 268–287. [CrossRef]
19. Bob, B.; Shi, J.; Feng, J.; Yang, Z.; Peng, B.; Zhao, S. Research progress of surfactant enhanced oil recovery based on microfluidics technology. *Acta Pet. Sin.* **2022**, *43*, 432–442+452.
20. Zhao, X.; Feng, Y.; Liao, G.; Liu, W. Visualizing in-situ emulsification in porous media during surfactant flooding: A microfluidic study. *J. Colloid Interface Sci.* **2020**, *578*, 629–640. [CrossRef]
21. Sugar, A.; Torrealba, V.; Buttner, U.; Hoteit, H. Assessment of Polymer-Induced Clogging Using Microfluidics. *SPE J.* **2020**, *26*, 3793–3804. [CrossRef]
22. Li, J. Enhanced Oil Recovery Using Bubbles in a Reservoir-on-a-Chip (ROC). Ph.D. Thesis, Shandong University, Jinan, China, 2021.
23. Kotdawala, R.R.; Kazantzis, N.; Thompson, R.W. Analysis of binary adsorption of polar and nonpolar molecules in narrow slit-pores by mean-field perturbation theory. *J. Chem. Phys.* **2005**, *123*, 244709. [CrossRef]
24. Li, Z.; Jin, Z.; Firoozabadi, A. Phase Behavior and Adsorption of Pure Substances and Mixtures and Characterization in Nanopore Structures by Density Functional Theory. *SPE J.* **2014**, *19*, 1096–1109. [CrossRef]
25. Travalloni, L.; Castier, M.; Tavares, F.W. Phase equilibrium of fluids confined in porous media from an extended Peng–Robinson equation of state. *Fluid Phase Equilibria* **2014**, *362*, 335–341. [CrossRef]
26. Derouane, E.G. On the physical state of molecules in microporous solids. *Microporous Mesoporous Mater.* **2007**, *104*, 46–51. [CrossRef]

27. Tan, S.P.; Piri, M. Equation-of-state modeling of confined-fluid phase equilibria in nanopores. *Fluid Phase Equilibria* **2015**, *393*, 48–63. [CrossRef]
28. Wang, F.; Wu, Y.; Nestler, B. Wetting Effect on Patterned Substrates. *Adv. Mater.* **2023**, *35*, e2210745. [CrossRef]
29. Snustad, I.; Røe, I.T.; Brunsvold, A.; Ervik, Å.; He, J.; Zhang, Z. A review on wetting and water condensation—Perspectives for $CO_2$ condensation. *Adv. Colloid Interface Sci.* **2018**, *256*, 291–304. [CrossRef]
30. Cassie, A.B.D.; Baxter, S. Wettability of porous surfaces. *Trans. Faraday Soc.* **1944**, *40*, 546–551. [CrossRef]
31. Zeng, Y.; Harrison, D.J. Self-Assembled Colloidal Arrays as Three-Dimensional Nanofluidic Sieves for Separation of Biomolecules on Microchips. *Anal. Chem.* **2007**, *79*, 2289–2295. [CrossRef]
32. Angelova, A.; Angelov, B.; Lesieur, S.; Mutafchieva, R.; Ollivon, M.; Bourgaux, C.; Willumeit, R.; Couvreur, P. Dynamic control of nanofluidic channels in protein drug delivery vehicles. *J. Drug Deliv. Sci. Technol.* **2008**, *18*, 41–45. [CrossRef]
33. Abgrall, P.; Nguyen, N.T. Nanofluidic Devices and Their Applications. *Anal. Chem.* **2008**, *80*, 2326–2341. [CrossRef]
34. Yang, H.Y.; Han, Z.J.; Yu, S.F.; Pey, K.L.; Ostrikov, K.; Karnik, R. Carbon nanotube membranes with ultrahigh specific adsorption capacity for water desalination and purification. *Nat. Commun.* **2013**, *4*, 2220. [CrossRef]
35. Morais, S.; Cario, A.; Liu, N.; Bernard, D.; Lecoutre, C.; Garrabos, Y.; Ranchou-Peyruse, A.; Dupraz, S.; Azaroual, M.; Hartman, R.L.; et al. Studying key processes related to $CO_2$ underground storage at the pore scale using high pressure micromodels. *React. Chem. Eng.* **2020**, *5*, 1156–1185. [CrossRef]
36. Hele-Shaw, H.S. Flow of water. *Nature* **1898**, *58*, 520. [CrossRef]
37. Engel, M.; Wunsch, B.H.; Neumann, R.F.; Giro, R.; Bryant, P.W.; Smith, J.T.; Steiner, M.B. Nanoscale Flow Chip Platform for Laboratory Evaluation of Enhanced Oil Recovery Materials. In Proceedings of the SPE Annual Technical Conference and Exhibition, San Antonio, TX, USA, 9–11 October 2017; D021S019R002.
38. Zhang, Y.; Zhou, C.; Qu, C.; Wei, M.; He, X.; Bai, B. Fabrication and verification of a glass–silicon–glass micro-/nanofluidic model for investigating multi-phase flow in shale-like unconventional dual-porosity tight porous media. *Lab A Chip* **2019**, *19*, 4071–4082. [CrossRef]
39. Lee, H.; Lee, S.G.; Doyle, P.S. Photopatterned oil-reservoir micromodels with tailored wetting properties. *Lab A Chip* **2015**, *15*, 3047–3055. [CrossRef]
40. Wegner, J.; Ganzer, L. Rock-on-a-Chip Devices for High p, T Conditions and Wettability Control for the Screening of EOR Chemicals. In Proceedings of the SPE Europec Featured at 79th EAGE Conference and Exhibition, Paris, France, 12–15 June 2017; D041S010R007.
41. Lignos, I.; Ow, H.; Lopez, J.P.; McCollum, D.L.; Zhang, H.; Imbrogno, J.; Shen, Y.; Chang, S.; Wang, W.; Jensen, K.F. Continuous Multistage Synthesis and Functionalization of Sub-100 nm Silica Nanoparticles in 3D-Printed Continuous Stirred-Tank Reactor Cascades. *ACS Appl. Mater. Interfaces* **2020**, *12*, 6699–6706. [CrossRef]
42. Jacobs, T. Reservoir-on-a-Chip Technology Opens a New Window Into Oilfield Chemistry. *J. Pet. Technol.* **2019**, *71*, 25–27. [CrossRef]
43. Wang, Y.; Wang, J.; Ma, C.; Qiao, W.; Ling, L. Fabrication of hierarchical carbon nanosheet-based networks for physical and chemical adsorption of $CO_2$. *J. Colloid Interface Sci.* **2018**, *534*, 72–80. [CrossRef]
44. Khansary, M.A.; Aroon, M.A.; Shirazian, S. Physical adsorption of $CO_2$ in biomass at atmospheric pressure and ambient temperature. *Environ. Chem. Lett.* **2020**, *18*, 1423–1431. [CrossRef]
45. Williamson, I.; Nelson, E.B.; Li, L. Carbon dioxide sorption in a nanoporous octahedral molecular sieve. *J. Phys. D Appl. Phys.* **2015**, *48*, 335304. [CrossRef]
46. Xie, L.; Jin, Z.; Dai, Z.; Zhou, T.; Zhang, X.; Chang, Y.; Jiang, X. Fabricating self-templated and N-doped hierarchical porous carbon spheres via microfluidic strategy for enhanced $CO_2$ capture. *Sep. Purif. Technol.* **2023**, *322*, 124267. [CrossRef]
47. Qi, Z.; Xu, L.; Xu, Y.; Zhong, J.; Abedini, A.; Cheng, X.; Sinton, D. Disposable silicon-glass microfluidic devices: Precise, robust and cheap. *Lab A Chip* **2018**, *18*, 3872–3880. [CrossRef] [PubMed]
48. Towler, B.F. *Fundamental Principles of Reservoir Engineering*; Society of Petroleum Engineers: Houston, TX, USA, 2002.
49. Nojabaei, B.; Johns, R.T.; Chu, L. Effect of Capillary Pressure on Phase Behavior in Tight Rocks and Shales. *SPE Reserv. Eval. Eng.* **2013**, *16*, 281–289. [CrossRef]
50. Alfi, M. Experimental Study of Confinement Effect on Phase Behavior of Hydrocarbons in Nano-Slit Channels Using Nanofluidic Devices. Ph.D. Thesis, Texas A&M University, College Station, TX, USA, 2019.
51. Liu, Y.; Li, H.A.; Okuno, R. Phase behavior of fluid mixtures in a partially confined space. In Proceedings of the SPE Annual Technical Conference and Exhibition, Dubai, United Arab Emirates, 27 September 2016; OnePetro: Richardson, TX, USA, 2016.
52. Ustinov, E.A.; Do, D.D. Modeling of adsorption in finite cylindrical pores by means of density functional theory. *Adsorption* **2005**, *11*, 455–477. [CrossRef]
53. Achour, S.H.; Okuno, R. Phase stability analysis for tight porous media by minimization of the Helmholtz free energy. *Fluid Phase Equilibria* **2020**, *520*, 112648. [CrossRef]
54. Nichita, D.V. Volume-based phase stability analysis including capillary pressure. *Fluid Phase Equilibria* **2019**, *492*, 145–160. [CrossRef]
55. Aljamaan, H.M. Multiscale and Multicomponent Flow and Storage Capacity Investigation of Unconventional Resources. Ph.D. Thesis, Stanford University, Stanford, CA, USA, 2017.

56. Konno, M.; Shibata, K.; Saito, S. Adsorption of light hydrocarbon mixtures on molecular sieving carbon MSC-5A. *J. Chem. Eng. Jpn.* **1985**, *18*, 394–398. [CrossRef]
57. Yun, J.H.; Düren, T.; Keil, F.J.; Seaton, N.A. Adsorption of methane, ethane, and their binary mixtures on MCM-41: Experimental evaluation of methods for the prediction of adsorption equilibrium. *Langmuir* **2002**, *18*, 2693–2701. [CrossRef]
58. Luo, S.; Lutkenhaus, J.L.; Nasrabadi, H. Experimental study of confinement effect on hydrocarbon phase behavior in nano-scale porous media using differential scanning calorimetry. In Proceedings of the SPE Annual Technical Conference and Exhibition, Houston, TX, USA, 28–30 September 2015; SPE: Kuala Lumpur, Malaysia, 2015; D031S043R003.
59. Jin, B.; Nasrabadi, H. Phase behavior of multi-component hydrocarbon systems in nano-pores using gauge-GCMC molecular simulation. *Fluid Phase Equilibria* **2016**, *425*, 324–334. [CrossRef]
60. Luo, S.; Lutkenhaus, J.L.; Nasrabadi, H. Use of differential scanning calorimetry to study phase behavior of hydrocarbon mixtures in nano-scale porous media. *J. Pet. Sci. Eng.* **2018**, *163*, 731–738. [CrossRef]
61. Liu, Y.; Jiang, L.; Song, Y.; Zhao, Y.; Zhang, Y.; Wang, D. Estimation of minimum miscibility pressure (MMP) of $CO_2$ and liquid n-alkane systems using an improved MRI technique. *Magn. Reson. Imaging* **2016**, *34*, 97–104. [CrossRef] [PubMed]
62. Rossi, A.; Piccinin, S.; Pellegrini, V.; de Gironcoli, S.; Tozzini, V. Nano-Scale Corrugations in Graphene: A Density Functional Theory Study of Structure, Electronic Properties and Hydrogenation. *J. Phys. Chem. C* **2015**, *119*, 7900–7910. [CrossRef]
63. Jin, B.; Nasrabadi, H. Phase Behavior in Shale Organic/Inorganic Nanopores from Molecular Simulation. *SPE Reserv. Eval. Eng.* **2018**, *21*, 626–637. [CrossRef]
64. Pitakbunkate, T.; Balbuena, P.B.; Moridis, G.J.; Blasingame, T.A. Effect of Confinement on Pressure/Volume/Temperature Properties of Hydrocarbons in Shale Reservoirs. *SPE J.* **2016**, *21*, 621–634. [CrossRef]
65. Jin, B.; Bi, R.; Nasrabadi, H. Molecular simulation of the pore size distribution effect on phase behavior of methane confined in nanopores. *Fluid Phase Equilibria* **2017**, *452*, 94–102. [CrossRef]
66. Luo, S.; Lutkenhaus, J.L.; Nasrabadi, H. Confinement-Induced Supercriticality and Phase Equilibria of Hydrocarbons in Nanopores. *Langmuir* **2016**, *32*, 11506–11513. [CrossRef]
67. Luo, S.; Lutkenhaus, J.L.; Nasrabadi, H. Effect of nano-scale pore size distribution on fluid phase behavior of gas IOR in shale reservoirs. In Proceedings of the SPE Improved Oil Recovery Conference, Tulsa, OK, USA, 14–18 April 2018; SPE: Kuala Lumpur, Malaysia, 2018; D041S019R003.
68. Liu, Y.; Jin, Z.; Li, H.A. Comparison of Peng-Robinson Equation of State WITH Capillary Pressure Model with Engineering Density-Functional Theory in Describing the Phase Behavior of Confined Hydrocarbons. *SPE J.* **2018**, *23*, 1784–1797. [CrossRef]
69. Salahshoor, S.; Fahes, M. Experimental Investigation of the Effect of Pore Size on Saturation Pressure for Gas Mixtures. In Proceedings of the Annual Technical Conference and Exhibition, Orlando, FL, USA, 7–10 May 2018; SPE: Kuala Lumpur, Malaysia, 2018; D011S006R003.
70. Molla, S.; Mostowfi, F. Microfluidic PVT--Saturation Pressure and Phase-Volume Measurement of Black Oils. *SPE Reserv. Eval. Eng.* **2016**, *20*, 233–239. [CrossRef]
71. Bao, B.; Qiu, J.; Liu, F.; Fan, Q.; Luo, W.; Zhao, S. Capillary trapping induced slow evaporation in nanochannels. *J. Pet. Sci. Eng.* **2020**, *196*, 108084. [CrossRef]
72. Wang, L.; Parsa, E.; Gao, Y.; Ok, J.T.; Neeves, K.; Yin, X.; Ozkan, E. Experimental study and modeling of the effect of nanoconfinement on hydrocarbon phase behavior in unconventional reservoirs. In Proceedings of the SPE Western Regional Meeting, Denver, CO, USA, 17–18 April 2014; SPE: Kuala Lumpur, Malaysia, 2014; SPE-169581-MS.
73. Zhong, J.; Zhao, Y.; Lu, C.; Xu, Y.; Jin, Z.; Mostowfi, F.; Sinton, D. Nanoscale Phase Measurement for the Shale Challenge: Multicomponent Fluids in Multiscale Volumes. *Langmuir* **2018**, *34*, 9927–9935. [CrossRef] [PubMed]
74. Zhong, J.; Zandavi, S.H.; Li, H.; Bao, B.; Persad, A.H.; Mostowfi, F.; Sinton, D. Condensation in One-Dimensional Dead-End Nanochannels. *ACS Nano* **2016**, *11*, 304–313. [CrossRef] [PubMed]
75. Yang, Q.; Jin, B.; Banerjee, D.; Nasrabadi, H. Direct visualization and molecular simulation of dewpoint pressure of a confined fluid in sub-10 nm slit pores. *Fuel* **2018**, *235*, 1216–1223. [CrossRef]
76. Zhang, K.; Jia, N.; Li, S.; Liu, L. Static and dynamic behavior of $CO_2$ enhanced oil recovery in shale reservoirs: Experimental nanofluidics and theoretical models with dual-scale nanopores. *Appl. Energy* **2019**, *255*, 113752. [CrossRef]
77. Jatukaran, A. *Visualization of Fluid Dynamics in Nanoporous Media for Unconventional Hydrocarbon Recovery*; University of Toronto: Toronto, ON, Canada, 2018.
78. Jatukaran, A.; Zhong, J.; Abedini, A.; Sherbatian, A.; Zhao, Y.; Jin, Z.; Mostowfi, F.; Sinton, D. Natural gas vaporization in a nanoscale throat connected model of shale: Multi-scale, multi-component and multi-phase. *Lab A Chip* **2018**, *19*, 272–280. [CrossRef]
79. Alfi, M.; Nasrabadi, H.; Banerjee, D. Effect of confinement on bubble point temperature shift of hydrocarbon mixtures: Experimental investigation using nanofluidic devices. In Proceedings of the Annual Technical Conference and Exhibition, Anaheim, CA, USA, 8–10 May 2017; SPE: Kuala Lumpur, Malaysia, 2017; D011S009R004.
80. Washburn, E.W. The Dynamics of Capillary Flow. *Phys. Rev. B* **1921**, *17*, 273–283. [CrossRef]
81. Song, Y.; Song, Z.; Liu, Y.; Guo, J.; Bai, B.; Hou, J.; Bai, M.; Song, K. Phase behavior and minimum miscibility pressure of confined fluids in organic nanopores. In Proceedings of the Improved Oil Recovery Conference, Tulsa, OK, USA, 31 August–4 September 2020; SPE: Kuala Lumpur, Malaysia, 2020; D021S035R002.

82. Li, H.; Zhong, J.; Pang, Y.; Zandavi, S.H.; Persad, A.H.; Xu, Y.; Mostowfi, F.; Sinton, D. Direct visualization of fluid dynamics in sub-10 nm nanochannels. *Nanoscale* **2017**, *9*, 9556–9561. [CrossRef]
83. Lu, H.; Xu, Y.; Duan, C.; Jiang, P.; Xu, R. Experimental Study on Capillary Imbibition of Shale Oil in Nanochannels. *Energy Fuels* **2022**, *36*, 5267–5275. [CrossRef]
84. Zuo, M.; Chen, H.; Xu, C.; Stephenraj, I.R.; Qi, X.; Yu, H.; Liu, X.Y. Study on Dynamic Variation Characteristics of Reservoir Fluid Phase Behavior During $CO_2$ Injection in $CO_2$ Based Enhanced Oil Recovery Process. In Proceedings of the IADC/SPE Asia Pacific Drilling Technology Conference and Exhibition, Bangkok, Thailand, 9–10 August 2022; SPE: Kuala Lumpur, Malaysia, 2022; D012S001R004.
85. Rezk, M.G.; Foroozesh, J. Phase behavior and fluid interactions of a $CO_2$-Light oil system at high pressures and temperatures. *Heliyon* **2019**, *5*, e02057. [CrossRef]
86. Fu, T. Microfluidics in $CO_2$ capture, sequestration, and applications. In *Advances in Microfluidics-New Applications in Biology, Energy, and Materials Sciences*; BoD—Books on Demand: Norderstedt, Germany, 2016; pp. 293–313.
87. Belyaev, A.V.; Dedov, A.V.; Krapivin, I.I.; Varava, A.N.; Jiang, P.; Xu, R. Study of Pressure Drops and Heat Transfer of Nonequilibrial Two-Phase Flows. *Water* **2021**, *13*, 2275. [CrossRef]
88. Du, L.; Liu, W.; Chen, X.; Qin, X.; Ren, X. Research Progress on the Diffusion of $CO_2$ in Crude Oil. *Oilfield Chem.* **2019**, *36*, 372–380.
89. Kuhn, S.; Jensen, K.F. A pH-sensitive laser-induced fluorescence technique to monitor mass transfer in multiphase flows in microfluidic devices. *Ind. Eng. Chem. Res.* **2012**, *51*, 8999–9006. [CrossRef]
90. Qiu, J.; Bao, B.; Zhao, S.; Lu, X. Microfluidics-based determination of diffusion coefficient for gas-liquid reaction system with hydrogen peroxide. *Chem. Eng. Sci.* **2020**, *231*, 116248. [CrossRef]
91. Hu, H.; Jiang, P.; Huang, F.; Xu, R. Role of trapped liquid in flow boiling inside micro-porous structures: Pore-scale visualization and heat transfer enhancement. *Sci. Bull.* **2021**, *66*, 1885–1894. [CrossRef] [PubMed]
92. Alfarge, D.; Alsaba, M.; Wei, M.; Bai, B. Miscible gases based EOR in unconventional liquids rich reservoirs: What we can learn. In Proceedings of the SPE International Heavy Oil Conference and Exhibition, Kuwait City, Kuwait, 10–12 December 2018; SPE: Kuala Lumpur, Malaysia, 2018; D022S034R002.
93. Cronin, M.; Emami-Meybodi, H.; Johns, R.T. Diffusion-Dominated Proxy Model for Solvent Injection in Ultratight Oil Reservoirs. *SPE J.* **2018**, *24*, 660–680. [CrossRef]
94. Bao, B.; Feng, J.; Qiu, J.; Zhao, S. Direct Measurement of Minimum Miscibility Pressure of Decane and $CO_2$ in Nanoconfined Channels. *ACS Omega* **2020**, *6*, 943–953. [CrossRef] [PubMed]
95. Flock, D.; Nouar, A. Parametric Analysis on the Determination of The Minimum Miscibility Pressure In Slim Tube Displacements. *J. Can. Pet. Technol.* **1984**, *23*, 05. [CrossRef]
96. Christiansen, R.L.; Haines, H.K. Rapid Measurement of Minimum Miscibility Pressure with the Rising-Bubble Apparatus. *SPE Reserv. Eng.* **1987**, *2*, 523–527. [CrossRef]
97. Rao, D.N. A new technique of vanishing interfacial tension for miscibility determination. *Fluid Phase Equilibria* **1997**, *139*, 311–324. [CrossRef]
98. Pereponov, D.; Tarkhov, M.; Dorhjie, D.B.; Rykov, A.; Filippov, I.; Zenova, E.; Krutko, V.; Cheremisin, A.; Shilov, E. Microfluidic Studies on Minimum Miscibility Pressure for n-Decane and $CO_2$. *Energies* **2023**, *16*, 4994. [CrossRef]
99. Zhang, K.; Jia, N.; Li, S.; Liu, L. Thermodynamic phase behaviour and miscibility of confined fluids in nanopores. *Chem. Eng. J.* **2018**, *351*, 1115–1128. [CrossRef]
100. Peng, D.-Y.; Robinson, D.B. *The Characterization of the Heptanes and Heavier Fractions for the GPA Peng-Robinson Programs*; GPA Research Report RR-28; Gas Processors Association: Tulsa, OK, USA, 1978.
101. van der Waals, J.D. Over de Continuiteit van der Gas-en Vloeistoftoestand. Ph.D. Thesis, Leiden University, Leiden, The Netherlands, 1873. (In Dutch)
102. Zhang, Y. *Fabrication of Micro-/Nanofluidic Models and Their Applications for Enhanced Oil Recovery Mechanism Study*; Missouri University of Science and Technology: Rolla, MO, USA, 2020.
103. Adyani Wan Razak, W.N.; Kechut, N.I. Advanced technology for rapid minimum miscibility pressure determination (part 1). In Proceedings of the SPE Asia Pacific Oil and Gas Conference and Exhibition, Jakarta, Indonesia, 30 October–1 November 2007; SPE: Kuala Lumpur, Malaysia, 2007; SPE-110265-MS.
104. Nguyen, P.; Mohaddes, D.; Riordon, J.; Fadaei, H.; Lele, P.; Sinton, D. Fast Fluorescence-Based Microfluidic Method for Measuring Minimum Miscibility Pressure of $CO_2$ in Crude Oils. *Anal. Chem.* **2015**, *87*, 3160–3164. [CrossRef] [PubMed]
105. Chen, C.; Balhoff, M.T.; Mohanty, K.K. Effect of Reservoir Heterogeneity on Primary Recovery and $CO_2$ Huff 'n' Puff Recovery in Shale-Oil Reservoirs. *SPE Reserv. Eval. Eng.* **2014**, *17*, 404–413. [CrossRef]
106. Yu, W.; Lashgari, H.R.; Wu, K.; Sepehrnoori, K. $CO_2$ injection for enhanced oil recovery in Bakken tight oil reservoirs. *Fuel* **2015**, *159*, 354–363. [CrossRef]
107. Alharthy, N.; Teklu, T.W.; Kazemi, H.; Graves, R.M.; Hawthorne, S.B.; Braunberger, J.; Kurtoglu, B. Enhanced Oil Recovery in Liquid–Rich Shale Reservoirs: Laboratory to Field. *SPE Reserv. Eval. Eng.* **2017**, *21*, 137–159. [CrossRef]
108. Hoffman, B.T. Comparison of various gases for enhanced recovery from shale oil reservoirs. In Proceedings of the SPE Improved Oil Recovery Symposium, Tulsa, OK, USA, 14–18 April 2012; OnePetro: Richardson, TX, USA, 2012.

109. Yu, Y.; Sheng, J.J. Experimental investigation of light oil recovery from fractured shale reservoirs by cyclic water injection. In Proceedings of the SPE Western Regional Meeting, Anchorage, AK, USA, 23–26 May 2016; SPE: Kuala Lumpur, Malaysia, 2016; SPE-180378-MS.
110. Wang, L.; Tian, Y.; Yu, X.; Wang, C.; Yao, B.; Wang, S.; Winterfeld, P.H.; Wang, X.; Yang, Z.; Wang, Y.; et al. Advances in improved/enhanced oil recovery technologies for tight and shale reservoirs. *Fuel* **2017**, *210*, 425–445. [CrossRef]
111. Luo, S. Experimental and Simulation Studies on Phase Behavior of Petroleum Fluids in Nanoporous Media. Ph.D. Thesis, Texas A&M University, College Station, TX, USA, 2018.
112. Jia, B.; Tsau, J.-S.; Barati, R. A review of the current progress of $CO_2$ injection EOR and carbon storage in shale oil reservoirs. *Fuel* **2018**, *236*, 404–427. [CrossRef]
113. Guo, Y.; Liu, F.; Qiu, J.; Xu, Z.; Bao, B. Microscopic transport and phase behaviors of $CO_2$ injection in heterogeneous formations using microfluidics. *Energy* **2022**, *256*, 124524. [CrossRef]
114. Al-Kindi, I.; Babadagli, T. Revisiting Kelvin equation and Peng–Robinson equation of state for accurate modeling of hydrocarbon phase behavior in nano capillaries. *Sci. Rep.* **2021**, *11*, 6573. [CrossRef]
115. Bao, B.; Zhao, S. A review of experimental nanofluidic studies on shale fluid phase and transport behaviors. *J. Nat. Gas Sci. Eng.* **2020**, *86*, 103745. [CrossRef]
116. Lu, H.; Huang, F.; Jiang, P.; Xu, R. Exsolution effects in $CO_2$ huff-n-puff enhanced oil recovery: Water-Oil-$CO_2$ three phase flow visualization and measurements by micro-PIV in micromodel. *Int. J. Greenh. Gas Control.* **2021**, *111*, 103445. [CrossRef]
117. Huang, F.; Xu, R.; Jiang, P.; Wang, C.; Wang, H.; Lun, Z. Pore-scale investigation of $CO_2$/oil exsolution in $CO_2$ huff-n-puff for enhanced oil recovery. *Phys. Fluids* **2020**. [CrossRef]
118. Jang, X.Z.J. Impacts of gettability on immiscible fluid flow pattern-Microfluidic chip experiment. *China Pet. Process. Petrochem. Technol.* **2019**, *21*, 80.
119. Guo, Y.; Zhang, L.; Yang, Y.; Xu, Z.; Bao, B. Pore-scale investigation of immiscible displacement in rough fractures. *J. Pet. Sci. Eng.* **2021**, *207*, 109107. [CrossRef]
120. Zhong, J.; Abedini, A.; Xu, L.; Xu, Y.; Qi, Z.; Mostowfi, F.; Sinton, D. Nanomodel visualization of fluid injections in tight formations. *Nanoscale* **2018**, *10*, 21994–22002. [CrossRef]
121. Liu, T.; Ma, X.; Diao, Y.; Jin, X.; Fu, J.; Zhang, C. Research status of $CO_2$ geological storage potential evaluation methods at home and abroad. *Geol. Surv. China* **2021**, *8*, 101–108.
122. Zheng, X.; Mahabadi, N.; Yun, T.S.; Jang, J. Effect of capillary and viscous force on $CO_2$ saturation and invasion pattern in the microfluidic chip. *J. Geophys. Res. Solid Earth* **2017**, *122*, 1634–1647. [CrossRef]
123. Morais, S.; Liu, N.; Diouf, A.; Bernard, D.; Lecoutre, C.; Garrabos, Y.; Marre, S. Monitoring $CO_2$ invasion processes at the pore scale using geological labs on chip. *Lab A Chip* **2016**, *16*, 3493–3502. [CrossRef]
124. Liu, N.; Aymonier, C.; Lecoutre, C.; Garrabos, Y.; Marre, S. Microfluidic approach for studying CO2 solubility in water and brine using confocal Raman spectroscopy. *Chem. Phys. Lett.* **2012**, *551*, 139–143. [CrossRef]
125. Hosseini, H.; Guo, F.; Ghahfarokhi, R.B.; Aryana, S.A. Microfluidic fabrication techniques for high-pressure testing of microscale supercritical $CO_2$ foam transport in fractured unconventional reservoirs. *JoVE J. Vis. Exp.* **2020**, *161*, e61369. [CrossRef]
126. McCourt, T.A.; Zhou, F.; Bianchi, V.; Pike, D.; Donovan, J. Chip-Firing on a Graph for Modelling Complex Geological Architecture in $CO_2$ Injection and Storage. *Transp. Porous Media* **2019**, *129*, 281–294. [CrossRef]
127. Diao, Y.; Zhang, S.; Guo, J.; Li, X.; Zhang, H. Geological safety evaluation method for $CO_2$ geological storage in deep saline aquifer. *Geol. China* **2011**, *38*, 786–792.
128. Gerami, A.; Alzahid, Y.; Mostaghimi, P.; Kashaninejad, N.; Kazemifar, F.; Amirian, T.; Mosavat, N.; Warkiani, M.E.; Armstrong, R.T. Microfluidics for Porous Systems: Fabrication, Microscopy and Applications. *Transp. Porous Media* **2018**, *130*, 277–304. [CrossRef]
129. de Lima, V.; Einloft, S.; Ketzer, J.M.; Jullien, M.; Bildstein, O.; Petronin, J.-C. $CO_2$ Geological storage in saline aquifers: Paraná Basin caprock and reservoir chemical reactivity. *Energy Procedia* **2011**, *4*, 5377–5384. [CrossRef]
130. Zirrahi, M.; Hassanzadeh, H.; Abedi, J. Modeling of $CO_2$ dissolution by static mixers using back flow mixing approach with application to geological storage. *Chem. Eng. Sci.* **2013**, *104*, 10–16. [CrossRef]
131. Jafari, M.; Cao, S.C.; Jung, J. Geological $CO_2$ sequestration in saline aquifers: Implication on potential solutions of China's power sector. *Resour. Conserv. Recycl.* **2017**, *121*, 137–155. [CrossRef]
132. Uemura, S.; Matsui, Y.; Noda, A.; Tsushima, S.; Hirai, S. Nanosized $CO_2$ Droplets Injection for Stable Geological Storage. *Energy Procedia* **2013**, *37*, 5596–5600. [CrossRef]
133. Ho, T.H.M.; Tsai, P.A. Microfluidic salt precipitation: Implications for geological $CO_2$ storage. *Lab A Chip* **2020**, *20*, 3806–3814. [CrossRef]
134. Tirapu-Azpiroz, J.; Ferreira, M.E.; Silva, A.F.; Ohta, R.L.; Ferreira, R.N.B.; Giro, R.; Wunsch, B.; Steiner, M.B. Advanced optical on-chip analysis of fluid flow for applications in carbon dioxide trapping. In Proceedings of the Microfluidics, BioMEMS, and Medical Microsystems XX, San Francisco, CA, USA, 22–27 January 2022; SPE: Kuala Lumpur, Malaysia, 2022; Volume 11955, pp. 34–45.

**Disclaimer/Publisher's Note:** The statements, opinions and data contained in all publications are solely those of the individual author(s) and contributor(s) and not of MDPI and/or the editor(s). MDPI and/or the editor(s) disclaim responsibility for any injury to people or property resulting from any ideas, methods, instructions or products referred to in the content.

*Review*

# CO₂ Mineralized Sequestration and Assistance by Microorganisms in Reservoirs: Development and Outlook

Shumin Ni [1,2], Weifeng Lv [2,3,*], Zemin Ji [2] and Kai Wang [1,2]

1. University of Chinese Academy of Sciences, Beijing 100049, China; nishumin22@mails.ucas.ac.cn (S.N.); wangkaiz@petrochina.com.cn (K.W.)
2. Institute of Porous Flow & Fluid Mechanics, Chinese Academy of Sciences, Langfang 065007, China; jizemin@petrochina.com.cn
3. State Key Laboratory of Enhanced Oil Recovery, Research Institute of Petroleum Exploration and Development, Beijing 100083, China
* Correspondence: lweifeng@petrochina.com.cn

**Abstract:** The goals of carbon neutrality and peak carbon have officially been proposed; consequently, carbon dioxide utilization and sequestration technology are now in the limelight. Injecting carbon dioxide into reservoirs and solidifying and sequestering it in the form of carbonates after a series of geochemical reactions not only reduces carbon emissions but also prevents carbon dioxide from leaking out of the formation. Carbon dioxide mineralization sequestration, which has good stability, has been considered the best choice for large-scale underground $CO_2$ sequestration. To provide a comprehensive exploration of the research and prospective advancements in $CO_2$ mineralization sequestration within Chinese oil and gas reservoirs, this paper undertakes a thorough review of the mechanisms involved in $CO_2$ mineralization and sequestration. Special attention is given to the advancing front of carbon dioxide mineralization, which is driven by microbial metabolic activities and the presence of carbonic anhydrase within oil and gas reservoirs. The paper presents an in-depth analysis of the catalytic mechanisms, site locations, and structural attributes of carbonic anhydrase that are crucial to the mineralization processes of carbon dioxide. Particular emphasis is placed on delineating the pivotal role of this enzyme in the catalysis of carbon dioxide hydration and the promotion of carbonate mineralization and, ultimately, in the facilitation of efficient, stable sequestration.

**Keywords:** carbon dioxide sequestration; mechanism of mineralization; microbial-assisted mineralization; carbonic anhydrase

## 1. Introduction

Excessive emissions of greenhouse gases such as carbon dioxide have caused global temperatures to rise [1]. According to six major international datasets integrated by the World Meteorological Organization (WMO), the world's average temperature in 2021 is about 1.11 (±0.13) °C above pre-industrial levels (1850–1900) [2–4]. During the meeting of the United Nations General Assembly in September 2020, the Chinese government outlined its 2030 "carbon peak" and 2060 "carbon neutrality" goals in response to the global climate crisis [5]. Carbon dioxide capture, sequestration, and utilization (CCUS) are recognized as critical and viable means of achieving carbon neutrality and mitigating the greenhouse effect [6–8], as illustrated in Figure 1. Carbon dioxide sequestration represents a pivotal aspect of CCUS technology; primarily, it involves the injection of captured carbon dioxide into deep geological formations and storing it in geological bodies such as saline aquifers [9], depleted oil and gas reservoirs, and unexploited coal seams [10,11] through geological structures, capillary force binding, dissolution, and mineralization [12–14]. Among the four main sequestration methods, mineralization sequestration involves reacting carbon dioxide with minerals in the rock layer to create a stable secondary mineral and then permanently

storing it in the form of solid carbonate. Mineralized sequestration is considered the most secure and stable method for carbon sequestration due to its durability and the safe and stable storage of the products. Achieving mineralized sequestration through natural forces alone typically requires at least a century [15], as shown in Figure 2. Consequently, how to accelerate the mining process with the assistance of microorganisms has become a research hotspot. This paper comprehensively reviews the metabolic mechanism of microbial-induced mineralization, the historical research, the action sites, and the mechanisms of the key enzyme, carbonic anhydrase, in mineralization induction. Additionally, it addresses the existing challenges and prospects.

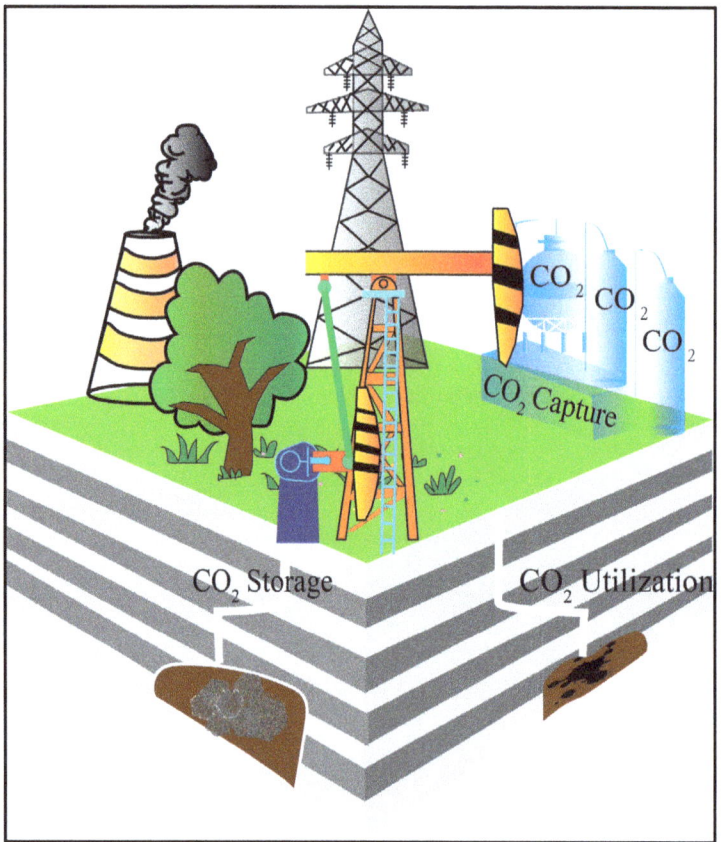

**Figure 1.** CCUS industrial concept diagram.

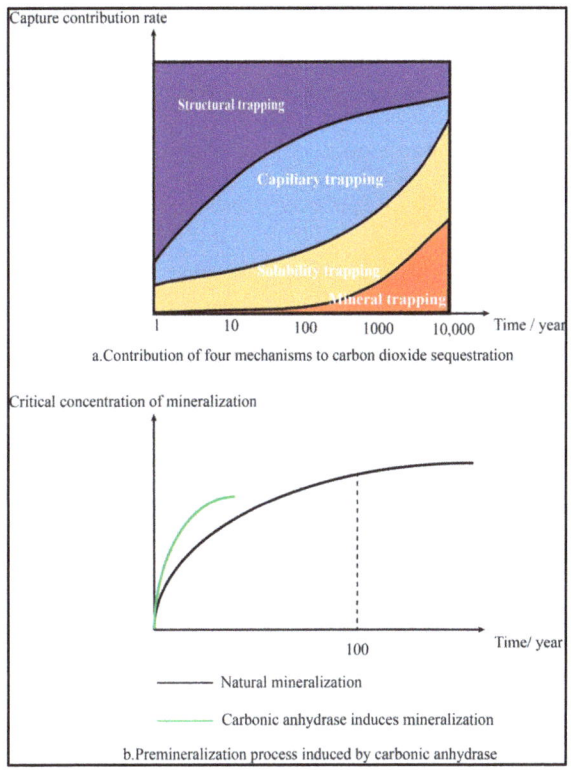

**Figure 2.** Carbonic anhydrase accelerates mineralization [16].

## 2. Sequestration Mechanism of Carbon Dioxide Mineralization

Currently, oil fields such as Daqing and Shengli are in the developmental phase; they are characterized by low permeability and a high water cut and present challenges in their operation [17–19]. Injecting carbon dioxide into oil reservoirs has been shown to enhance the recovery rate while securely sequestering carbon dioxide [20–22]. Figure 3 illustrates the schematic diagram of carbon dioxide geological sequestration. This technology has a double effect in fostering economic development and contributing to environmental protection. Therefore, $CO_2$ mineralization in an underground reservoir environment is considered to be one of the most promising and safe methods for reducing $CO_2$ emissions.

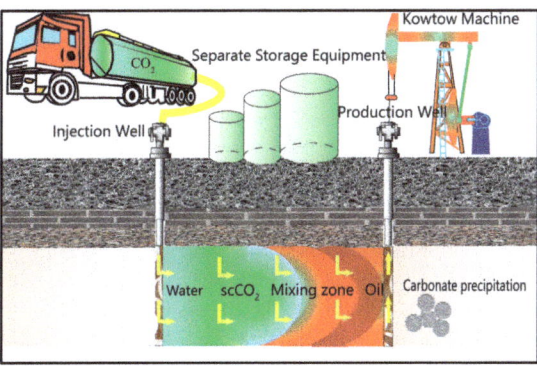

**Figure 3.** Schematic diagram of carbon dioxide geological sequestration.

The concept of mineralized utilization of carbon dioxide emerged in the 1990s; it primarily involves the carbonation of natural silicate minerals [23,24]. Carbonation can be broadly categorized into dry carbonation and wet carbonation [25]. In wet carbonation, water primarily serves as a medium to facilitate mineral dissolution and carbon dioxide dissolution.

Carbon dioxide mineralization in the reservoir belongs to in situ wet carbonation. During the process of carbon dioxide injection into the target layer (usually at a depth of 800–3000 m), both the temperature and the pressure exceed the critical point of the carbon dioxide phase diagram (31.1 °C and 73.9 bars), as shown in Figure 4, and the carbon dioxide is in a supercritical state [26].

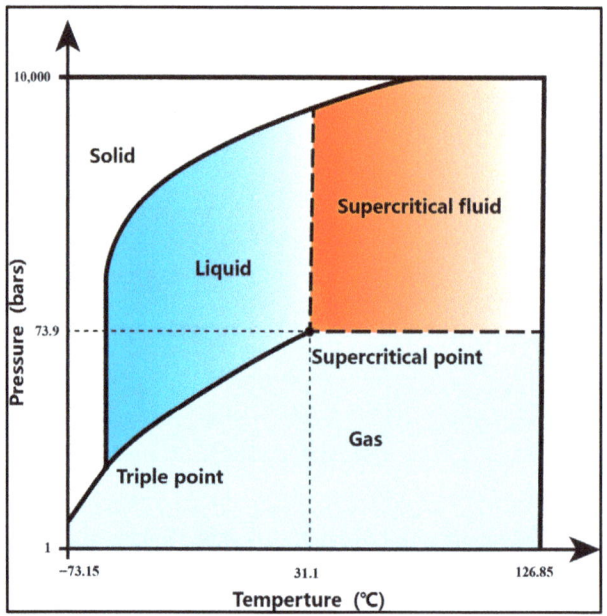

**Figure 4.** Carbon dioxide phase diagram (Copyright © 1999 Chemical Logic Corporation, 99 South Bedford Street, Suite 207, Burlington, MA 01803 USA. All rights reserved).

Supercritical carbon dioxide is introduced into the deep reservoir, following the principle of similar phase dissolution. During this process, a portion of the carbon dioxide dissolves in the crude oil while another portion dissolves in the formation water. Notably, the solubility of carbon dioxide in crude oil is significantly higher than in water. A fraction of the $CO_2$ undergoes hydration within the formation, resulting in the formation of carbonic acid, which subsequently ionizes into $CO_3^{2-}$ and $HCO_3^-$. Biogeochemical processes eventually convert some of the $CO_2$ into solid carbonate, while a minute amount remains in a free state. The different occurrence states of $CO_2$ within the reservoir are illustrated in Figure 5 below.

Supercritical carbon dioxide undergoes a reaction with water, resulting in the production of carbonic acid. This carbonic acid, in turn, engages in distinct dissolution reactions with various types of minerals [28–30]. Leaching cations ($Ca^{2+}$, $Mg^{2+}$, $Fe^{2+}$, etc.) combine with carbonic anions produced by carbonic acid ionization to produce carbonate precipitation. Under specific conditions, this combination leads to the precipitation of carbonates, facilitating the enduring sequestration of carbon dioxide in a solid state. The whole process can be roughly divided into three stages: carbon dioxide dissolution and ionization, mineral dissolution, and carbonate mineral deposition.

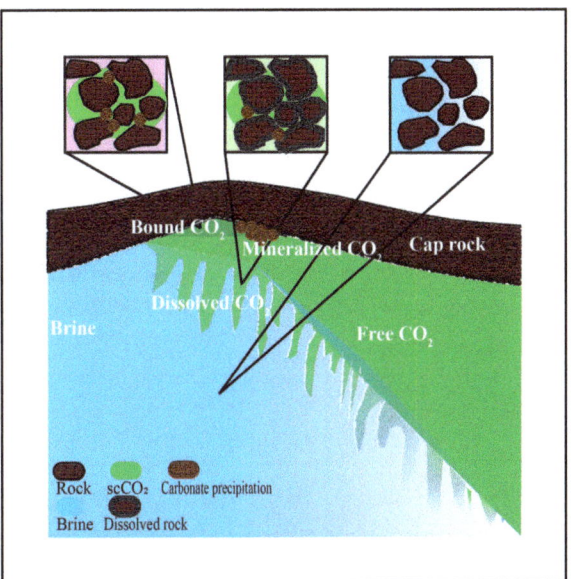

**Figure 5.** Carbon dioxide in the reservoir occurrence state [27].

(1) Dissolution and ionization of carbon dioxide

Upon the entry of supercritical carbon dioxide (scCO$_2$) into the formation, a density disparity between the scCO$_2$ and the formation water is evident. This prompts the initiation of a gravity-driven flow in the longitudinal direction, causing some scCO$_2$ to ascend and react with the cap rock. Due to the higher density of the scCO$_2$ flow compared to the formation of water, a significant portion of the scCO$_2$ settles, resulting in natural convection between the scCO$_2$ and the brine aquifer during this phase [31]. Subsequently, as the gravity flow reaches the interface of the scCO$_2$–brine phase, the scCO$_2$ undergoes molecular diffusion and dissolves into the brine aquifer due to concentration differences. Concurrently, at the phase interface, the CO$_2$ undergoes a hydration reaction, forming carbonic acid water, and continues to engage in ionization reactions, generating free protons. The hydration reaction of carbon dioxide exhibits a pH dependence, presenting roughly two reaction forms in different pH ranges. The reaction formulas are as follows:

$$CO_2 + H_2O \leftrightarrow H_2CO_3 \leftrightarrow H^+ + HCO_3^- \text{ (pH < 8)} \tag{1}$$

$$CO_2 + OH^- \rightarrow HCO_3^- \text{ (pH > 10)} \tag{2}$$

Considering that the pH of reservoir groundwater typically remains within a neutral or weakly alkaline range within the oil–gas reservoir environment, the carbon dioxide hydration reaction aligns with Equation (1). The kinetic constant of the reaction ranges from 0.026 to 0.044 s$^{-1}$ at a room temperature of 25 °C [32,33]. The hydration rate of carbon dioxide is relatively slow. To expedite the hydration reaction of carbon dioxide, domestic and international researchers primarily employ catalysts such as phosphate buffers, metal nanoparticles (e.g., nickel nanoparticles or NiNPs), and carbonic anhydrase [34–36].

(2) Mineral corrosion

The surfaces of clay minerals, carbonate minerals, and feldspar mineral particles are enveloped by a carbonate water film. This film leads to the leaching of alkaline earth metal ions such as calcium, magnesium, and iron, resulting in the acquisition of Ca$^{2+}$, Mg$^{2+}$, Fe$^{2+}$,

$OH^-$, $HCO_3^-$, $CO_3^{2-}$, and Si-O groups [37]. The mineral dissolution process is depicted in Figure 6, and the reaction equations are presented below:

$$M_xSi_yO_{x+2y}(s) + 2H^+(aq) \rightarrow M^{2+}(aq) + SiO_2(s) + H_2O(l) \tag{3}$$

$$Calcite + H^+ \leftrightarrow Ca^{2+} + HCO_3^- \tag{4}$$

$$Dolomite + H^+ \leftrightarrow Ca^{2+} + Mg^{2+} + HCO_3^- \tag{5}$$

$$Siderite + H^+ \leftrightarrow Fe^{2+} + HCO_3^- \tag{6}$$

$$Magnesite + H^+ \leftrightarrow Mg^{2+} + HCO_3^- \tag{7}$$

$$Ca - felspar + 2H^+ + H_2O \leftrightarrow Ca^{2+} + Al_2Si_2O_5(OH)_4 \tag{8}$$

$$Na - felspar + 2H^+ + H_2O \leftrightarrow 2Na^+ + Al_2Si_2O_5(OH)_4 + 4SiO_2(aq) \tag{9}$$

$$K - feldspar + 2H^+ + H_2O \leftrightarrow 2K^+ + Al_2Si_2O_5(OH)_4 + 4SiO_2(aq) \tag{10}$$

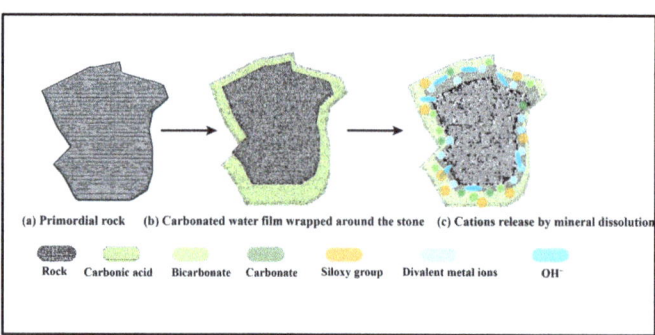

**Figure 6.** Mineral dissolution process.

Through laboratory static reaction experiments, domestic and foreign researchers have found that carbonate minerals represented by dolomite, feldspar minerals represented by potassium albite, and clay minerals such as chlorite show different degrees of dissolution in the early stage of the reaction, indicating that the degree of dissolution of minerals depends on their chemical structure [38–41]. Due to the limited presence of carbonate rocks and the comparatively low reactivity of feldspar minerals in domestic reservoir minerals, the dissolution of minerals is generally considered to be the rate-limiting step in the mineral carbonation process. Literature indicates that factors such as reservoir temperature, pH [42], and mineral surface area influence the dissolution rate [37,43]. In general, higher temperatures, lower pH values, and the larger specific surface areas of minerals lead to faster dissolution rates and higher degrees of mineral dissolution within the same time frame. Furthermore, research has shown that the addition of a salt solution (e.g., a mixed solution of sodium chloride and sodium bicarbonate or a mixed solution of potassium nitrate and sodium nitrate) during the reaction can act as a catalyst, accelerating the leaching of alkaline earth metal ions to a certain extent [44,45].

(3) Carbonate mineral deposition

As the minerals continue to dissolve and consume protons, the pH of the environment gradually rises, reaching a slightly alkaline level. In this alkaline environment, when carbon anions (carbonates and bicarbonates) within the environment reach the supersaturated state, they combine with cations such as $Ca^{2+}$, $Mg^{2+}$, $Fe^{2+}$, and others and form the respective carbonate precipitates. The equations are listed as follows:

$$M^{2+}(aq) + CO_3^{2-}(aq) \rightarrow MCO_3(s) \tag{11}$$

$$Mg_2SiO_4 + 2CO_2 + 2H_2O \rightarrow 2MgCO_3 + H_4SiO_4 \tag{12}$$

$$Mg_3Si_2O_5(OH)_4 + 3CO_2 + 2H_2O \rightarrow 3MgCO_3 + 2H_4SiO_4 \tag{13}$$

$$CaSiO_3 + CO_2 + H_2O \rightarrow CaCO_3 + H_2SiO_3 \tag{14}$$

Indoor experiments involving carbon dioxide–water–rock interactions have been conducted to detect and analyze sedimentary minerals. These experiments have revealed that the sedimentary minerals mainly consist of carbonate rocks, predominantly calcite and dolomite. Additionally, a smaller proportion consists of silicate rocks, including kaolin, dolomite, and quartz [40,46,47].

Regarding the three phases of geochemical reactions, the literature has consolidated the key factors influencing mineral deposition efficiency, encompassing the concentration of carbon anions in the solution, the concentration of divalent metal cations like calcium and magnesium, and the availability of mineral nucleation sites [48]. From a reaction kinetics perspective, the rate of carbonate deposition is directly influenced by the concentrations of carbon anions and divalent metal cations such as calcium, magnesium, and iron in the solution. Divalent metal cations primarily stem from groundwater composition and the dissolution and weathering of rock minerals [49]. The existing literature data demonstrate marked distinctions in the mineral composition of reservoir rocks between those of typical domestic regions and those of foreign regions, as outlined in Table 1. Among these minerals, the dissolution of both carbonate minerals and clay minerals can supply divalent metal ions. However, the dissolution and the re-precipitation of carbonate minerals do not significantly contribute to carbon dioxide sequestration. Therefore, the content of clay minerals in reservoir rock holds substantial importance in the context of carbon dioxide mineralization and sequestration.

Most of the reservoirs in China are terrestrial and are characterized by quartz clastic minerals and carbonate minerals, which are their principal components. In contrast, clay minerals constitute a relatively minor portion compared to foreign marine-phase reservoirs [50,51]. Among these minerals, the dissolution of both carbonate minerals and clay minerals can supply divalent metal ions. However, the dissolution and the re-precipitation of carbonate minerals do not significantly contribute to carbon dioxide sequestration. Therefore, the content of clay minerals in reservoir rock holds substantial importance in the context of carbon dioxide mineralization and sequestration.

To maximize the sequestration of $CO_2$ in the form of solid carbonates, it is essential for the reservoir rock to possess a relatively high clay mineral content, as this indicates richness in silicate minerals (e.g., chlorite, illite, montmorillonite, kaolinite, etc.). These silicate minerals neutralize acids and provide the necessary raw materials for precipitation by providing $Ca^{2+}$, $Mg^{2+}$, and $Fe^{2+}$. Consequently, achieving effective carbonation of natural silicate minerals is most feasible in porous minerals rich in divalent metal cations, such as basalts and mantle peridotites [61–63]. An exemplar of implementation is the Carbon Fix project in Iceland, where carbon dioxide was injected into the ground in the form of carbonated water. Remarkably, over 95% of the carbon dioxide mineralized into carbonate minerals within 2 years and completed in situ mineralization [64].

Constrained by reservoir conditions, mineralized carbon dioxide sequestration in China's oil reservoirs faces significant challenges, which are primarily characterized by a slow natural mineralization rate and a lengthy cycle. Furthermore, the indoor in situ static simulation experiments conducted thus far have predominantly remained at the dissolution stage of rock minerals. To expedite the mineralization process, both domestic and international researchers commonly opt for the introduction of microorganisms.

Table 1. Rock mineral composition of typical regions at home and abroad.

| Region | Felsic Mineral Content/% | Carbonate Mineral Content/% | Clay Mineral Content/% | Reference |
|---|---|---|---|---|
| Cambrian Yurtus Formation, Tarim Basin, China | 21.2–94.8<br>57.8<br>(Mean) | 0.1–69.6<br>16.9<br>(Mean) | 0.8–48.5<br>15.7<br>(Mean) | [52] |
| Cretaceous Qingshankou Formation, Songliao Basin, China | 3.7–73.4<br>49.4<br>(Mean) | 0.1–93.5<br>13.1<br>(Mean) | 2.5–49.6<br>33.6<br>(Mean) | [53] |
| Permian Luchaogou Formation, Junggar Basin, China | 0.1–73.5<br>49.4<br>(Mean) | 0.1–91.5<br>32.8<br>(Mean) | 0.8–48.5<br>11.9<br>(Mean) | [54] |
| Paleogene Kongdian Formation, Cangdong Sag, Bohai Bay Basin, China | 0.1–56.0<br>36.7<br>(Mean) | 0.1–95.0<br>32.8<br>(Mean) | 0.8–48.5<br>15.7<br>(Mean) | [55] |
| Permian Longtan Formation, Southeast Xiang-tan Depression, China | 9.0–43.0<br>27.0<br>(Mean) | 23.0–50.0<br>38.0<br>(Mean) | 0.8–48.5<br>15.7<br>(Mean) | [56] |
| Devonian–Mississippian of the Western Canadian Basin | 21.2–94.8<br>57.8<br>(Mean) | 0.1–85.4<br>4.7<br>(Mean) | 0.8–48.5<br>15.7<br>(Mean) | [57] |
| Barnett Shale, Fort Worth Basin, USA | 51.9 | 8.1 | 35.0 | [58] |
| Upper Jurassic Haynesville Shale, Gulf of Mexico Basin, USA | 31.8 | 22.7 | 45.5 | [59] |
| West Philippine Sea | 22.7–75.4<br>66.3<br>(Mean) | 14.6–41.5<br>28.0<br>(Mean) | 5.3–54.3<br>21.5<br>(Mean) | [60] |

## 3. Mechanism of Microbial-Assisted Mineralization

Microorganisms play critical roles in elemental cycling, as well as in the transformation of metals and minerals, decomposition, and the formation of soil and sediment [65]. Reservoirs act as natural geological bioreactors [66], harboring a diverse array of microorganisms, which primarily include fermenting bacteria, sulfate-reducing bacteria, nitrate-reducing bacteria, iron-reducing bacteria, methanogens, etc. [67]. These microorganisms are directly or indirectly involved in the nucleation, crystallization, and growth processes of minerals during mineral formation. Academics classify the microbial involvement in mineralization into two main categories based on the mode of microbial action: microbial-induced mineralization and microbial-controlled mineralization. Microbial-controlled mineralization refers to the process in which microbial cells and metabolic products are directly involved in the carbonation process, resulting in microbial minerals with distinctive properties. On the other hand, microbial-induced mineralization involves microbes altering the local microenvironment through metabolic activities, creating physicochemical conditions that are conducive to mineral precipitation and thus lead to the formation of induced microbial minerals. The process of microbial-assisted carbon dioxide mineralization in the in situ environment of oil reservoirs is considered a microbial-induced process. As early as 1973, Boquest et al. isolated 210 strains of Proteus vulgaris from soil microorganisms, demonstrating the ability to induce calcium carbonate precipitation generation through metabolic activities [68]. Subsequently, numerous experiments have highlighted the fact that microbial-induced mineralization is mainly achieved through metabolic activities [69], such as urea hydrolysis [70,71], denitrification [72,73], trivalent iron reduction [74], sulfate reduction [75], and organic matter degradation. Table 2 lists the main microorganisms involved in mineralization.

Table 2. Major microorganisms that assist mineralization.

| Types of Microorganisms that Assist Mineralization | Typical Microbial Representative | Main Mechanism | Environment | Reference |
|---|---|---|---|---|
| Sulfate-reducing bacteria | Acinetobacter calcoaceticus SRB4 | Consumption of specific electron donors, forming a metal sulfide precipitate | Anaerobic environment rich in organic matter, calcium, and sulfate; can survive in oil reservoirs | [76] |
| Iron-reducing bacteria | Shewanella oneidensis MR-4 | Consumption of specific electron donors, adjusting Eh value, promoting siderite precipitation | Anaerobic environment; most of them are thermophilic bacteria, and a few can survive in oil reservoirs | [77] |
| Urea-decomposing bacteria | Thermoanaerobacterium | Decomposition of urea | Aerobic environment | [78] |
| Denitrifying bacteria | Pseudomonas stutzeri | Consumption of specific electron donors | Anaerobic environment; can survive in oil reservoirs | [79] |
| Methanogenic bacteria | Methanococcales | Oxidization of methane, producing carbon anions | Anaerobic environment; can survive in oil reservoirs | [55,80] |
| Photosynthetic microorganisms | Cyanobacteria | Consumption of $CO_2$, promoting carbon anion generation | Aerobic environment, light conditions | [81] |
| Microorganisms producing carbonic anhydrase | Sporosarcina Kluyver | Accelerating $CO_2$ hydration, increasing carbon anion concentration | Aerobic environment | [82] |

(1) Reservoir microbial function in carbon dioxide mineralization

Reservoir microorganisms play a pivotal role in $CO_2$ mineralization and sequestration; this role is primarily manifested in two key aspects: the regulation of environmental factors (e.g., pH, Eh value) and the adherence to minerals to offer nucleation sites.

Reservoir microorganisms engage in metabolic activities that utilize crude oil components in the environment, humic substances in minerals, and sulfates in the brine aquifer, resulting in the generation of organic acids (formic acid, acetic acid, citric acid, etc.) and $H_2S$, among other acidic substances. These acidic substances can regulate the environmental pH by hydrolyzing and producing protons. Concurrently, the generated protons can react with feldspar minerals (e.g., potassium feldspar and sodium feldspar) and carbonate rock minerals (e.g., calcite and dolomite) to liberate additional free metal ions (e.g., $Ca^{2+}$, $Mg^{2+}$, $Fe^{2+}$, etc.), providing raw materials for the formation of carbonate mineral precipitates. Moreover, microorganisms in oil reservoirs encompass a significant population of electron acceptor-reducing bacteria. These bacteria, which are categorized based on their electron acceptor redox potential from high to low, include nitrate-reducing bacteria, iron-reducing bacteria, and sulfate-reducing bacteria. Among them, nitrate-reducing bacteria are responsible for reducing nitrate to nitrogen or ammonium, and some have the ability to metabolize monosaccharides, polysaccharides, and volatile fatty acids into small organic acids and gases such as nitrogen and carbon dioxide. Iron-reducing bacteria utilize hydrogen, carbon dioxide, and certain small organic acids as electron donors. These bacteria utilize extracellular Fe(III) as the terminal electron acceptor, reducing Fe(III) to Fe(II) through the oxidation of organic matter. The presence of iron-reducing bacteria facilitates the reduction of Fe(III) to Fe(II), which regulates the environmental redox value (Eh) and creates a conducive environment for the generation of rhodochrosite precipitation. The potential metabolic pathways of reservoir mineralization and precipitation are depicted in Figure 7.

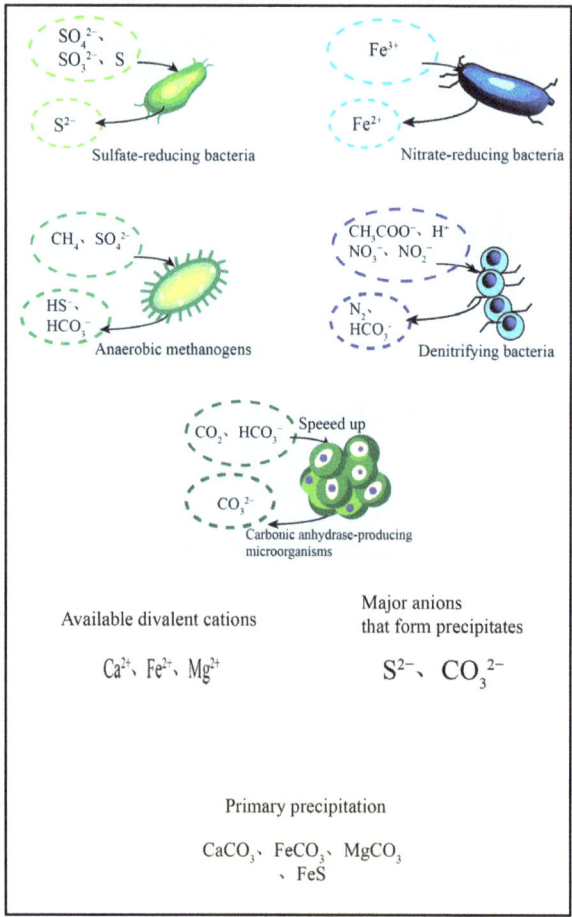

**Figure 7.** Possible metabolic pathways of reservoir mineralization and precipitation.

More importantly, microorganisms create a favorable microenvironment for precipitation, and their metabolism produces extracellular polymers (EPSs) that can provide the nucleation sites necessary for nucleation and accelerate the growth of carbonate crystals, as shown in Figure 8. Reservoir microorganisms predominantly exist in the form of microbial communities within the scCO$_2$–water–rock system, which can synergistically resist extreme environments and further develop into biofilms [65]. In the scCO$_2$–water–rock system, where sandstone serves as a porous medium, microbial retention and attachment are highly favorable. Microorganisms proliferate on the moist surface of the porous reservoir medium, secreting viscous extracellular polymeric substances (EPSs) that culminate in biofilm formation. Due to the presence of various organic functional groups, such as carboxyl and hydroxyl, on the biofilm surface, it exhibits a certain degree of hydrophobicity [83–85]. The interplay of the electrostatic force, the van der Waals force, and hydrophobicity leads to a robust and stable adhesion between microorganisms and porous media. Additionally, the extracellular polymers typically consist of polar amino acids such as metabolically produced glutamic acid and aspartic acid. These amino acids are highly prone to proton dissociation under alkaline conditions, which results in an overall negative charge on the biofilm surface. This charge facilitates the adsorption of cations such as calcium, magnesium, and iron, providing essential nucleation sites for the formation of minerals like dolomite, calcite, and ferro-dolomite [61,86,87]. In parallel with extracellular mineralization, intracellular bio-mineralization occurs in certain prokaryotic bacteria, algae, etc. Benzerara et al. found that amorphous calcium carbonate can be formed by some cyanobacterial cells [88]. Yan Huaxiao and other researchers utilized Bacillus subtilis Daniel-1 to study biomineralization and found that intracellular and extracellular mineralization can occur simultaneously and that the two have a synergistic effect under certain conditions [89].

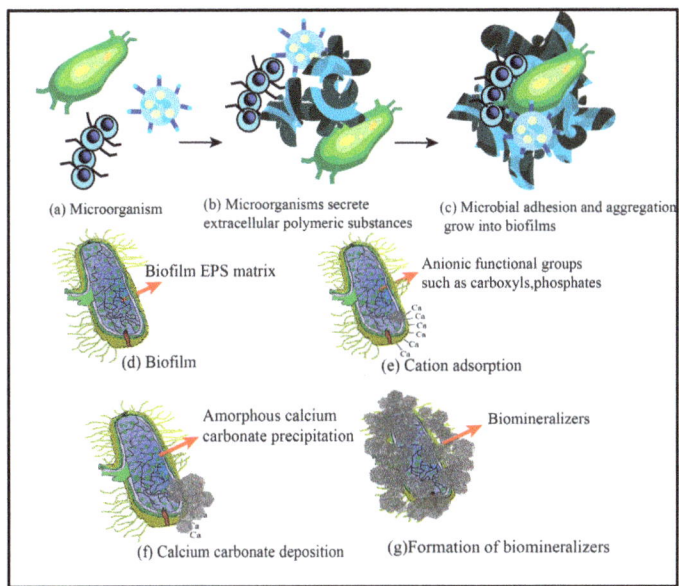

**Figure 8.** Microorganisms provide nucleation sites [90].

(2) Key microbial enzymes in $CO_2$ mineralization

Currently, the critical enzymes implicated in mineralization and sequestration primarily encompass urease and carbonic anhydrase, as demonstrated in Table 3. Urease chiefly elevates the environmental pH by decomposing urea, whereas carbonic anhydrase augments the $CO_3^{2-}$ concentration by expediting $CO_2$ hydration. However, most urea-producing and urea-decomposing bacteria are aerobic bacteria, which do not exist in the oil reservoir environment. Furthermore, the ammonia generated during urea decomposition poses new environmental risks. Consequently, this paper places emphasis on carbonic anhydrase as the auxiliary enzyme for $CO_2$ mineralization in the reservoir. The core function of carbonic anhydrase in accelerating $CO_2$ mineralization in reservoirs is to achieve rapid and efficient conversion between $CO_2$ and bicarbonate and carbonate. Notably, $CO_2$ hydration stands as a pivotal component of this conversion and represents one of the critical rate-limiting steps in the mineralization process, alongside mineral dissolution. The reaction rates of both processes are somewhat dependent on the ambient pH value. The mineralization of carbonate deposition occurs in weakly alkaline environments, and the kinetics of $CO_2$ hydration in mildly alkaline solutions are notably sluggish, merely 0.026–0.044 $s^{-1}$ at room temperature [91], during which the rate of carbonic anhydrase-catalyzed hydration is dramatically increased by the addition of carbonic anhydrase, with turnover numbers of up to $1.4 \times 10^6$ $s^{-1}$ at 25 °C and pH 9, surpassing the natural rate by $10^8$ times [92,93]. Thus, carbonic anhydrase plays an important role as a key enzyme catalyzing the $CO_2$ hydration reaction during $CO_2$ mineralization.

Table 3. Key enzymes involved in carbon dioxide mineralization.

| Key Enzyme for Carbon Dioxide Mineralization | Main Role | Catalytic Rate | Maintain Active Environment | Application | Reference |
|---|---|---|---|---|---|
| Urease | Decomposes urea and increases the pH | - | pH 7.0, 40 °C catalyzed the conversion of urea to ammonium carbonate; the optimal pH is 7.4. | Ecological restoration, soil reinforcement | [78] |
| Carbonic anhydrase | Accelerates $CO_2$ hydration and increases $CO_3^{2-}$ ion concentration | $K_{cat}$ $10^4$–$10^6$/s | The pH value between 4.0 and 9.0 and the temperature below 65 °C can maintain high activity and stability. | Fixed $CO_2$, biological monitoring | [94] |

### 3.1. Catalytic Mechanism of Carbonic Anhydrase

Carbonic anhydrase (CA) constitutes a class of metalloenzymes that utilize metal ions such as $Zn^{2+}$, $Co^{2+}$, $Fe^{2+}$, $Cd^{2+}$ ($\zeta$ – CAcontaining), $Cu^{2+}$, and other related metal ions; the active sites are responsible for the participation in the reversible hydration of carbon dioxide [91,95]. The discovery of carbonic anhydrase traces back to the 1930s, when scientists like Meldrum, in their investigations into carbon dioxide transport in the blood, identified a protein in the blood that was responsible for converting carbon dioxide into carbonic acid [96]. Since then, scientists have progressively detected carbonic anhydrase in various mammals, plants, and microorganisms [97–99]. Carbonic anhydrases can be broadly categorized into eight major groups based on amino acid sequence differences; these are α, β, γ, δ, ζ, η, θ, and ᴄ types [100–104], of which α, β, and γ are widely distributed in plants and animals, algae, bacteria, and other organisms, with α-carbonic anhydrase alone exhibiting the presence of at least a dozen isozymes.

Currently, the most detailed structural analysis of carbonic anhydrase is that of the isoenzyme CAII within α-CA. Due to its robust catalytic activity, the predominant carbonic anhydrase sequestered by carbon dioxide mineralization in reservoirs belongs to the α type. The structure of α-CA is depicted in Figure 9 [105]. The active region of carbonic anhydrase comprises multiple peptide chains shaped by β-folding, α-helix, and other structural elements within the cavity. The catalytic center of the cavity is anchored by a $Zn^{2+}$ positioned approximately 15Å ($1.5 \times 10^{-9}$ m) from the bottom of the cavity. Simultaneously, four coordination bonds extend around the zinc ion, connecting three amino acid residues (His-94, 96, 119) and a water molecule or $OH^-$, which together form a nearly symmetric tetrahedral coordination geometry [106,107]. When the environmental pH becomes neutral or weakly basic, the ligand $H_2O$ attached to the site at the time deprotonates, resulting in the formation of a hydroxide ligand (-$OH^-$). The genesis of this tetrahedral structure primarily hinges on the binding force between the zinc ions and the nitrogen atoms in the imidazole group of the side chain of the three histidine residues, as well as the oxygen atoms in the water molecule. Additionally, the cavity environment near the catalytic site of $Zn^{2+}$ can be roughly divided into hydrophobic and hydrophilic regions. The hydrophobic side chain is composed of branched amino acids such as valine (Val143,121), tryptophan (Trp-209), and leucine (Leu-198), which surround the active site to form a hydrophobic pocket responsible for $CO_2$ fixation and act as a catalytic binding site for $CO_2$ [108]. On the other hand, the hydrophilic portion consists mainly of histidine (His-64), threonine (Thr-199), glutamic acid (Glu-106), and water molecules, forming a proton channel in which histidine plays the role of a proton shuttle, which provides a chain of water molecules to remotely control the E·Zn$H_2O$ deprotonation process [109].

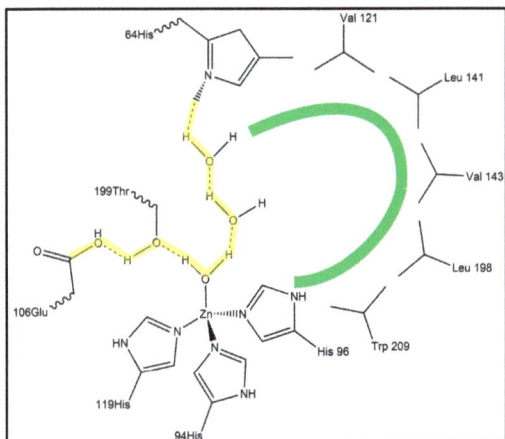

**Figure 9.** Active region structure of carbonic anhydrase [105]. Yellow represents the proton transport chain and green represents the hydrophobic pocket.

A large number of studies have revealed the structure of α-CA with zinc as the active center and have provided insights into its catalytic mechanism [95,97,99,110]. During its active state, carbonic anhydrase readily binds to $OH^-$ to form $E \cdot ZnOH^-$. Subsequently, $E \cdot ZnOH^-$ launches a nucleophilic attack on the carbon atom of the carbon dioxide molecule surrounded by a hydrophobic pocket and combines to form $E \cdot ZnHCO_3^-$. Later, the water molecule reacts with the zinc bicarbonate, $E \cdot ZnHCO_3^-$. Once the bicarbonate is displaced into the solution, the enzyme is inactivated as it combines with water to produce $E \cdot ZnH_2O$. The catalytic process of carbonic anhydrase is illustrated in Figure 10.

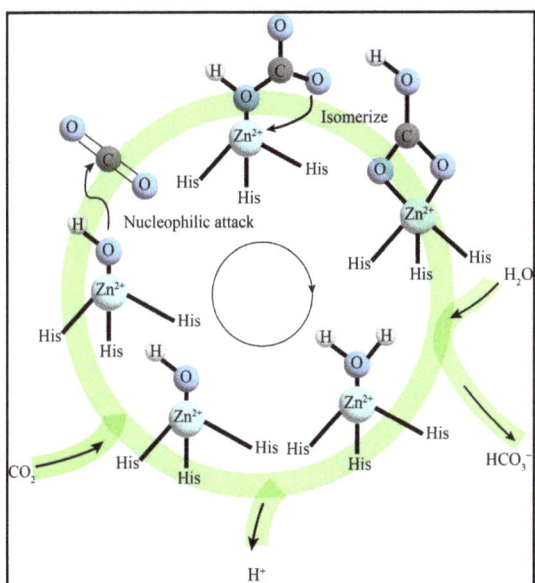

**Figure 10.** Catalytic process of carbonic anhydrase [111].

To restore carbonic anhydride enzyme activity, the deprotonation of $E \cdot ZnH_2O$ has to be completed [95,112,113]. The $E \cdot ZnH_2O$ deprotonation process, during which the proton transfer rate is only $10^6$ s$^{-1}$, represents the key rate-limiting step in the carbonic anhydrase catalytic process and is crucial for ensuring the sustained stability of the catalytic process. The proton is transferred to His64 in three steps along the hydrogen-bonded water molecule chain. This protonated His64 side chain undergoes rotation to maximize its exposure to the solvent outside the enzyme. Upon the release of $H^+$, His64 rotates back to its initial position. The literature suggests that the proton exchanger during deprotonation may be a buffer in the solution, in addition to water, as depicted in Equation (20). However, the contribution of carbonic anhydrase to the catalytic efficiency of the deprotonation process diminishes, resulting in a reduced catalytic effect [110]. The equations describing the carbonate precipitation process induced by carbonic anhydrase are as follows:

$$E \cdot ZnH_2O \leftrightarrow E \cdot ZnOH^- + H^+ \tag{15}$$

$$E \cdot ZnOH^- + CO_2 \leftrightarrow E \cdot ZnHCO_3^- \tag{16}$$

$$E \cdot ZnHCO_3^- + H_2O \leftrightarrow E \cdot ZnH_2O + HCO_3^- \tag{17}$$

$$HCO_3^- + OH^- \leftrightarrow CO_3^{2-} + H_2O \tag{18}$$

$$Cell + Ca^{2+} \leftrightarrow Cell - Ca^{2+} \tag{19}$$

$$E \cdot ZnH_2O + buffer \rightarrow E \cdot ZnOH^- + bufferH^+ \tag{20}$$

Overall, carbonic anhydrase-producing bacteria grow and multiply in the reservoir environment and secrete carbonic anhydrase to catalyze carbon dioxide hydration, thus promoting the conversion between carbon dioxide and bicarbonate and carbonate and increasing the $CO_3^{2-}$ concentration in the formation water. Meanwhile, carbonic acid formed by carbon dioxide hydration dissolves minerals such as silicates, releasing divalent cations, which are adhered to the surface of microbial membranes through electrostatic forces. Eventually, the cations combine with locally oversaturated carbonate ions, culminating in carbonate precipitation, and thereby achieve the solid-state mineralization of carbon dioxide. The process of carbonate deposition induced by carbonic anhydrase-producing microorganisms is shown in Figure 11.

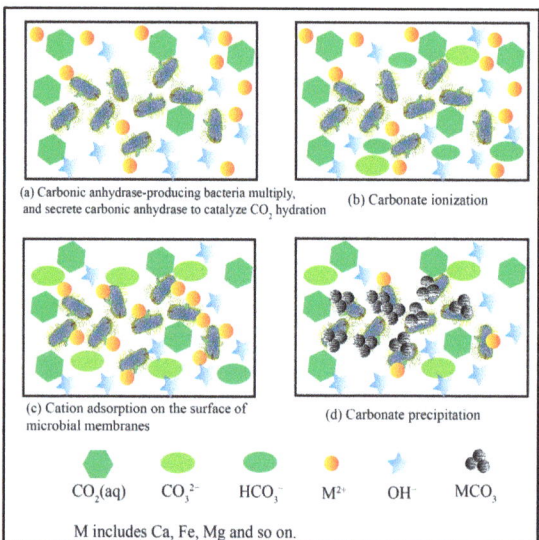

**Figure 11.** Mechanistic modeling of carbonate deposition induced by carbonic anhydrase-producing microorganisms.

*3.2. Carbonic Anhydrase Catalytic Activity and Stability Factors*

Carbonic anhydrase (CA), which is renowned as an environmentally friendly and highly efficient catalyst for carbon dioxide hydration, is widely used in various industrial processes for carbon dioxide abatement. To align with the technical requirements of industrial applications, CA must endure severe environments, including, but not limited to, those with high-temperature and high-salinity conditions. Thus, a key research focus is on the preservation of its catalytic activity and stability under extreme conditions [114,115]. Some studies have underscored that the catalytic activity and thermal stability of carbonic anhydrase are largely influenced by the enzyme structure [104,116].

The function and catalytic activity of proteases are fundamentally controlled by the sequence of amino acids comprising the protease (enzyme primary structure). The peptide chain adopts stable secondary structures such as α-helix and β-folding, forming a foundation upon which the peptide chain further folds to attain a three-dimensional or four-dimensional protease structure. Most carbonic anhydrases exist in forms like dimers, trimers, and tetramers. Upon the three-dimensional structure analysis of several isozymes (I, II, III, IV, V, XII, XIII, and XIV), a notable structural resemblance is observed. They all exhibit a tetrahedral-like structure, with the metal ions located within the interior of the tetrahedral cavity, the three conserved histidine residues, and a water molecule or hydroxyl as the four ligands of the tetrahedral structure. The tetrahedral cavity structure can be roughly divided into two parts according to its distribution of amino acid sequences; the hydrophobic part is mostly composed of nonpolar amino acids such as alanine (Ala121, 135), valine (Val 207), phenylalanine (Phe91), leucine (Leu131, 138, 146, 109), and proline (Pro201, 202), whereas the hydrophilic portion of the cavity is mainly composed of polar amino acids such as histidine (His64, 67, 200), aspartic acid (Asn69), glutamine (Gln92), threonine (Thr199), and tyrosine (Tyr7). The hydrophobicity of these residues plays a crucial role in catalysis. Consequently, some researchers have utilized amino acid modification to reinforce the hydrophobic pocket to minimize the exposure of hydrophobic groups in aqueous solutions, thus enhancing the structural stability and enzymatic activity. Furthermore, the arrangement of amino acids and the presence of disulfide bonding also influence the enhancement of stability and catalytic activity. In CAII, which is characterized by high catalytic activity, the histidine residue His64 primarily governs the conversion of $Zn^{2+}$-bound water molecules into hydroxide ions before catalysis. This histidine residue is a pivotal component of the proton shuttle process. The replacement of this histidine residue with other residues, such as Phe66, Tyr64, Glu207, etc., significantly impacts the catalytic activity of the CA enzyme. Additionally, the formation of a disulfide bond between the two conserved cysteine residues can reduce the degree of freedom for conformational changes in the protease, enhancing thermal stability. Consequently, researchers have explored methods to enhance protease surface densification by introducing disulfide bonds, among other approaches, to bolster rigidity and thermal stability.

*3.3. Application of Carbonic Anhydrase Immobilization in $CO_2$ Mineralization*

Carbon dioxide geological sequestration is a crucial strategy for meeting carbon emission reduction targets [8]. Carbon dioxide in geological formations exists in various states, including dissolved, free, bound, and mineralized states, among which the mineralized state in the form of carbonate is the most secure, stable, and sustainable. The mineralization process is facilitated by carbonic anhydrase (CA), an efficient and environmentally friendly catalyst that accelerates carbonate deposition by catalyzing $CO_2$ hydration. However, its protease activity is affected by a variety of factors, including temperature, pressure, pH, ionic strength, and the presence of metal ions [117,118]. Even under optimal conditions, the enzyme's catalytic activity gradually decreases as the reaction progresses [119]. To maintain the tolerance and catalytic activity of carbonic anhydrase in the anaerobic environment at the high temperature and pressure in oil and gas reservoirs [114], domestic and foreign researchers have pursued immobilization techniques, binding the enzyme in a specialized phase to enhance its suitability within oil reservoirs.

Common enzyme immobilization methods mainly include adsorption, embedding, covalent binding, and cross-linking [117,120–122], which are shown in Figure 12. Various types of carrier materials are used for immobilization and encompass organic and inorganic materials such as polyacrylamide gel [123], polyurethane foam, diatomaceous earth [124], and silica [125]. Additionally, newer materials like magnetic nanoparticles and hollow nanofiber membranes are being explored [126]. Scientists, such as B. Ray and others, have embedded and immobilized carbonic anhydrase on polyacrylamide gel to enhance the thermal stability; carbonic anhydrase (CA) immobilized on polyacrylamide gel improved the tolerance to sulfonamide inhibitors [123]. Another study immobilized carbonic anhydrase on surfactant-modified silylated chitosan (SMSC) to induce bionic-driven carbonate precipitation. The results demonstrated improved stability, prolonged enzyme activity, and increased carbonate production compared to the free enzyme under the same conditions [127]. Building upon this research, Vinoba et al. modified the porous nanomaterials and immobilized carbonic anhydrase on the modified materials, and the experimental results indicated that the cross-linking method for immobilizing the CA enzyme resulted in catalytic activity comparable to that of the free enzyme [128]. Utilizing biomimetic principles, the synthesis of biomimetic materials emulating the activity of natural carbonic anhydrase and the subsequent immobilization through material modification using chemical reagents has shown promise. This method can solve not only the problem of the slow rate of carbon dioxide absorption and hydration but also the problem of the poor stability of the natural carbonic anhydrase. Furthermore, it allows controllable adjustments in the molecular structure, catalytic rate, and other parameters, enhancing its potential for practical applications. Simultaneously, domestic and international researchers have conducted numerical simulations to predict carbon dioxide sequestration on a 10,000-year scale. The consensus is that natural conditions will facilitate carbon dioxide mineral sequestration within approximately one hundred years. Currently, by employing bionic carbonic anhydrase immobilization technology, the introduction of carbonic anhydrase coagulation nuclei has proven effective in reducing the critical concentration for induced carbonate precipitation. This advancement enables a mineralization process with controllable speed and the ability to regulate the size of carbonate precipitation particles.

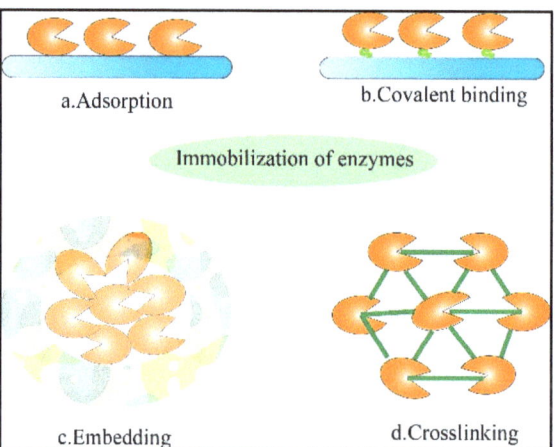

Figure 12. Common enzyme immobilization methods [120].

## 4. Comparisons of the Natural and Microbial-Catalyzed Sequestration Processes

Based on the natural mineralization mechanism, microbial-assisted mineralization integrates the characteristics of accelerated hydration by carbonic anhydrase and those of the surface nucleation site provided by microorganisms. Compared with the natural mineralization process, it has more advantages in terms of storage time, storage cost, efficiency, and derivative application value.

According to previous studies [16,129–131], the storage processes of sedimentary basins within 1000 years primarily consist of structural storage, residual storage, and dissolution storage, as seen in Figure 2. Mineralization storage is limited due to the comparatively low reactivity of sedimentary rock minerals and the low content of bivalent metals producing carbonate minerals, and it may take thousands of years to accomplish in situ formation mineralization storage. Currently, most of the research on in situ mineralized carbon sequestration is conducted in laboratories, with only a few countries carrying out pilot test projects. Unlike the traditional carbon dioxide storage of sandstone reservoirs, dissolution storage and mineralization storage play a major role in the in situ carbon sequestration projects of basalt reservoirs; this, in a sense, is equivalent to advancing the mineralization process and achieving the mineralization effect in a short time, thus enhancing the project's safety. In the case of the supercritical $CO_2$ mineralization sequestration project in Wallula, USA, 60% of the injected $CO_2$ (293 tons/year) was fixed as a carbonate mineral within two years [132,133]. Similarly, the CarbFix dissolved $CO_2$ mineralization sequestration project in Iceland mineralized 95% of the injected $CO_2$ (57–75 tons/year) within two years [64,134]. The latter sequestration time is advanced in part because the injection of dissolved $CO_2$ reduces the time for $CO_2$ hydration. In the natural state, the hydration rate of $CO_2$ is slow; the microbial carbonic anhydrase catalyzes the hydration, and the conversion number is increased by about $10^8$ times that of the natural hydration, thus significantly shortening the hydration time and accelerating the mineralization process. At present, the research on microbial-assisted mineralization mainly remains at the laboratory research stage, and relevant engineering test data are temporarily lacking. Combined with the results of the laboratory experiments, it can be seen that microbial-assisted mineralization can significantly improve the mineralization rate and is expected to advance the time scale to several decades.

Carbon dioxide is turned into stable carbonate minerals by injecting captured carbon dioxide into geological formations such as oil reservoirs; this results in long-term carbon fixation and emissions reduction while enhancing oil and gas recovery. In terms of environmental benefits, in situ mineral carbonation achieves storage by fixing carbon dioxide in the reservoir minerals as stable carbonates, successfully capturing carbon dioxide and lowering the danger of carbon dioxide leakage. Mineralization is accelerated by microbial-assisted in situ mineralization with no additional energy use. From the perspective of economic benefits, even though carbon dioxide storage accounts for the smallest proportion of investment in the entire CCUS industry chain, the cost of carbon dioxide storage is expected to be CNY 40–50/ton by 2030 [135]. However, due to the high capture and transport and operating costs of CCUS projects, oil companies must increase oil and gas production to break even. In recent years, China has carried out $CO_2$ flooding demonstration projects for low-permeability and ultra-low-permeability reservoirs, covering about 100 million tons of reserves and improving the recovery rate by 6–20% [21]. Carbon dioxide injection into oil and gas reservoirs has both the economic benefits of enhanced oil recovery and the environmental benefits of storage and emissions reduction. At present, microbial-assisted mineralization for the porous media of oil and gas reservoirs is still at the laboratory research stage, and the mineralization rate and effect are affected by many factors; therefore, the economic benefits and emission reduction efficiency of reservoir mineralization storage are difficult to predict.

In terms of mineralization storage derivative applications, microbial-induced mineralization is primarily employed in soil reinforcement [136,137], micro-crack repair [138], cement bonding [139], environmental remediation, and other domains [140]. Carbonate formed by carbon dioxide mineralization in reservoirs can be employed to plug dominant seepage pathways, thereby improving volumetric sweep efficiency and oil recovery. The activity and stability of microorganisms and their enzymes will continue to improve in the future as the fields of extreme-environment microorganism mining, protein transformation engineering, and enzyme immobilization technology develop. This will broaden the scope of applications and increase the likelihood of industrial application.

## 5. Outlook

In the context of the carbon peak and the commitment to carbon neutrality, carbon capture, utilization, and storage (CCUS) technology holds significant potential for application and development. The Annual Report of China on Carbon Dioxide Capture, Utilization, and Storage (CCUS) in 2021 highlights that CCUS technology stands as the sole technological option for achieving the decarbonization of fossil energy. Additionally, it underscores that China possesses a substantial geological sequestration potential for $CO_2$, which is estimated to range from 12.1 to 41.3 million tons [141]. $CO_2$ mineralization and fixation represent effective technologies for mitigating the greenhouse effect and decelerating global warming. These approaches offer new avenues for the study of the multiphase reactions of $CO_2$ with water and rocks, as well as the exploration of the application of microbial carbonic anhydrase in multicomponent reactions.

Overall, $CO_2$ mineralization and sequestration are still primarily at the laboratory stage, with only a few pilot demonstration projects. Currently, several key challenges are being faced in the realm of $CO_2$ mineralization sequestration.

The storage potential of supercritical carbon dioxide in the subsurface is influenced by various complex environmental factors, including the stratum temperature, pressure, mineralization level of the formation water, and mineral composition of the stratum. However, there is currently a lack of established methods to measure and assess the specific impact of each factor. It is crucial to introduce evaluation indices to quantitatively measure the influence of each factor on the rate of mineralization and the amount of sequestered carbon dioxide.

In contrast to the extensive research conducted by foreign scholars on carbon dioxide sequestration mechanisms in typical marine sedimentary minerals, China boasts abundant deep saline aquifers with terrestrial sedimentation. Given the significant disparities in rock types and mineral compositions between marine and terrestrial sedimentary strata, the mechanisms of mineralization and sequestration differ. Currently, there is a dearth of research on the $CO_2$–water–rock interaction mechanisms specific to $CO_2$ mineralization and storage in the deep saline aquifers associated with terrestrial deposition. Future research should focus on investigating the intricate mechanisms of $CO_2$ mineralization and storage. Particular attention should be given to the reaction mechanisms between the typical terrestrial sedimentary minerals and $CO_2$ in saline solutions. This research will facilitate the evaluation of suitable geological conditions and storage technologies and ultimately enhance the efficiency and safety of $CO_2$ storage.

Investigation into accelerated mineralization induced by microorganisms is currently confined to indoor simulations and lacks on-site samples. Numerous studies have been conducted at ambient temperature and pressure and have primarily employed exogenously cultivated CA-producing bacteria instead of reservoir-originated microorganisms. The adaptability of microorganisms to the extreme environment of mineralization and sequestration in oil reservoirs and deep saline aquifers has not been thoroughly assessed. Future research should prioritize the targeted evolution, screening, and cultivation of microorganisms that are highly adaptable to extreme environments and should aim to obtain carbonic anhydrase with superior catalytic activity from native microorganisms. Furthermore, the research should delve into the catalytic mechanism and active sites of key enzymes like carbonic anhydrase in microbial-assisted mineralization. Structural alterations and immobilization should be explored for enzyme modifications, with the aim of enhancing enzyme activity and stability in extreme environments.

Additionally, most microorganisms are adversely affected by high temperature, high pressure, and $scCO_2$, resulting in a significant reduction in their activity and biomass within a specific time frame. There is a need to identify viable alternatives to microorganisms to assist with mineralization. Simultaneously, chemo-mimicry is vital to the enhancement of the adaptability for engineering applications, given the impact of extreme environments on microbial activity. In ongoing studies related to the synthesis and preparation of biomimetic enzymes, broader focus should be placed on evaluating enzyme-induced mineralization performance in simulated in situ reservoir core replacement experiments. Moreover, it is critical to investigate the factors influencing the activity and stability of bionic CA enzymes. Establishing a comprehensive evaluation system for bionic carbonic anhydrase, including reference comparisons with biological enzymes, will optimize the function of bionic enzymes and provide a scientific basis for the subsequent engineering applications of bionic carbonic anhydride-induced mineralization.

## 6. Conclusions

Carbon dioxide mineralization sequestration is regarded as the safest and most effective method for $CO_2$ geological sequestration; it has a crucial role to play in the achievement of carbon neutrality goals. Through the utilization of the calcium- and magnesium-based resources present in reservoir rocks, captured industrial $CO_2$ is injected into reservoirs. Through the carbonation reaction between silicate minerals and $CO_2$ within the reservoir, a permanent fixation of $CO_2$ is achieved, which is in alignment with the objective of carbon neutrality.

Drawing upon the relevant literature on $CO_2$ mineralization sequestration and microbial-induced mineralization, several fundamental conclusions can be outlined:

(1) The geophysical–chemical process of carbon dioxide mineralization in reservoirs primarily encompasses three stages: carbon dioxide dissolution ionization, mineral dissolution, and carbonate mineral precipitation.
(2) The rate of $CO_2$ mineralization sequestration in alkaline environments is influenced by factors such as carbon anion concentrations; concentrations of divalent metal cations like calcium, magnesium, and iron; and the availability of mineral nucleation sites.
(3) Microbial induction can expedite the mineralization process by enhancing the precipitation environment and providing nucleation sites. Additionally, the carbonic anhydrase produced during microbial metabolism is a pivotal enzyme in the process of $CO_2$ mineralization induced by bacteria.
(4) Carbonic anhydrase primarily catalyzes the carbon dioxide hydration reaction, influencing the $CO_2$ mineralization process. Its enzyme activity and stability are impacted by factors like the structure of the active region, the environmental temperature, and the pressure. Utilizing covalent bonding and other methods to prepare immobilized carbonic anhydrase by chemo-mimicry can augment enzyme catalytic activity, thereby enhancing the rate of bio-induced mineralization.

**Author Contributions:** S.N.: investigation, writing—original draft, formal analysis, data curation; W.L.: writing—review and editing, validation, formal analysis, resources; Z.J.: formal analysis, validation, data curation, writing—review and editing; K.W.: investigation, formal analysis, writing—review and editing, software. All authors have read and agreed to the published version of the manuscript.

**Funding:** This research was funded by the Major Science and Technology project of the CNPC in China (grant No. 2021ZZ05).

**Data Availability Statement:** Not applicable.

**Acknowledgments:** The authors thank two anonymous reviewers for their constructive feedback.

**Conflicts of Interest:** The authors declare no conflict of interest.

## References

1. Soeder, D.J. Greenhouse gas sources and mitigation strategies from a geosciences perspective. *Adv. Geo Energy Res.* **2021**, *5*, 274–285. [CrossRef]
2. IPCC, E. *IPCC Special Report on Carbon Dioxide Capture and Storage*; Cambridge University Press: New York, NY, USA, 2005; p. 442.

3. Tapia, J.F.D.; Lee, J.-Y.; Ooi, R.E.; Foo, D.C.; Tan, R.R. A review of optimization and decision-making models for the planning of $CO_2$ capture, utilization and storage (CCUS) systems. *Sustain. Prod. Consum.* **2018**, *13*, 1–15. [CrossRef]
4. Zhou, C.; Wu, S.; Yang, Z.; Pan, S.; Wang, G.; Jiang, X.; Guan, M.; Yu, C.; Yu, Z.; Shen, Y. Progress, challenge and significance of building carbon industrial system in the context of carbon neutral strategy. *Pet. Explor. Dev.* **2023**, *50*, 190–205.
5. Jin, L.; Zhou, D.; Liu, R. Research on ecological protection compensation mechanism for adapting to carbon peak and carbon neutrality—Based on the perspective of carbon sink value. *Proc. Chin. Acad. Sci. USA* **2022**, *37*, 1623–1634.
6. Liu, H.; Were, P.; Li, Q.; Gou, Y.; Hou, Z. Worldwide status of CCUS technologies and their development and challenges in China. *Geofluids* **2017**, *2017*, 6126505. [CrossRef]
7. Núñez-López, V.; Gil-Egui, R.; Hosseini, S.A. Environmental and operational performance of $CO_2$-EOR as a CCUS technology: A Cranfield example with dynamic LCA considerations. *Energies* **2019**, *12*, 448. [CrossRef]
8. Xu, T.; Tian, H.; Zhu, H.; Cai, J. China actively promotes $CO_2$ capture, utilization and storage research to achieve carbon peak and carbon neutrality. *Adv. Geo Energy Res.* **2022**, *6*, 1–3. [CrossRef]
9. Cui, G.; Hu, Z.; Ning, F.; Jiang, S.; Wang, R. A review of salt precipitation during $CO_2$ injection into saline aquifers and its potential impact on carbon sequestration projects in China. *Fuel* **2023**, *334*, 126615. [CrossRef]
10. Gunter, W.; Bachu, S.; Law, D.-S.; Marwaha, V.; Drysdale, D.; MacDonald, D.; McCann, T. Technical and economic feasibility of $CO_2$ disposal in aquifers within the Alberta Sedimentary Basin, Canada. *Energy Convers. Manag.* **1996**, *37*, 1135–1142. [CrossRef]
11. Shukla, R.; Ranjith, P.; Haque, A.; Choi, X. A review of studies on $CO_2$ sequestration and caprock integrity. *Fuel* **2010**, *89*, 2651–2664. [CrossRef]
12. Emami-Meybodi, H.; Hassanzadeh, H.; Green, C.P.; Ennis-King, J. Convective dissolution of $CO_2$ in saline aquifers: Progress in modeling and experiments. *Int. J. Greenh. Gas Control* **2015**, *40*, 238–266.
13. Iglauer, S. *Dissolution Trapping of Carbon Dioxide in Reservoir Formation Brine—A Carbon Storage Mechanism*; INTECH Open Access Publisher: London, UK, 2011; pp. 233–262.
14. Li, L.; Zhao, N.; Wei, W.; Sun, Y. A review of research progress on $CO_2$ capture, storage, and utilization in Chinese Academy of Sciences. *Fuel* **2013**, *108*, 112–130. [CrossRef]
15. Zhang, L.; Soong, Y.; Dilmore, R.; Lopano, C. Numerical simulation of porosity and permeability evolution of Mount Simon sandstone under geological carbon sequestration conditions. *Chem. Geol.* **2015**, *403*, 1–12. [CrossRef]
16. Song, Y.; Sung, W.; Jang, Y.; Jung, W. Application of an artificial neural network in predicting the effectiveness of trapping mechanisms on $CO_2$ sequestration in saline aquifers. *Int. J. Greenh. Gas Control* **2020**, *98*, 103042. [CrossRef]
17. Liu, C.; Zhou, W.; Jiang, J.; Shang, F.; He, H.; Wang, S. Remaining oil distribution and development strategy for offshore unconsolidated sandstone reservoir at ultrahigh water-cut stage. *Geofluids* **2022**, *2022*, 6856298. [CrossRef]
18. Huang, S.; Wu, Y.; Meng, X.; Liu, L.; Ji, W. Recent advances on microscopic pore characteristics of low permeability sandstone reservoirs. *Adv. Geo Energy Res.* **2018**, *2*, 122–134. [CrossRef]
19. Shiyi, Y.; Qiang, W. New progress and prospect of oilfields development technologies in China. *Pet. Explor. Dev.* **2018**, *45*, 698–711.
20. Guo, H.; Wang, Z.; Wang, B.; Zhang, Y.; Meng, H.; Sui, H. Molecular dynamics simulations of oil recovery from dolomite slit nanopores enhanced by $CO_2$ and $N_2$ injection. *Adv. Geo Energy Res.* **2022**, *6*, 306–313. [CrossRef]
21. Xinmin, S.; Feng, W.; Desheng, M.; Ming, G.; Zhang, Y. Progress and prospect of carbon dioxide capture, utilization and storage in CNPC oilfields. *Pet. Explor. Dev.* **2023**, *50*, 229–244.
22. Yang, B.; Shao, C.; Hu, X.; Ngata, M.R.; Aminu, M.D. Advances in Carbon Dioxide Storage Projects: Assessment and Perspectives. *Energy Fuels* **2023**, *37*, 1757–1776. [CrossRef]
23. Lackner, K.S.; Wendt, C.H.; Butt, D.P.; Joyce, E.L., Jr.; Sharp, D.H. Carbon dioxide disposal in carbonate minerals. *Energy* **1995**, *20*, 1153–1170. [CrossRef]
24. Seifritz, W. $CO_2$ disposal by means of silicates. *Nature* **1990**, *345*, 486. [CrossRef]
25. Saito, T.; Sakai, E.; Morioka, M.; Otsuki, N. Carbonation of $\gamma$-$Ca_2SiO_4$ and the Mechanism of Vaterite Formation. *J. Adv. Concr. Technol.* **2010**, *8*, 273–280. [CrossRef]
26. Bachu, S. Sequestration of $CO_2$ in geological media: Criteria and approach for site selection in response to climate change. *Energy Convers. Manag.* **2000**, *41*, 953–970. [CrossRef]
27. Xu, R.; Li, R.; Ma, J.; He, D.; Jiang, P. Effect of mineral dissolution/precipitation and $CO_2$ exsolution on $CO_2$ transport in geological carbon storage. *Acc. Chem. Res.* **2017**, *50*, 2056–2066. [CrossRef] [PubMed]
28. Raza, A.; Glatz, G.; Gholami, R.; Mahmoud, M.; Alafnan, S. Carbon mineralization and geological storage of $CO_2$ in basalt: Mechanisms and technical challenges. *Earth Sci. Rev.* **2022**, *229*, 104036. [CrossRef]
29. Kou, Z.; Wang, H.; Alvarado, V.; Nye, C.; Bagdonas, D.A.; McLaughlin, J.F.; Quillinan, S.A. Effects of carbonic acid-rock interactions on $CO_2$/Brine multiphase flow properties in the upper minnelusa sandstones. *SPE J.* **2023**, *28*, 754–767. [CrossRef]
30. Campbell, J.S.; Foteinis, S.; Furey, V.; Hawrot, O.; Pike, D.; Aeschlimann, S.; Maesano, C.N.; Reginato, P.L.; Goodwin, D.R.; Looger, L.L. Geochemical negative emissions technologies: Part I. Review. *Front. Clim.* **2022**, *4*, 879133. [CrossRef]
31. Khosrokhavar, R.; Elsinga, G.; Farajzadeh, R.; Bruining, H. Visualization and investigation of natural convection flow of $CO_2$ in aqueous and oleic systems. *J. Pet. Sci. Eng.* **2014**, *122*, 230–239. [CrossRef]
32. Gibbons, B.H.; Edsall, J.T. Rate of hydration of carbon dioxide and dehydration of carbonic acid at 25. *J. Biol. Chem.* **1963**, *238*, 3502–3507. [CrossRef]

33. Ho, C.; Sturtevant, J.M. The kinetics of the hydration of carbon dioxide at 25. *J. Biol. Chem* **1963**, *238*, 3499–3501. [CrossRef] [PubMed]
34. Dennard, A.; Williams, R. The catalysis of the reaction between carbon dioxide and water. *J. Chem. Soc. A Inorg. Phys. Theor.* **1966**, 812–816. [CrossRef]
35. Bhaduri, G.A. Catalytic Enhancement of Hydration of $CO_2$ Using Nickel Nanoparticles for Carbon Capture and Storage. Ph.D. Thesis, Newcastle University, Newcastle upon Tyne, UK, 2018.
36. Verma, M.; Bhaduri, G.A.; Phani Kumar, V.S.; Deshpande, P.A. Biomimetic catalysis of $CO_2$ hydration: A materials perspective. *Ind. Eng. Chem. Res.* **2021**, *60*, 4777–4793. [CrossRef]
37. Zhang, L.; Wang, Y.; Miao, X.; Gan, M.; Li, X. Geochemistry in geologic $CO_2$ utilization and storage: A brief review. *Adv. Geo Energy Res.* **2019**, *3*, 304–313. [CrossRef]
38. André, L.; Audigane, P.; Azaroual, M.; Menjoz, A. Numerical modeling of fluid–rock chemical interactions at the supercritical $CO_2$–liquid interface during $CO_2$ injection into a carbonate reservoir, the Dogger aquifer (Paris Basin, France). *Energy Convers. Manag.* **2007**, *48*, 1782–1797. [CrossRef]
39. Ketzer, J.; Iglesias, R.; Einloft, S.; Dullius, J.; Ligabue, R.; De Lima, V. Water–rock–$CO_2$ interactions in saline aquifers aimed for carbon dioxide storage: Experimental and numerical modeling studies of the Rio Bonito Formation (Permian), Southern Brazil. *Appl. Geochem.* **2009**, *24*, 760–767. [CrossRef]
40. Zhang, R.; Winterfeld, P.H.; Yin, X.; Xiong, Y.; Wu, Y.-S. Sequentially coupled THMC model for $CO_2$ geological sequestration into a 2D heterogeneous saline aquifer. *J. Nat. Gas Sci. Eng.* **2015**, *27*, 579–615. [CrossRef]
41. Khudhur, F.W.; MacDonald, J.M.; Macente, A.; Daly, L. The utilization of alkaline wastes in passive carbon capture and sequestration: Promises, challenges and environmental aspects. *Sci. Total Environ.* **2022**, *823*, 153553. [CrossRef]
42. Jorat, M.E.; Goddard, M.A.; Manning, P.; Lau, H.K.; Ngeow, S.; Sohi, S.P.; Manning, D.A. Passive $CO_2$ removal in urban soils: Evidence from brownfield sites. *Sci. Total Environ.* **2020**, *703*, 135573. [CrossRef]
43. Ali, M.; Jha, N.K.; Pal, N.; Keshavarz, A.; Hoteit, H.; Sarmadivaleh, M. Recent advances in carbon dioxide geological storage, experimental procedures, influencing parameters, and future outlook. *Earth Sci. Rev.* **2022**, *225*, 103895. [CrossRef]
44. Gerdemann, S.J.; O'Connor, W.K.; Dahlin, D.C.; Penner, L.R.; Rush, H. Ex situ aqueous mineral carbonation. *Environ. Sci. Technol.* **2007**, *41*, 2587–2593. [CrossRef] [PubMed]
45. Prigiobbe, V.; Costa, G.; Baciocchi, R.; Hänchen, M.; Mazzotti, M. The effect of $CO_2$ and salinity on olivine dissolution kinetics at 120 °C. *Chem. Eng. Sci.* **2009**, *64*, 3510–3515. [CrossRef]
46. Wigand, M.; Carey, J.; Schütt, H.; Spangenberg, E.; Erzinger, J. Geochemical effects of $CO_2$ sequestration in sandstones under simulated in situ conditions of deep saline aquifers. *Appl. Geochem.* **2008**, *23*, 2735–2745. [CrossRef]
47. Bénézeth, P.; Palmer, D.A.; Anovitz, L.M.; Horita, J. Dawsonite synthesis and reevaluation of its thermodynamic properties from solubility measurements: Implications for mineral trapping of $CO_2$. *Geochim. Cosmochim. Acta* **2007**, *71*, 4438–4455. [CrossRef]
48. Goldberg, P.; Chen, Z.-Y.; O'Connor, W.; Walters, R.; Ziock, H. $CO_2$ Mineral Sequestration Studies in the US. In Proceedings of the National Conference on Carbon Sequestration, Washington, DC, USA, 15–17 May 2001; National Energy Technology Laboratory: Pittsburgh, PA, USA; Washington, DC, USA, 2001.
49. Manning, D.A.; Renforth, P. Passive sequestration of atmospheric $CO_2$ through coupled plant-mineral reactions in urban soils. *Environ. Sci. Technol.* **2013**, *47*, 135–141. [CrossRef]
50. Xi, K.; Cao, Y.; Haile, B.G.; Zhu, N.; Liu, K.; Wu, S.; Hellevang, H. Diagenetic variations with respect to sediment composition and paleo-fluids evolution in conglomerate reservoirs: A case study of the Triassic Baikouquan Formation in Mahu Sag, Junggar Basin, Northwestern China. *J. Pet. Sci. Eng.* **2021**, *197*, 107943. [CrossRef]
51. Liu, B.; Fu, X.; Li, Z. Impacts of $CO_2$-brine-rock interaction on sealing efficiency of sand caprock: A case study of Shihezi formation in Ordos basin. *Adv. Geo Energy Res.* **2018**, *2*, 380–392. [CrossRef]
52. Zhang, P.; Chen, Z.; Xue, L.; Bao, Y.; Fang, Y. Differential diagenetic evolution of the Lower Cambrian black rock system at the northwestern margin of the Tarim Basin and its influencing factors. *J. Petrol.* **2020**, *36*, 3463–3476.
53. Li, S.Z.; Yang, J.G.; Liu, B.; Yao, Y.L.; Xiao, F.; Bai, L.H.; Huang, Y.M.; Li, A.; Zhang, L.Y. Petrologic characterization and petrographic delineation of mud shale in a section of Qingshankou Formation, Sanzhao Depression, Songliao Basin—An example from Songpai Oil Well No. 3. *Geol. Resour.* **2021**, *30*, 317–324.
54. Alonso-Zarza, A.M.; Tanner, L.H. *Carbonates in Continental Settings: Facies, Environments, and Processes*; Elsevier: Amsterdam, The Netherlands, 2009; p. 371.
55. Schieber, J.; Southard, J.; Thaisen, K. Accretion of mudstone beds from migrating floccule ripples. *Science* **2007**, *318*, 1760–1763. [CrossRef]
56. Yan, J.H.; Deng, Y.; Pu, X.G.; Zhou, L.H.; Chen, S.Y.; Jiao, Y.X. Characteristics and controlling factors of fine-grained mixed sedimentary rocks in the Kong II section of the Cangdong Depression, Bohai Bay Basin. *Oil Gas Geol.* **2017**, *38*, 98–109.
57. Liu, Y.Q.; Jiao, X.; Li, H.; Yuan, M.S.; Zhou, X.H.; Liang, H.; Zhou, D.W.; Zheng, C.Y.; Sun, Q. Permian mantle hydrothermal jet-type primary dolomite from Yuejin Gully, Santang Lake, Xinjiang. *Chin. Sci. Earth Sci.* **2011**, *41*, 1862–1871.
58. Manger, K.C.; Curtis, J. Geologic influences on location and production of Antrim Shale gas. *Devonian Gas Shales Technol. Rev. GRI* **1991**, *7*, 5–16.
59. Curtis, J.B. Fractured shale-gas systems. *AAPG Bull.* **2002**, *86*, 1921–1938.

60. Shi, X.F.; Chen, L.R.; Li, K.Y.; Yang, H.L. Mineral assemblages of West Philippine Sea sediments and their geologic significance. *Ocean. Lakes* **1994**, *25*, 328–335.
61. Snæbjörnsdóttir, S.Ó.; Sigfússon, B.; Marieni, C.; Goldberg, D.; Gislason, S.R.; Oelkers, E.H. Carbon dioxide storage through mineral carbonation. *Nat. Rev. Earth Environ.* **2020**, *1*, 90–102. [CrossRef]
62. Al Kalbani, M.; Serati, M.; Hofmann, H.; Bore, T. A comprehensive review of enhanced in-situ $CO_2$ mineralisation in Australia and New Zealand. *Int. J. Coal Geol.* **2023**, *276*, 104316. [CrossRef]
63. Stokreef, S.; Sadri, F.; Stokreef, A.; Ghahreman, A. Mineral carbonation of ultramafic tailings: A review of reaction mechanisms and kinetics, industry case studies, and modelling. *Clean. Eng. Technol.* **2022**, *8*, 100491. [CrossRef]
64. Matter, J.M.; Stute, M.; Snæbjörnsdottir, S.Ó.; Oelkers, E.H.; Gislason, S.R.; Aradottir, E.S.; Sigfusson, B.; Gunnarsson, I.; Sigurdardottir, H.; Gunnlaugsson, E. Rapid carbon mineralization for permanent disposal of anthropogenic carbon dioxide emissions. *Science* **2016**, *352*, 1312–1314. [CrossRef]
65. Ebigbo, A.; Gregory, S.P. The relevance of microbial processes in geo-energy applications. *Adv. Geo Energy Res.* **2021**, *5*, 5–7. [CrossRef]
66. Herrmann, G.; Jayamani, E.; Mai, G.; Buckel, W. Energy conservation via electron-transferring flavoprotein in anaerobic bacteria. *J. Bacteriol.* **2008**, *190*, 784–791. [CrossRef]
67. Jones, D.; Head, I.; Gray, N.; Adams, J.; Rowan, A.; Aitken, C.; Bennett, B.; Huang, H.; Brown, A.; Bowler, B. Crude-oil biodegradation via methanogenesis in subsurface petroleum reservoirs. *Nature* **2008**, *451*, 176–180. [CrossRef] [PubMed]
68. Boquet, E.; Boronat, A.; Ramos-Cormenzana, A. Production of calcite (calcium carbonate) crystals by soil bacteria is a general phenomenon. *Nature* **1973**, *246*, 527–529. [CrossRef]
69. Chen, J.; Liu, B.; Zhong, M.; Jing, C.; Guo, B. Research status and development of microbial induced calcium carbonate mineralization technology. *PLoS ONE* **2022**, *17*, e0271761. [CrossRef] [PubMed]
70. Mirjafari, P. *Sequestration of Carbon Dioxide: A Biological Approach to Mineralization of Carbon Dioxide in Depleted Oil and Gas Reservoirs*; University of Regina: Regina, SK, Canada, 2006.
71. De Muynck, W.; De Belie, N.; Verstraete, W. Microbial carbonate precipitation in construction materials: A review. *Ecol. Eng.* **2010**, *36*, 118–136. [CrossRef]
72. Wang, Z.; Zhang, N.; Cai, G.; Jin, Y.; Ding, N.; Shen, D. Review of ground improvement using microbial induced carbonate precipitation (MICP). *Mar. Georesources Geotechnol.* **2017**, *35*, 1135–1146. [CrossRef]
73. Jain, S.; Fang, C.; Achal, V. A critical review on microbial carbonate precipitation via denitrification process in building materials. *Bioengineered* **2021**, *12*, 7529–7551. [CrossRef]
74. Semple, K.; Westlake, D. Characterization of iron-reducing Alteromonas putrefaciens strains from oil field fluids. *Can. J. Microbiol.* **1987**, *33*, 366–371. [CrossRef]
75. Personna, Y.R.; Ntarlagiannis, D.; Slater, L.; Yee, N.; O'Brien, M.; Hubbard, S. Spectral induced polarization and electrodic potential monitoring of microbially mediated iron sulfide transformations. *J. Geophys. Res. Biogeosciences* **2008**, *113*. [CrossRef]
76. Han, Z.; Zhao, Y.; Yan, H.; Zhao, H.; Han, M.; Sun, B.; Sun, X.; Hou, F.; Sun, H.; Han, L. Struvite precipitation induced by a novel sulfate-reducing bacterium Acinetobacter calcoaceticus SRB4 isolated from river sediment. *Geomicrobiol. J.* **2015**, *32*, 868–877. [CrossRef]
77. Wang, F.; Zheng, S.; Qiu, H.; Cao, C.; Tang, X.; Hao, L.; Liu, F.; Li, J. The process by which the iron-reducing bacterium Shewanella oneidensis MR-4 induces hydrated iron oxide to form cyanite. *J. Microbiol. Biotechnol.* **2018**, *58*, 573–583.
78. Mobley, H.L. Urease. In *Helicobacter Pylori: Physiology and Genetics*; John Wiley & Sons, Inc.: Hoboken, NJ, USA, 2001; pp. 177–191.
79. Lalucat, J.; Bennasar, A.; Bosch, R.; García-Valdés, E.; Palleroni, N.J. Biology of Pseudomonas stutzeri. *Microbiol. Mol. Biol. Rev.* **2006**, *70*, 510–547. [CrossRef]
80. Zeikus, J. The biology of methanogenic bacteria. *Bacteriol. Rev.* **1977**, *41*, 514–541. [CrossRef] [PubMed]
81. Fernandes, B.D.; Mota, A.; Teixeira, J.A.; Vicente, A.A. Continuous cultivation of photosynthetic microorganisms: Approaches, applications and future trends. *Biotechnol. Adv.* **2015**, *33*, 1228–1245. [CrossRef]
82. Kwon, S.-W.; Kim, B.-Y.; Song, J.; Weon, H.-Y.; Schumann, P.; Tindall, B.J.; Stackebrandt, E.; Fritze, D. Sporosarcina koreensis sp. nov. and Sporosarcina soli sp. nov., isolated from soil in Korea. *Int. J. Syst. Evol. Microbiol.* **2007**, *57*, 1694–1698. [CrossRef] [PubMed]
83. Xiao, R.; Zheng, Y. Overview of microalgal extracellular polymeric substances (EPS) and their applications. *Biotechnol. Adv.* **2016**, *34*, 1225–1244. [CrossRef]
84. Rajkhowa, S.; Sarma, J. Biosurfactant: An Alternative Towards Sustainability. In *Innovative Bio-Based Technologies for Environmental Remediation*; CRC Press: Boca Raton, FL, USA, 2022; pp. 377–402.
85. Thanigaivelan, R.; Prakash, S.; Maniraj, S. Surface Modification Techniques for Bio-Materials: An Overview. *Adv. Manuf. Tech. Eng. Eng. Mater.* **2022**, 42–60. [CrossRef]
86. Sánchez-Román, M.; Fernández-Remolar, D.; Amils, R.; Sánchez-Navas, A.; Schmid, T.; Martin-Uriz, P.S.; Rodríguez, N.; McKenzie, J.A.; Vasconcelos, C. Microbial mediated formation of Fe-carbonate minerals under extreme acidic conditions. *Sci. Rep.* **2014**, *4*, 4767. [CrossRef]
87. Jun, Y.-S.; Zhu, Y.; Wang, Y.; Ghim, D.; Wu, X.; Kim, D.; Jung, H. Classical and nonclassical nucleation and growth mechanisms for nanoparticle formation. *Annu. Rev. Phys. Chem.* **2022**, *73*, 453–477. [CrossRef]

88. Benzerara, K.; Skouri-Panet, F.; Li, J.; Férard, C.; Gugger, M.; Laurent, T.; Couradeau, E.; Ragon, M.; Cosmidis, J.; Menguy, N. Intracellular Ca-carbonate biomineralization is widespread in cyanobacteria. *Proc. Natl. Acad. Sci. USA* **2014**, *111*, 10933–10938. [CrossRef]
89. Yan, H.; Owusu, D.C.; Han, Z.; Zhao, H.; Ji, B.; Zhao, Y.; Tucker, M.E.; Zhao, Y. Extracellular, surface, and intracellular biomineralization of Bacillus subtilis Daniel-1 bacteria. *Geomicrobiol. J.* **2021**, *38*, 698–708. [CrossRef]
90. Hammes, F.; Boon, N.; de Villiers, J.; Verstraete, W.; Siciliano, S.D. Strain-specific ureolytic microbial calcium carbonate precipitation. *Appl. Environ. Microbiol.* **2003**, *69*, 4901–4909. [CrossRef] [PubMed]
91. Larachi, F. Kinetic model for the reversible hydration of carbon dioxide catalyzed by human carbonic anhydrase II. *Ind. Eng. Chem. Res.* **2010**, *49*, 9095–9104. [CrossRef]
92. Khajepour, H.; Mahmoodi, M.; Biria, D.; Ayatollahi, S. Investigation of wettability alteration through relative permeability measurement during MEOR process: A micromodel study. *J. Pet. Sci. Eng.* **2014**, *120*, 10–17. [CrossRef]
93. Khalifah, R.G. The carbon dioxide hydration activity of carbonic anhydrase: I. Stop-flow kinetic studies on the native human isoenzymes B and C. *J. Biol. Chem.* **1971**, *246*, 2561–2573. [CrossRef] [PubMed]
94. Cuesta-Seijo, J.A.; Borchert, M.S.; Navarro-Poulsen, J.-C.; Schnorr, K.M.; Mortensen, S.B.; Leggio, L.L. Structure of a dimeric fungal α-type carbonic anhydrase. *FEBS Lett.* **2011**, *585*, 1042–1048. [CrossRef] [PubMed]
95. Lindskog, S. Structure and mechanism of carbonic anhydrase. *Pharmacol. Ther.* **1997**, *74*, 1–20. [CrossRef]
96. Meldrum, N.U.; Roughton, F.J. Carbonic anhydrase. Its preparation and properties. *J. Physiol.* **1933**, *80*, 113. [CrossRef]
97. Hassan, M.I.; Shajee, B.; Waheed, A.; Ahmad, F.; Sly, W.S. Structure, function and applications of carbonic anhydrase isozymes. *Bioorganic Med. Chem.* **2013**, *21*, 1570–1582. [CrossRef]
98. Rowlett, R.S. Structure and catalytic mechanism of the β-carbonic anhydrases. *Biochim. Biophys. Acta Proteins Proteom.* **2010**, *1804*, 362–373. [CrossRef]
99. Supuran, C.T. Structure and function of carbonic anhydrases. *Biochem. J.* **2016**, *473*, 2023–2032. [CrossRef] [PubMed]
100. Bose, H.; Satyanarayana, T. Microbial carbonic anhydrases in biomimetic carbon sequestration for mitigating global warming: Prospects and perspectives. *Front. Microbiol.* **2017**, *8*, 1615. [CrossRef] [PubMed]
101. Del Prete, S.; Vullo, D.; Fisher, G.M.; Andrews, K.T.; Poulsen, S.-A.; Capasso, C.; Supuran, C.T. Discovery of a new family of carbonic anhydrases in the malaria pathogen Plasmodium falciparum—The η-carbonic anhydrases. *Bioorganic Med. Chem. Lett.* **2014**, *24*, 4389–4396. [CrossRef] [PubMed]
102. Lane, T.W.; Morel, F.M. A biological function for cadmium in marine diatoms. *Proc. Natl. Acad. Sci. USA* **2000**, *97*, 4627–4631. [CrossRef]
103. Lapointe, M.; MacKenzie, T.D.; Morse, D. An external δ-carbonic anhydrase in a free-living marine dinoflagellate may circumvent diffusion-limited carbon acquisition. *Plant Physiol.* **2008**, *147*, 1427–1436. [CrossRef]
104. Jensen, E.L.; Maberly, S.C.; Gontero, B. Insights on the functions and ecophysiological relevance of the diverse carbonic anhydrases in microalgae. *Int. J. Mol. Sci.* **2020**, *21*, 2922. [CrossRef]
105. Boone, C.D.; Habibzadegan, A.; Tu, C.; Silverman, D.N.; McKenna, R. Structural and catalytic characterization of a thermally stable and acid-stable variant of human carbonic anhydrase II containing an engineered disulfide bond. *Acta Crystallogr. Sect. D Biol. Crystallogr.* **2013**, *69*, 1414–1422. [CrossRef]
106. An, H.; Tu, C.; Duda, D.; Montanez-Clemente, I.; Math, K.; Laipis, P.J.; McKenna, R.; Silverman, D.N. Chemical rescue in catalysis by human carbonic anhydrases II and III. *Biochemistry* **2002**, *41*, 3235–3242. [CrossRef]
107. Tu, C.; Silverman, D.N.; Forsman, C.; Jonsson, B.H.; Lindskog, S. Role of histidine 64 in the catalytic mechanism of human carbonic anhydrase II studied with a site-specific mutant. *Biochemistry* **1989**, *28*, 7913–7918. [CrossRef]
108. Jönsson, B.; Håkansson, K.; Liljas, A. The structure of human carbonic anhydrase II in complex with bromide and azide. *FEBS Lett.* **1993**, *322*, 186–190. [CrossRef]
109. Thoms, S. Hydrogen bonds and the catalytic mechanism of human carbonic anhydrase II. *J. Theor. Biol.* **2002**, *215*, 399–404. [CrossRef]
110. Silverman, D.N.; Lindskog, S. The catalytic mechanism of carbonic anhydrase: Implications of a rate-limiting protolysis of water. *Acc. Chem. Res.* **1988**, *21*, 30–36. [CrossRef]
111. Park, D.; Lee, M.S. Kinetic study of catalytic $CO_2$ hydration by metal-substituted biomimetic carbonic anhydrase model complexes. *R. Soc. Open Sci.* **2019**, *6*, 190407. [CrossRef] [PubMed]
112. Boone, C.D.; Rasi, V.; Tu, C.; McKenna, R. Structural and catalytic effects of proline substitution and surface loop deletion in the extended active site of human carbonic anhydrase II. *FEBS J.* **2015**, *282*, 1445–1457. [CrossRef]
113. Lindskog, S.; Coleman, J.E. The catalytic mechanism of carbonic anhydrase. *Proc. Natl. Acad. Sci. USA* **1973**, *70*, 2505–2508. [CrossRef] [PubMed]
114. Liszka, M.J.; Clark, M.E.; Schneider, E.; Clark, D.S. Nature versus nurture: Developing enzymes that function under extreme conditions. *Annu. Rev. Chem. Biomol. Eng.* **2012**, *3*, 77–102. [CrossRef] [PubMed]
115. Gomes, J.; Steiner, W. The biocatalytic potential of extremophiles and extremozymes. *Food Technol. Biotechnol.* **2004**, *42*, 223–225.
116. Sharma, T.; Sharma, S.; Kamyab, H.; Kumar, A. Energizing the $CO_2$ utilization by chemo-enzymatic approaches and potentiality of carbonic anhydrases: A review. *J. Clean. Prod.* **2020**, *247*, 119138. [CrossRef]
117. Rasouli, H.; Nguyen, K.; Iliuta, M.C. Recent advancements in carbonic anhydrase immobilization and its implementation in $CO_2$ capture technologies: A review. *Sep. Purif. Technol.* **2022**, *296*, 121299. [CrossRef]

118. de Oliveira Maciel, A.; Christakopoulos, P.; Rova, U.; Antonopoulou, I. Carbonic anhydrase to boost $CO_2$ sequestration: Improving carbon capture utilization and storage (CCUS). *Chemosphere* **2022**, *299*, 134419. [CrossRef]
119. Ozdemir, E. Biomimetic $CO_2$ sequestration: 1. Immobilization of carbonic anhydrase within polyurethane foam. *Energy Fuels* **2009**, *23*, 5725–5730. [CrossRef]
120. Molina-Fernández, C.; Luis, P. Immobilization of carbonic anhydrase for $CO_2$ capture and its industrial implementation: A review. *J. CO2 Util.* **2021**, *47*, 101475. [CrossRef]
121. Yuan, Y.; Wang, F.; Li, H.; Su, S.; Gao, H.; Han, X.; Ren, S. Potential application of the immobilization of carbonic anhydrase based on metal organic framework supports. *Process Biochem.* **2022**, *122*, 214–223. [CrossRef]
122. Rodriguez, L.C.; Restrepo-Sánchez, N.; Pelaez, C.; Bernal, C. Enhancement of the catalytic activity of Carbonic Anhydrase by covalent immobilization on Magnetic Cellulose Crystals. *Bioresour. Technol. Rep.* **2023**, *21*, 101380. [CrossRef]
123. Ray, B. Purification and immobilization of human carbonic anhydrase B by using polyacrylamide gel. *Experientia* **1977**, *33*, 1439–1440. [CrossRef]
124. Li, J.; Zhang, Y.; Yang, Y. Characterization of the diatomite binding domain in the ribosomal protein L2 from E. coli and functions as an affinity tag. *Appl. Microbiol. Biotechnol.* **2013**, *97*, 2541–2549. [CrossRef]
125. Chen, Q.; Kenausis, G.L.; Heller, A. Stability of oxidases immobilized in silica gels. *J. Am. Chem. Soc.* **1998**, *120*, 4582–4585. [CrossRef]
126. Zhang, S.; Du, M.; Shao, P.; Wang, L.; Ye, J.; Chen, J.; Chen, J. Carbonic anhydrase enzyme-MOFs composite with a superior catalytic performance to promote $CO_2$ absorption into tertiary amine solution. *Environ. Sci. Technol.* **2018**, *52*, 12708–12716. [CrossRef] [PubMed]
127. Yadav, R.; Wanjari, S.; Prabhu, C.; Kumar, V.; Labhsetwar, N.; Satyanarayanan, T.; Kotwal, S.; Rayalu, S. Immobilized carbonic anhydrase for the biomimetic carbonation reaction. *Energy Fuels* **2010**, *24*, 6198–6207. [CrossRef]
128. Vinoba, M.; Kim, D.H.; Lim, K.S.; Jeong, S.K.; Lee, S.W.; Alagar, M. Biomimetic sequestration of $CO_2$ and reformation to $CaCO_3$ using bovine carbonic anhydrase immobilized on SBA-15. *Energy Fuels* **2011**, *25*, 438–445. [CrossRef]
129. Krevor, S.; De Coninck, H.; Gasda, S.E.; Ghaleigh, N.S.; de Gooyert, V.; Hajibeygi, H.; Juanes, R.; Neufeld, J.; Roberts, J.J.; Swennenhuis, F. Subsurface carbon dioxide and hydrogen storage for a sustainable energy future. *Nat. Rev. Earth Environ.* **2023**, *4*, 102–118. [CrossRef]
130. Joseph, E.T.; Ashok, A.; Singh, D.; Ranganathan, A.; Pandey, G.; Bhan, U.; Singh, Y. Carbon sequestration: Capture, storage & utilization of $CO_2$ emissions from anthropogenic sources. *AIP Conf. Proc.* **2023**, *2521*, 030022.
131. Mwakipunda, G.C.; Ngata, M.R.; Mgimba, M.M.; Yu, L. Carbon Dioxide Sequestration in Low Porosity and Permeability Deep Saline Aquifer: Numerical Simulation Method. *J. Energy Resour. Technol.* **2023**, *145*, 073401. [CrossRef]
132. White, S.K.; Spane, F.A.; Schaef, H.T.; Miller, Q.R.; White, M.D.; Horner, J.A.; McGrail, B.P. Quantification of $CO_2$ mineralization at the Wallula basalt pilot project. *Environ. Sci. Technol.* **2020**, *54*, 14609–14616. [CrossRef]
133. Polites, E.G.; Schaef, H.T.; Horner, J.A.; Owen, A.T.; Holliman, J.E., Jr.; McGrail, B.P.; Miller, Q.R. Exotic carbonate mineralization recovered from a deep basalt carbon storage demonstration. *Environ. Sci. Technol.* **2022**, *56*, 14713–14722. [CrossRef]
134. Pogge von Strandmann, P.A.; Burton, K.W.; Snæbjörnsdóttir, S.O.; Sigfússon, B.; Aradóttir, E.S.; Gunnarsson, I.; Alfredsson, H.A.; Mesfin, K.G.; Oelkers, E.H.; Gislason, S.R. Rapid $CO_2$ mineralisation into calcite at the CarbFix storage site quantified using calcium isotopes. *Nat. Commun.* **2019**, *10*, 1983. [CrossRef]
135. Luo, J.; Xie, Y.; Hou, M.Z.; Xiong, Y.; Wu, X.; Lüddeke, C.T.; Huang, L. Advances in subsea carbon dioxide utilization and storage. *Energy Rev.* **2023**, *2*, 100016. [CrossRef]
136. Xu, F.; Wang, D. Review on Soil Solidification and Heavy Metal Stabilization by Microbial-Induced Carbonate Precipitation (MICP) Technology. *Geomicrobiol. J.* **2023**, *40*, 503–518. [CrossRef]
137. Khodabandeh, M.A.; Nagy, M.; Török, Á. Stabilization of collapsible soils with nanomaterials, fibers, polymers, industrial waste, and microbes: Current trends. *Constr. Build. Mater.* **2023**, *368*, 130463. [CrossRef]
138. Lu, C.; Ge, H.; Li, Z.; Zheng, Y. Effect evaluation of microbial mineralization for repairing load-induced crack in concrete with a cyclic injection-immersion process. *Case Stud. Constr. Mater.* **2022**, *17*, e01702. [CrossRef]
139. Carter, M.S.; Tuttle, M.J.; Mancini, J.A.; Martineau, R.; Hung, C.-S.; Gupta, M.K. Microbially induced calcium carbonate precipitation by sporosarcina pasteurii: A case study in optimizing biological $CaCO_3$ precipitation. *Appl. Environ. Microbiol.* **2023**, *89*, e01794-22. [CrossRef] [PubMed]
140. Seifan, M.; Berenjian, A. Microbially induced calcium carbonate precipitation: A widespread phenomenon in the biological world. *Appl. Microbiol. Biotechnol.* **2019**, *103*, 4693–4708. [CrossRef] [PubMed]
141. Cai, B.F.; Li, Q.; Zhang, X. *China Carbon Dioxide Capture, Utilization and Storage (CCUS) Annual Report (2021)—A Study of CCUS Pathways in China*; Environmental Planning Institute of the Ministry of Ecology and Environment, Wuhan Institute of Geotechnics, Chinese Academy of Sciences: Wuhan, China, 2021.

**Disclaimer/Publisher's Note:** The statements, opinions and data contained in all publications are solely those of the individual author(s) and contributor(s) and not of MDPI and/or the editor(s). MDPI and/or the editor(s) disclaim responsibility for any injury to people or property resulting from any ideas, methods, instructions or products referred to in the content.

Article

# Effect of H₂O Content on the Corrosion Behavior of X52 Steel in Supercritical CO₂ Streams Containing O₂, H₂S, SO₂ and NO₂ Impurities

Jia Liu [1,*], Dengzun Yao [2], Kai Chen [1], Chao Wang [2], Chong Sun [3,*], Huailiang Pan [1], Fanpeng Meng [1], Bin Chen [4] and Lili Wang [1]

1. China Petroleum Pipeline Engineering Corporation, Langfang 065000, China; cppechenkai@cnpc.com.cn (K.C.); panhuailiang@cnpc.com.cn (H.P.); meng-fanpeng@cnpc.com.cn (F.M.); shanqi861@sina.com (L.W.)
2. China Petroleum Pipeline Research Institute Co., Ltd., Langfang 065000, China; yaodengzun@cnpc.com.cn (D.Y.); chaowang03@cnpc.com.cn (C.W.)
3. School of Materials Science and Engineering, China University of Petroleum (East China), Qingdao 266580, China
4. China Petroleum Pipeline No.2 Engineering Company, Xuzhou 221000, China; gd2_chenbin@cnpc.com.cn
* Correspondence: cppeliujia@cnpc.com.cn (J.L.); sunchong@upc.edu.cn (C.S.); Tel.: +86-13070845753 (C.S.)

**Abstract:** In this study, the corrosion behavior of X52 pipeline steel affected by $H_2O$ content in supercritical $CO_2$ streams containing $O_2$, $H_2S$, $SO_2$ and $NO_2$ impurities was investigated by the weight loss test and surface characterization. The corrosion differences of the steel in impure supercritical $CO_2$ streams containing different $H_2O$ contents were analyzed. The influence of the variation of $H_2O$ content on the corrosion mechanism of steel in the complex impurity-containing supercritical $CO_2$ streams was discussed. The results show that the $H_2O$ content limit is 100 ppmv in supercritical $CO_2$ streams containing 200 ppmv $O_2$, 200 ppmv $H_2S$, 200 ppmv $SO_2$ and 200 ppmv $NO_2$ at 10 MPa and 50 °C. The impurities and their interactions significantly promote the formation of corrosive aqueous phase, thereby exacerbating the corrosion of X52 steel. The corrosion process of X52 steel in the environment with a low $H_2O$ content is controlled by the products of impurity reactions, whereas the impurities and the products of impurity reactions jointly control the corrosion process of the steel in the environment with a high $H_2O$ content.

**Keywords:** supercritical $CO_2$; $H_2O$ content; impurity; corrosion; corrosion product film

## 1. Introduction

Carbon capture, utilization and storage (CCUS) is an important technique to combat climate change [1,2]. A large number of CCUS projects have been successfully applied worldwide, and the success of these projects relies heavily on the safety, reliability, and cost effectiveness of the CCUS process [3,4]. In CCUS projects, $CO_2$ pipeline, as the primary transport manner of $CO_2$, is a key component in ensuring the safe and efficient transport of $CO_2$ from the capture site to its destination [5]. However, the inevitable presence of corrosive impurities such as $H_2O$, $O_2$, $SO_2$, $NO_2$ and $H_2S$ in the supercritical $CO_2$ streams transported by pipelines can pose a serious threat to the service safety of carbon steel pipelines [2,6–13].

It is generally accepted that $CO_2$ itself is not corrosive and does not cause the corrosion of pipeline steel when transported as pure $CO_2$ [14]. However, the presence of impurity components in the captured gas is inevitable. For the particular system of supercritical $CO_2$ transportation pipelines, $H_2O$ in supercritical $CO_2$ streams usually exists in a dissolved state, and this form of $H_2O$ is largely unlikely to cause corrosion of pipeline steel [2,15–17]. However, the presence of other impurity gases, such as $O_2$, $SO_2$, $NO_2$ and $H_2S$, can significantly reduce the solubility of $H_2O$ in supercritical $CO_2$ and induce the precipitation of free aqueous phase [8,17–20]. In addition, there are also complex chemical reactions among these impurity components, which can not only react to generate $H_2O$ but also form strong corrosive substances such as $H_2SO_4$ and $HNO_3$, thereby exacerbating the corrosion of pipeline steel [6]. Therefore, for the application of CCUS, it is of great interest to clarify the threshold of $H_2O$ content that does not cause significant corrosion for impurity-laden supercritical $CO_2$ transport environments. At present, numerous studies have been carried out to explore the possible threshold of $H_2O$ content in $CO_2$ transportation environments [6,8,21,22]. The $H_2O$ content thresholds obtained vary in the range of 40 ppmv to 2500 ppmv, which is closely related to the types of impurities, the concentration of impurities and the pipeline operating parameters. Although some progress has been achieved, the knowledge on the influence of $H_2O$ content on corrosion is still very limited, and most of the research work only involves a small number of impurity combinations. However, in the actual operation of the pipelines, it is inevitable that there will be a complex corrosive environment in which various impurities such as $O_2$, $H_2S$, $SO_2$ and $NO_2$ coexist. In this case, it not only raises a higher requirement for the limitation of $H_2O$ content, but also causes great changes in the corrosion law and mechanism of pipeline steel. Therefore, it is currently difficult for us to determine the $H_2O$ content limit in $CO_2$ transportation environments with the coexistence of all possibly corrosive impurities of $O_2$, $H_2S$, $SO_2$ and $NO_2$.

In view of this, this study investigated the effect of $H_2O$ content variation on the corrosion rate, corrosion morphology and corrosion film characteristics of X52 pipeline steel in a supercritical $CO_2$ environment where $H_2O$, $O_2$, $H_2S$, $SO_2$ and $NO_2$ impurities coexisted, and explored the influence mechanism of multiple impurity interactions on the corrosion of X52 steel under different $H_2O$ content conditions. Compared with the knowledge achieved in this domain, this study is expected to provide novel insights into the corrosion mechanism of pipeline steel in a complex impurity-containing $CO_2$ transportation environment, and also provide a scientific basis for the limitation of $H_2O$ content when the impurities of $O_2$, $H_2S$, $SO_2$ and $NO_2$ coexist in supercritical $CO_2$ transportation pipelines.

## 2. Materials and Methods

The specimen used for the corrosion test was machined to a size of 40 mm × 15 mm × 3 mm from a commercial X52 pipeline steel. Four parallel specimens were prepared for each group of tests to ensure the repeatability of test results. The chemical compositions of X52 steel were 0.10% C, 0.17% Si, 1.10% Mn, 0.011% P, 0.006% S, 0.006% Mo, 0.020% Cr, 0.009% Ni, 0.062% Al, 0.016% Cu, 0.002% V and Fe balance. Figure 1 shows the microstructure of X52 steel, which consisted of ferrite and pearlite. Prior to the tests, the surface of the specimen was successively ground with 240, 600, 800 and 1000 grit SiC paper to ensure the same surface roughness. After that, the specimen was cleaned with deionized (DI) water, dewatered with alcohol and dried with cold air. The original weight ($W_1$) of the prepared specimen was measured by an electronic balance with a metering precision of 0.0001 g.

**Figure 1.** Microstructure of X52 pipeline steel.

Corrosion simulation tests were carried out in a high-temperature and high-pressure reactor with a 3 L volume. The test device was described in a previous study [23]. The conditions of corrosion simulation tests are shown in Table 1, where the $H_2O$ content of 4333 ppmv was the saturation solubility of $H_2O$ in supercritical $CO_2$ at 50 °C and 10 MPa (at a $H_2O$ content of up to 4333 ppmv, $H_2O$ will be completely dissolved in supercritical $CO_2$, i.e., there is no free water phase in the initial corrosion environment). The concentrations of $O_2$, $H_2S$, $SO_2$ and $NO_2$ impurities were determined based on the commonly used $CO_2$ quality specifications [21,24–26]. Among various impurities, the maximum concentration of $H_2S$ was limited to 200 ppmv due to health and safety considerations [21,24]. Therefore, a concentration of 200 ppmv was selected in this study mainly according to the limitation on $H_2S$. In order to keep a consistent concentration of each impurity, the concentrations of $O_2$, $SO_2$ and $NO_2$ impurities were also fixed at 200 ppmv. Prior to the test, the specimens were placed on a Teflon fixture and the required amount of $H_2O$ (de-oxygenated DI water) was added to the reactor. In order to remove the air left in the reactor during the installation, a continuous flow of high-purity $CO_2$ was introduced into the reactor for 2 h after closing the reactor. The reactor was preferentially heated to 50 °C. The impurities of $O_2$, $H_2S$, $SO_2$ and $NO_2$ were added to the reactor in their respective required concentrations. Finally, $CO_2$ was added to reach 10 MPa. The test duration was 72 h.

**Table 1.** Test conditions.

| Temperature (°C) | $CO_2$ (MPa) | $O_2$ (ppmv) | $H_2S$ (ppmv) | $SO_2$ (ppmv) | $NO_2$ (ppmv) | $H_2O$ (ppmv) |
|---|---|---|---|---|---|---|
| 50 | 10 | 200 | 200 | 200 | 200 | 20, 100, 500 1000, 2000, 4333 |

At the end of the test, the specimens were taken out, photographed using a digital camera and then placed in a vacuum drying chamber for natural dehydration. A total of 1 L pickling solution was prepared using 100 mL of hydrochloric acid (the density was 1.19 g/mL), 5 g of hexamethyltetramine and DI water [27]. Three corroded specimens were placed in the above solution to remove the corrosion products. The weight ($W_2$) of each specimen was measured again after drying. The corrosion rate of the specimen was calculated using the weight loss method [27]:

$$V_{CR} = \frac{8.76 \times 10^4 (W_1 - W_2)}{S \rho t} \quad (1)$$

where $V_{CR}$ is the corrosion rate, mm/y; $W_1$ and $W_2$ are the weight of the specimen before and after corrosion, g; S is the exposed area of the specimen, $cm^2$; $\rho$ is the density of the specimen, $g/cm^3$; t is the corrosion time, h; $8.76 \times 10^4$ is the unit conversion constant. The corrosion rates with error bars reported in this study were the average values calculated from three parallel specimens.

The surface and cross-sectional morphologies of the corroded specimens were observed using scanning electron microscopy (SEM). The elemental compositions of the corrosion products and the elemental distributions in the cross-section of corrosion product film were analyzed by energy dispersive spectroscopy (EDS). The phase compositions of the corrosion products were determined by X-ray diffraction (XRD, Cu target, 40 kV, 40 mA). The chemical valences of the corrosion products were analyzed by X-ray photoelectron spectroscopy (XPS, Al target, $h\nu$ = 1486.6 eV).

## 3. Results and Discussion

*3.1. Corrosion Rate*

Figure 2 exhibits the variation of corrosion rate of X52 steel with $H_2O$ content exposed to supercritical $CO_2$ steams containing the impurities of $O_2$, $H_2S$, $SO_2$ and $NO_2$ at 10 MPa and 50 °C. At a $H_2O$ content of 20 ppmv, the corrosion rate of X52 steel is 0.0199 mm/y. The rate increases to 0.0234 mm/y when the $H_2O$ content reaches 100 ppmv. However, in the range of 100–2000 ppmv $H_2O$ content, the corrosion rate of X52 steel dramatically increases as the $H_2O$ content rises. Upon reaching the 2000 ppmv $H_2O$ content, X52 steel has a corrosion rate of 0.2671 mm/y. After the $H_2O$ content exceeds 2000 ppmv, the increment trend of corrosion rate slows down, reaching a value of 0.2838 mm/y at the saturated solubility of 4333 ppmv. Obviously, the corrosion rate of X52 steel is strongly associated with the variation of $H_2O$ content. Given that the remarkable increase in corrosion rate when the $H_2O$ content exceeds 100 ppmv, it is reasonably inferred that the $H_2O$ content limit should be 100 ppmv in supercritical $CO_2$ streams containing 200 ppmv $O_2$, 200 ppmv $H_2S$, 200 ppmv $SO_2$ and 200 ppmv $NO_2$ at 10 MPa and 50 °C.

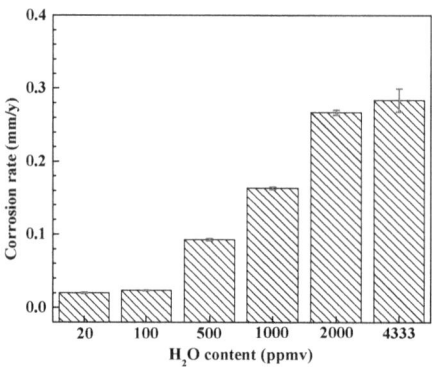

**Figure 2.** Variations of corrosion rate of X52 steel with $H_2O$ content exposed to supercritical $CO_2$ steams containing the impurities of $O_2$, $H_2S$, $SO_2$ and $NO_2$ at 10 MPa and 50 °C for 72 h.

*3.2. Morphological Observation of Corrosion Product Film*

3.2.1. $H_2O$ Content of between 20 and 100 ppmv

Figure 3 shows the macroscopic and SEM surface morphology, cross-sectional backscattered electron images and elemental distributions in cross-section of X52 steel in supercritical $CO_2$-$H_2O$-impurity environment with $H_2O$ content ranging from 20 to 100 ppmv, where the results of EDS analysis of corrosion products at regions A and B are shown in Table 2. It can be seen that at a $H_2O$ content of 20 ppmv, a thin layer of yellowish corrosion products is deposited unevenly on the surface of the specimen, and the polishing traces due to the pretreatment are still observed on the local areas of the specimen. A small amount of spherical corrosion products is distributed on the surface of the specimen, and the matrix below is slightly corroded. With the increase in $H_2O$ content, the coverage of the corrosion products on the specimen surface increases significantly, and the corrosion product film exhibits an obvious oxidation color. It can be seen from SEM images that a large number of spherical products and a small number of worm-like products are alternately distributed on the specimen surface. The results of the EDS analysis show that in the range of 20–100 ppmv, the corrosion products are mainly composed of Fe and O, indicating that the corrosion products are mainly oxygen-containing compounds. In addition, the increase in the coverage of corrosion product film and the corrosion rate of the steel also shows that even when the $H_2O$ content is far less than the saturation solubility (4333 ppmv) of $H_2O$ in supercritical $CO_2$ streams, the free aqueous phase can still precipitate from the $CO_2$ streams, thereby resulting in the corrosion of X52 steel. The increased corrosion rate strongly supports that the precipitation of the free aqueous phase increases with the increase in the $H_2O$ content in impure supercritical $CO_2$ streams.

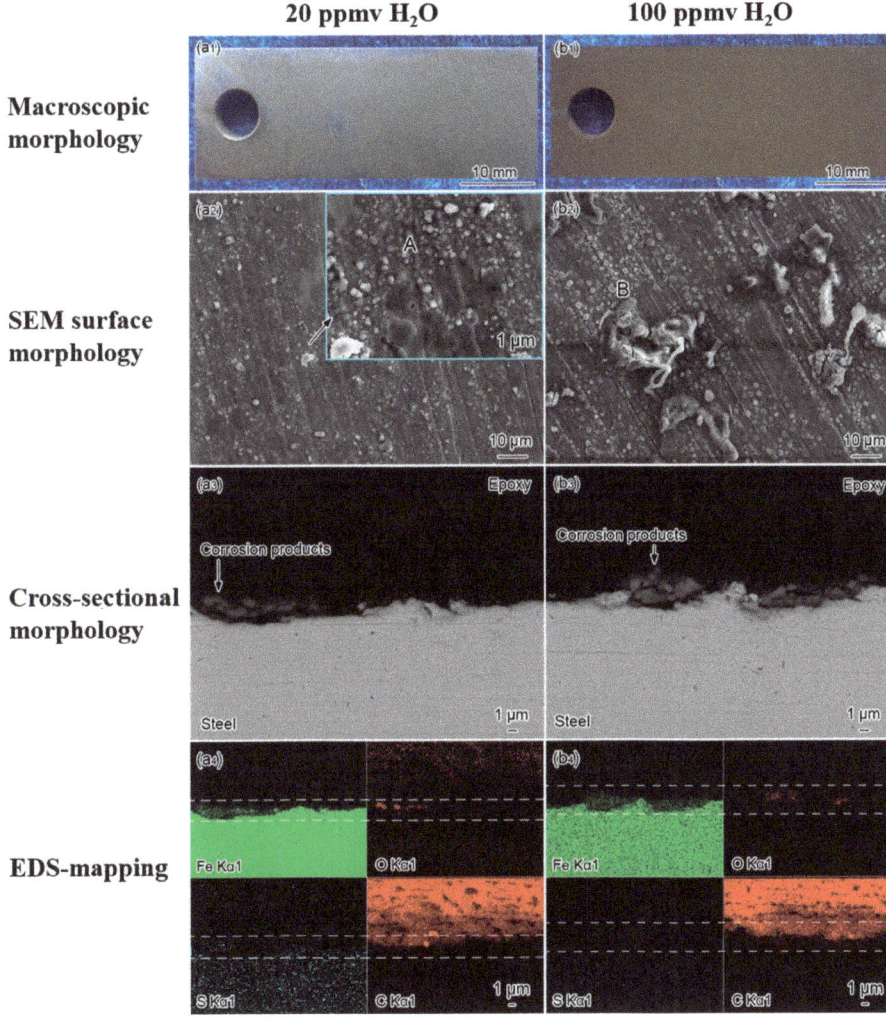

**Figure 3.** ($a_1,b_1$) Macroscopic morphologies, ($a_2,b_2$) SEM surface morphologies, ($a_3,b_3$) cross-sectional backscattered electron images and ($a_4,b_4$) distributions of main elements in cross-section of X52 steels after corrosion for 72 h in supercritical $CO_2$-$H_2O$-$O_2$-$H_2S$-$SO_2$-$NO_2$ environment with different $H_2O$ contents at 10 MPa and 50 °C: ($a_1$–$a_4$) 20 ppmv $H_2O$; ($b_1$–$b_4$) 100 ppmv $H_2O$. (A and B are the regions where EDS analysis is performed, the blue box region is the magnified image of the region denoted by the arrow, and the corrosion products are located in the region between the dotted lines.)

**Table 2.** Main elements in corrosion products on X52 steel denoted by A-B in Figure 3 (at%).

| Region | Fe | O | S |
|---|---|---|---|
| A | 27.2 | 65.1 | 4.6 |
| B | 33.7 | 64.6 | 1.1 |

### 3.2.2. $H_2O$ Content of between 500 and 2000 ppmv

Figure 4 shows the macroscopic and SEM surface morphology, cross-sectional backscattered electron images and elemental distributions in cross-section of X52 steel in the environment with a $H_2O$ content of 500 to 2000 ppmv. It can be seen that when the $H_2O$ content is 500 ppmv, the surface of the specimen is uniformly covered with a layer of spherical corrosion products with a thickness of about 1 μm. When the $H_2O$ content increases to 1000 ppmv, a large number of worm-like corrosion products are formed on the specimen surface, and the thickness of the corrosion product film reaches about 10 μm. However, when the $H_2O$ content increases to 2000 ppmv, the macro- and micro-morphologies of the corrosion product film change significantly. The corrosion products gradually become grey-black and present a mud-like morphology. Moreover, the cracks caused by dehydration can be observed. Compared to the uniform deposition of corrosion products at 500 and 1000 ppmv $H_2O$ contents, the corrosion product film at 2000 ppmv $H_2O$ content shows varying degrees of bulging and more pronounced corrosion of the matrix beneath it. The EDS results show that at the $H_2O$ content of 500 ppmv, the corrosion products are mainly composed of Fe and O elements, while at the $H_2O$ content of 1000 ppmv or 2000 ppmv, the corrosion products mainly contain Fe, O and S elements. Apparently, the increase in $H_2O$ content significantly promotes the formation of sulfur-containing products in the corrosion product film. This implies that the increase in $H_2O$ content causes the changes not only in corrosion rate and corrosion morphology, but also possibly in corrosion mechanism, i.e., the influence of sulfur-containing substances (e.g., $H_2S$ and $SO_2$) on the corrosion of X52 steel intensifies with the increase in $H_2O$ content.

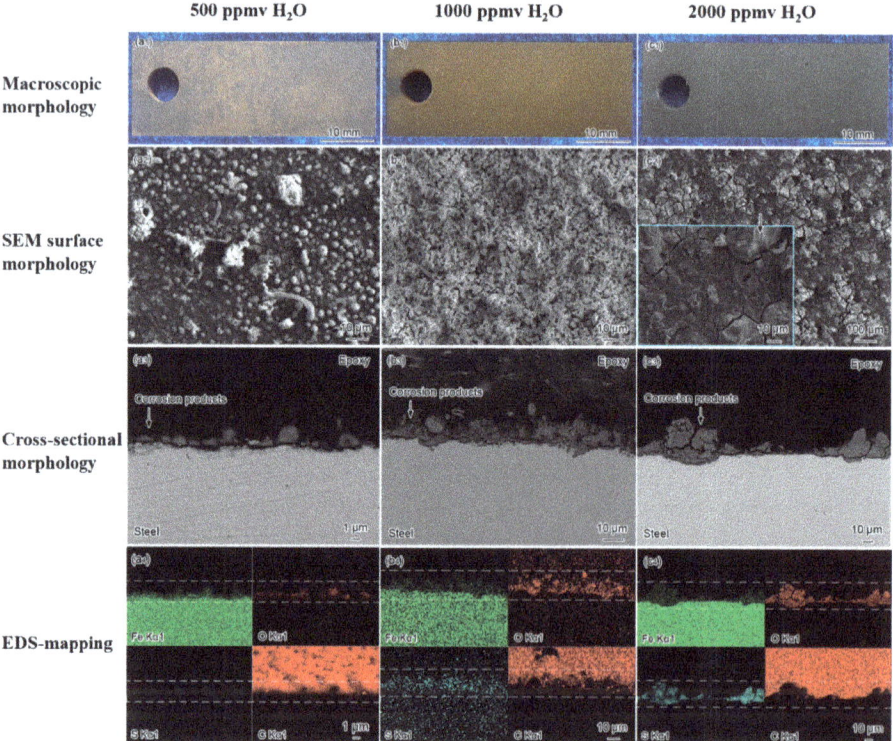

**Figure 4.** ($a_1$–$c_1$) Macroscopic morphologies, ($a_2$–$c_2$) SEM surface morphologies, ($a_3$–$c_3$) cross-sectional backscattered electron images and ($a_4$–$c_4$) distributions of main elements in cross-section of X52 steels after corrosion for 72 h in supercritical $CO_2$-$H_2O$-$O_2$-$H_2S$-$SO_2$-$NO_2$ environment with different $H_2O$ contents at 10 MPa and 50 °C: ($a_1$–$a_4$) 500 ppmv $H_2O$; ($b_1$–$b_4$) 1000 ppmv $H_2O$; ($c_1$–$c_4$) 2000 ppmv $H_2O$. (The blue box region is the magnified image of the region denoted by the arrow, and t the corrosion products are located in the region between the dotted lines.)

### 3.2.3. $H_2O$ Content of over 2000 ppmv

Figure 5 shows the macroscopic and SEM surface morphology, cross-sectional backscattered electron images and elemental distributions in cross-section of X52 steel in the environment with a $H_2O$ content of 4333 ppmv. It can be seen that at the saturation solubility (4333 ppmv) of $H_2O$ in supercritical $CO_2$ streams, the corrosion product film on the surface of the specimen is black in color and the corrosion products with droplet-shaped protrusions can be observed in local areas. SEM morphology shows that the raised areas of the corrosion products exhibit the same characteristics as the overall corrosion product film, all of which present the mud-like morphology. In addition, corrosion is more evident in the matrix below the raised areas of the corrosion products. The results of EDS line scanning analysis show that the corrosion products in different areas are mainly composed of Fe, O and S elements. Therefore, the formation of localized droplet-like corrosion products on the surface of the corrosion product film may be related to the deposition state of the aqueous phase on the surface of the specimen at this $H_2O$ content.

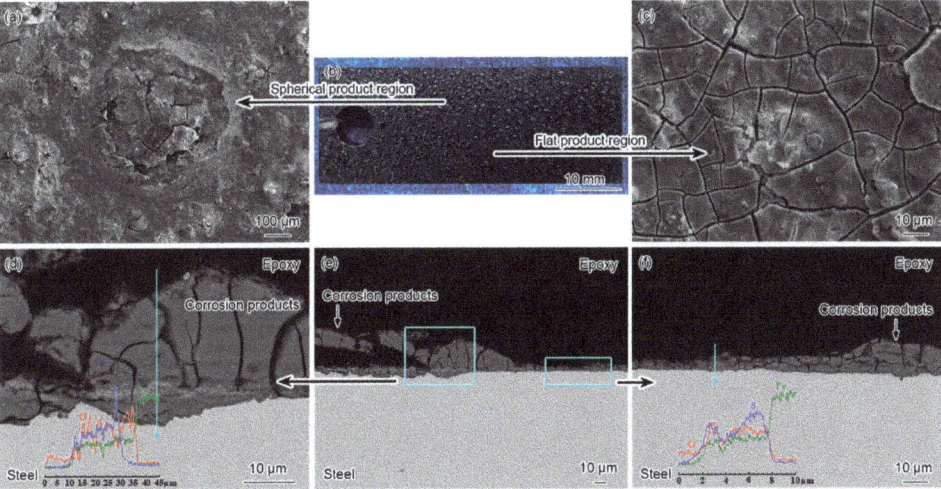

**Figure 5.** (**a,c**) SEM surface morphologies, (**b**) macroscopic morphologies, and (**d–f**) cross-sectional backscattered electron images and EDS line scanning analysis along blue arrow in cross-section of corrosion products on X52 steels after corrosion for 72 h in supercritical $CO_2$-$H_2O$-$O_2$-$H_2S$-$SO_2$-$NO_2$ environment with 4333 ppmv $H_2O$ content at 10 MPa and 50 °C. ((**d,f**) are the magnified images of the regions denoted by the blue frame and arrow in (**e**).)

Obviously, the increased $H_2O$ content can not only cause the change in corrosion rate, but also lead to the significant change in the characteristics of corrosion product film on the specimen surface. At a high $H_2O$ content, the content of S element in the corrosion products increases significantly, and the corrosion product film on the surface of X52 steel gradually transforms from Fe-O product-dominated film to Fe-O-S mixed product film.

### 3.3. Analysis of Corrosion Products
#### 3.3.1. XRD Analysis

Figure 6 shows the XRD patterns of the corrosion product film on the surface of X52 steel in supercritical $CO_2$-$H_2O$-impurity environment with different $H_2O$ contents. The diffraction peak of Fe is detected in the XRD pattern at a $H_2O$ content of 500 ppmv. As the average thickness of the corrosion product film formed on the surface of X52 steel is only about 2–3 μm (Figure 4($a_3$)), X-rays can penetrate this thin corrosion product film and excite the diffraction peaks of Fe in the steel matrix, which in turn masks the diffraction peaks of the corrosion products. The thickness of the corrosion film formed on the surface of X52 steel increases with the increase in $H_2O$ content. Correspondingly, the intensity of the diffraction peak of the corrosion products increases and appears in the XRD pattern. At $H_2O$ contents of 1000 and 2000 ppmv, XRD results show that the corrosion products are both mainly composed of FeOOH and $FeSO_4$. Combined with the previous EDS results, it can be deduced that the corrosion products have a low $FeSO_4$ content at 1000 ppmv $H_2O$ content and a high $FeSO_4$ content at 2000 and 4333 ppmv $H_2O$ contents.

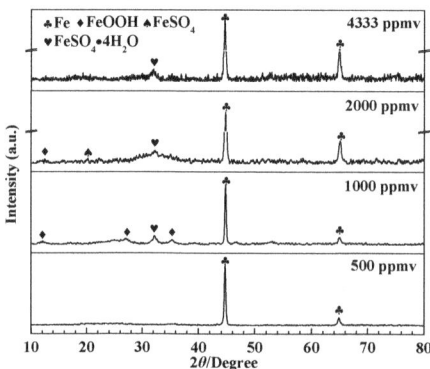

**Figure 6.** XRD patterns of corrosion products on X52 steel after corrosion for 72 h in supercritical $CO_2$-$H_2O$-$O_2$-$H_2S$-$SO_2$-$NO_2$ environment with different $H_2O$ contents at 10 MPa and 50 °C.

3.3.2. XPS Analysis

When the $H_2O$ content is below 500 ppmv, there are obvious corrosion products on the surface of the specimen (Figure 3). However, it is difficult to detect these products by XRD due to the fact that the depth that X-rays can penetrate is generally between ten and dozens of micrometers. For the thin corrosion product film on the surface of the specimen, the diffraction peaks of the corrosion products are easily masked by the high-intensity diffraction peaks of Fe derived from the matrix [28]. When the $H_2O$ content is higher than 500 ppmv, the thickness of the corrosion product film on the surface of the specimen increases significantly, and the products of FeOOH and $FeSO_4 \cdot 4H_2O$ can be detected by XRD, but the high-intensity diffraction peak of Fe from the matrix still masks some details of the composition of the corrosion products. In addition, some amorphous products are also difficult to characterize with XRD [14]. Different from XRD, XPS, as a high-resolution surface analysis technology, can extract the chemical information from 0 to 10 nm on the surface of the material, and amorphous or low-content corrosion products are easily detected by XPS [8]. Therefore, XPS was used to further determine the chemical state of the corrosion products on the surface of X52 steel formed at different $H_2O$ contents in order to eliminate the influence of the matrix.

Based on the results of EDS and XRD analyses, it can be concluded that the corrosion products mainly contain Fe, O and S elements. Therefore, these elements were selected for analysis in the XPS test. Figure 7 shows the high-resolution XPS spectra of the different elements in the corrosion products for $H_2O$ contents of 20 ppmv, 1000 ppmv and 4333 ppmv, respectively. The high-resolution XPS spectra of Fe 2p in Figure 7a show that the main peaks of Fe $2p_{3/2}$ and Fe $2p_{1/2}$ are located at the binding energies of 711.3 eV and 725.1 eV, respectively, indicating that Fe is in the oxidation states ($Fe^{2+}$ and/or $Fe^{3+}$) [14,29]. As the Fe-related compounds have similar binding energies, it is difficult to determine their specific compositions based on Fe 2p spectra alone. Therefore, this study determines the corrosion products mainly based on the fitting results of the split peak of O 1s and S 2p spectra. As shown in Figure 7b, the O 1s peak can be decomposed into three characteristic peaks at different $H_2O$ contents, the characteristic peaks at 530.1 eV and 531.6 eV correspond to hydroxyl oxides and the characteristic peak at 532.2 eV corresponds to sulphates [14,29,30].

According to the peak splitting results of S 2p spectra (Figure 7c), one S $2p_{3/2}$ peak at 168.6 eV indicates the presence of sulfate at a $H_2O$ content of 20 ppmv [14,30]; two S $2p_{3/2}$ peaks are present at 164.0 eV and 168.6 eV, which are ascribed to elemental sulfur and sulfate, respectively, at a $H_2O$ content of 1000 ppmv [14,30]; and at a $H_2O$ content of 4333 ppmv, three S 2p3/2 characteristic peaks are present at 164.0 eV, 166.8 eV and 168.6 eV, corresponding to elemental sulfur, sulfite and sulfate, respectively [14,30]. Taking into account the chemical state of each element, it can be determined that at a $H_2O$ content of 20 ppmv, the corrosion products are mainly FeOOH and a small amount of $FeSO_4$; at a $H_2O$ content of 1000 ppmv, the corrosion products are mainly FeOOH, $FeSO_4$ and S; at a $H_2O$ content of 4333 ppmv, the corrosion products are mainly FeOOH, $FeSO_3$, $FeSO_4$ and S. According to the results of EDS, XRD and XPS analyses, it can be concluded that the sulfur-containing products in the corrosion product film gradually increase as the $H_2O$ content increases.

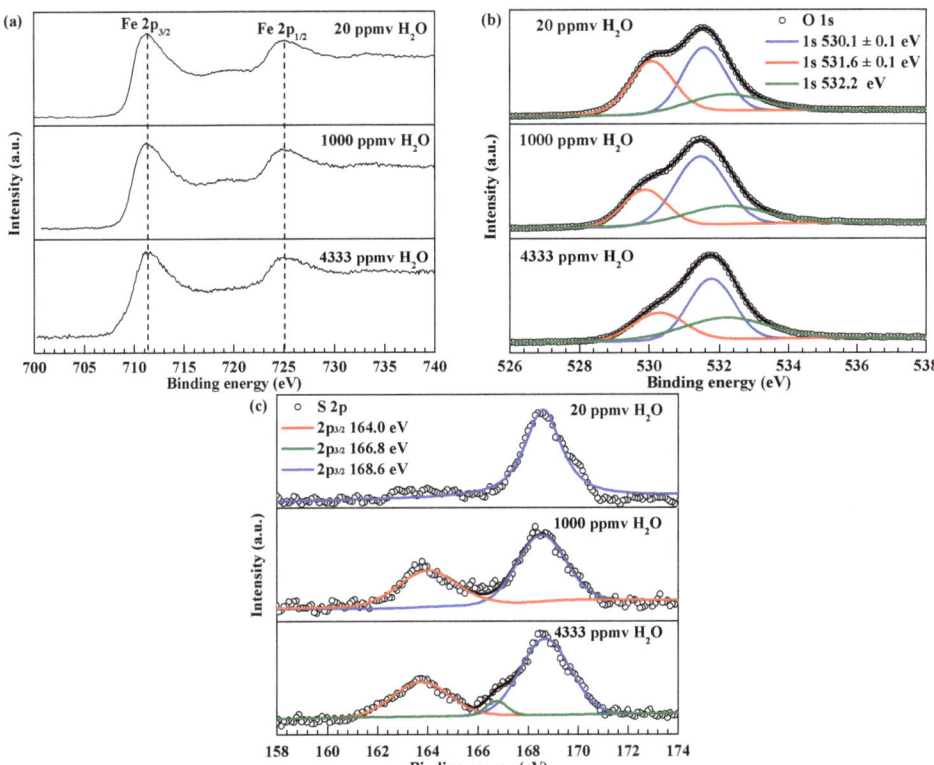

**Figure 7.** XPS spectra of (**a**) Fe 2p, (**b**) O 1s and (**c**) S 2p for the corrosion products on X52 steels after corrosion for 72 h in supercritical $CO_2$-$H_2O$-$O_2$-$H_2S$-$SO_2$-$NO_2$ environment with different $H_2O$ contents at 10 MPa and 50 °C.

### 3.4. Analysis of Corrosion Mechanism

In the environment of supercritical $CO_2$ transportation containing impurities, although supercritical $CO_2$ is the main body, the previous analysis results show that the chemical composition of the corrosion product film on the surface of X52 steel under different $H_2O$ contents is mainly FeOOH and $FeSO_4$ and a small amount of S or $FeSO_3$, without detecting the typical product of $CO_2$ corrosion ($FeCO_3$). This phenomenon indicates that the corrosion process and film formation of X52 steel are mainly controlled by impurity components. Related studies have shown that impurity components in the supercritical $CO_2$ streams can reduce the solubility of $H_2O$ in supercritical $CO_2$ and promote the formation of the aqueous phase [8,17–20]. Furthermore, the complex chemical reactions among various impurity components such as $O_2$, $H_2S$, $SO_2$ and $NO_2$ can generate additional corrosive substances such as $H_2SO_4$, $HNO_3$, elemental S and $H_2O$ [6]. Therefore, the obvious corrosion of X52 steel at conditions well below the saturation solubility of $H_2O$ is probably related to the additional corrosive substances, which are generated by the reaction between the impurities and exert influence on the amount of aqueous phase formation and the chemical environment of the aqueous phase. To further prove the above inference, hydrochemical simulations of supercritical $CO_2$ streams containing impurities were carried out with the aid of the Stream Analyzer module of the OLI Analyzer Studio software. The basic compositions of the streams used for the above calculation are 990 g $CO_2$ and 10 g $H_2O$, with $O_2$, $H_2S$, $H_2SO_3$, $H_2SO_4$ and $HNO_3$, with the content (mass fraction %) ranging from 0 to 0.03%, at a temperature of 50 °C and pressure of 10 MPa, respectively, the results of which are shown in Figure 8.

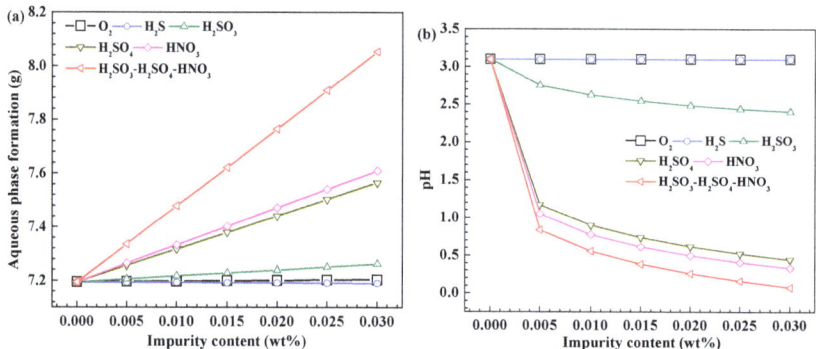

**Figure 8.** Influence of impurities on aqueous phase formation and pH value of aqueous phase formed in supercritical $CO_2$ streams at 10 MPa and 50 °C (calculated by OLI Analyzer Studio).

As shown in Figure 8a, an increase in $O_2$ or $H_2S$ content does not have a significant effect on the amount of aqueous phase formed in supercritical $CO_2$ streams. However, the amount of aqueous phase formed in supercritical $CO_2$ streams increases with an increase in $H_2SO_3$ (associated with $SO_2$), $H_2SO_4$ (associated with $O_2$ and $SO_2$) or $HNO_3$ (associated with $NO_2$). For supercritical $CO_2$ streams containing $H_2SO_3$, $H_2SO_4$ or $HNO_3$, the amount of aqueous phase formation increases significantly with an increase in their concentrations. It also implies that the presence of even a small amount of impurities in the environment where multiple impurities exist can significantly promote the formation of aqueous phase, leading to the formation of acid-rich aqueous phases in low-$H_2O$-content environments. Whereas these acid-rich aqueous phases usually have lower pH values (Figure 8b), the higher $H^+$ concentration in the aqueous phase promotes hydrogen evolution reactions at the cathode, which in turn intensifies the corrosion of the steel. This should be an important reason why the corrosion rate of up to 0.0199 mm/y can be achieved at the $H_2O$ content of only 20 ppmv.

When the $H_2O$ content is in the range of 20–100 ppmv, the corrosion products of X52 steel are mainly FeOOH and a small amount of $FeSO_4$. This suggests that the reaction between impurities is able to form at least the corrosive $H_2SO_4$ [31,32], thus producing the characteristic product of $FeSO_4$. Based on the results of this study, it is impossible to determine the exact FeOOH formation pathway. However, related studies have shown that FeOOH is a common corrosion product in environments containing strong oxidizing impurities of $O_2$ or $NO_2$ [31,33,34]. The amount of aqueous phase formed in low-$H_2O$-content environments is relatively small compared to high-$H_2O$-content environments. This is because the corrosion products formed on the steel surface can be oxidized in full contact with oxidants in supercritical $CO_2$ streams. For example, $FeSO_4$ can be oxidized by $O_2$ into FeOOH [8,35]:

$$4FeSO_4 + 6H_2O + O_2 \rightarrow 4FeOOH + H_2SO_4 \qquad (2)$$

As a result, the corrosion product film on the surface of X52 steel in the low-$H_2O$-content environment shows an obvious oxidation color (Figure 3($a_1$,$b_1$)), with the corrosion products being dominated by iron oxides. It is worth noting that the oxidation reaction between the primary products of corrosion and impurities also results in the cyclic regeneration of corrosive substances, which continues to cause corrosion of the steel matrix [8,35]. This may also be one of the reasons for the high corrosion rate of X52 steel in an extremely low-$H_2O$-content environment.

When the $H_2O$ content is in the range of 500–2000 ppmv, the amount of aqueous phase condensed on the surface of X52 steel under the interaction of impurities increases considerably due to the increased $H_2O$ content in the corrosion system, which also provides more electrolytes for the corrosion reaction, resulting in a significant increase in the corrosion rate. Moreover, the deposition of the aqueous phase on the corrosion products can prevent to some extent the oxidation of the corrosion products caused by the oxidizing agent in the supercritical $CO_2$ streams. This also leads to a gradual increase in the content of $FeSO_4$ in the corrosion products within this $H_2O$ content range (Figure 4($a_4$–$c_4$)), which also corresponds to a shift in the macroscopic morphological characteristics of X52 steel (Figure 4($a_1$–$c_1$)).

When the $H_2O$ content is above 2000 ppmv, a small amount of $FeSO_3$ can be detected in the corrosion products of X52 steel in addition to FeOOH and $FeSO_4$ (Figure 7), indicating that $SO_2$ has been involved in the film formation reaction. This shows that in addition to the chemical reaction products of impurities involved in the corrosion process, the impurities themselves can also be involved in the corrosion process. However, although the results of this study demonstrate that the reaction between impurities can form elemental S, it cannot be proved whether elemental S is involved in the corrosion process of X52 steel. But related studies have shown that elemental S formed in the supercritical $CO_2$ transport environment containing impurities can cause elemental sulfur corrosion of pipeline steel, which is one of the important reasons for the aggravation of pipeline steel corrosion in the environment where multiple impurities coexist [23]. As a result, X52 steel corrodes more severely in a high-$H_2O$-content environment under the combined effect of impurities and chemical reaction products between impurities, with corrosion rates exceeding 0.25 mm/y (Figure 2).

## 4. Conclusions

In summary, we explored the effect of $H_2O$ content in changing the corrosion behavior of X52 steel in a supercritical $CO_2$ environment containing the impurities of 200 ppmv $O_2$, 200 ppmv $H_2S$, 200 ppmv $SO_2$ and 200 ppmv $NO_2$ at 10 MPa and 50 °C. The main conclusions are drawn as follows:

(1) The corrosion rate of X52 steel increases from 0.0199 mm/y to 0.2838 mm/y as the $H_2O$ content increases from 20 ppmv to saturation solubility (4333 ppmv), while the critical $H_2O$ content that causes a significant change in the corrosion rate is 100 ppmv.

(2) $O_2$, $H_2S$, $SO_2$ and $NO_2$ impurities and their interactions jointly promote the formation of corrosive aqueous phases and aggravate the corrosion of X52 steel. With the increase in $H_2O$ content, the corrosion product film of X52 steel gradually changes from FeOOH-dominated film to $FeSO_4$ and FeOOH mixed film. Correspondingly, the corrosion process of X52 steel, which is controlled by the products of impurity reactions at a low $H_2O$ content, is transformed to be under the joint control of impurities and the products of impurity reactions at a high $H_2O$ content.

(3) The change in the corrosion rate of X52 steel strongly depends on the amount of aqueous phase precipitation in the environment and the amount of corrosive aqueous

phase generated by the chemical reactions between impurities. The increase in $H_2O$ content and the direct participation of impurities in the corrosion process greatly aggravate the corrosion of X52 steel.

**Author Contributions:** Conceptualization, J.L. and D.Y.; Investigation, J.L., K.C., C.W. and C.S.; Writing—original draft preparation, J.L. and C.S.; Writing—review and editing, C.S., H.P. and F.M.; Supervision, B.C. and L.W. All authors have read and agreed to the published version of the manuscript.

**Funding:** This work is supported by the PetroChina Scientific Research and Technology Development Project (2021ZZ01-02) and the National Natural Science Foundation of China (No. 52001328).

**Data Availability Statement:** The raw/processed data required to reproduce these findings cannot be shared at this time due to technical and time limitations. Concurrently, the data also form part of an ongoing study.

**Conflicts of Interest:** The authors declare no conflict of interest.

## References

1. Jones, A.C.; Lawson, A.J. *Carbon Capture and Sequestration (CCS) in the United States*; Congressional Research Service: Washington, DC, USA, 2021.
2. Kairy, S.K.; Zhou, S.; Turnbull, A.; Hinds, G. Corrosion of pipeline steel in dense phase $CO_2$ containing impurities: A critical review of test methodologies. *Corros. Sci.* **2023**, *214*, 110986. [CrossRef]
3. Yao, J.; Han, H.; Yang, Y.; Song, Y.; Li, G. A review of recent progress of carbon capture, utilization, and storage (CCUS) in China. *Appl. Sci.* **2023**, *13*, 1169. [CrossRef]
4. Peletiri, S.P.; Rahmanian, N.; Mujtaba, I.M. $CO_2$ Pipeline design: A review. *Energies* **2018**, *11*, 2184. [CrossRef]
5. Leung, D.Y.C.; Garamanna, G.; Maroto-Valer, M.M. An overview of current status of carbon dioxide capture and storage technologies. *Renew. Sustain. Energy Rev.* **2014**, *39*, 426–443. [CrossRef]
6. Barker, R.; Hua, Y.; Neville, A. Internal corrosion of carbon steel pipelines for dense-phase $CO_2$ transport in carbon capture and storage (CCS)—A review. *Int. Mater. Rev.* **2017**, *62*, 1–31. [CrossRef]
7. Halseid, M.; Dugstad, A.; Morland, B. Corrosion and bulk phase reactions in $CO_2$ transport pipelines with impurities: Review of recent published studies. *Energy Procedia* **2014**, *63*, 2557–2569. [CrossRef]
8. Sun, C.; Sun, J.B.; Liu, S.B.; Wang, Y. Effect of water content on the corrosion behavior of X65 pipeline steel in supercritical $CO_2$-$H_2O$-$O_2$-$H_2S$-$SO_2$ environment as relevant to CCS application. *Corros. Sci.* **2018**, *137*, 151–162. [CrossRef]
9. Morland, B.H.; Tadesse, A.; Svenningsen, G.; Springer, R.D.; Anderko, A. Nitric and sulfuric acid solubility in dense phase $CO_2$. *Ind. Eng. Chem. Res.* **2019**, *58*, 22924–22933. [CrossRef]
10. Xiang, Y.; Xu, M.H.; Choi, Y.S. State-of-the-art overview of pipeline steel corrosion in impure dense $CO_2$ for CCS transportation: Mechanisms and models. *Corros. Eng. Sci. Technol.* **2017**, *52*, 485–509. [CrossRef]
11. Sun, C.; Wang, Y.; Sun, J.B.; Lin, X.Q.; Li, X.D.; Liu, H.F.; Cheng, X.K. Effect of impurity on the corrosion behavior of X65 steel in water-saturated supercritical $CO_2$ system. *J. Supercrit. Fluids* **2016**, *116*, 70–82. [CrossRef]
12. Zhao, G.X.; Wang, Y.C.; Zhang, S.Q.; Song, Y. Influence mechanism of $H_2S$/$CO_2$-charging on corrosion of J55 steel in an artificial solution. *J. Chin. Soc. Corros. Prot.* **2022**, *42*, 785–790.
13. Sun, C.; Liu, J.X.; Sun, J.B.; Li, H.; Zhao, Z.; Lin, X.Q.; Wang, Y. Corrosion behaviors of X65 steel in gaseous $CO_2$ environment containing impurities. *J. China Univ. Petrol.* **2022**, *43*, 129–139.
14. Li, C.; Xiang, Y.; Song, C.C.; Ji, Z.L. Assessing the corrosion product scale formation characteristics of X80 steel in supercritical $CO_2$-$H_2O$ binary systems with flue gas and NaCl impurities relevant to CCUS technology. *J. Supercrit. Fluids* **2019**, *146*, 107–119. [CrossRef]
15. Cui, G.; Yang, J.G.; Liu, J.G.; Li, Z.L. A comprehensive review of metal corrosion in a supercritical $CO_2$ environment. *Int. J. Greenh. Gas. Control* **2019**, *90*, 102814. [CrossRef]
16. Jiang, X.; Qu, D.R.; Song, X.L.; Liu, X.H.; Zhang, Y.L. Critical water content for corrosion of X65 mild steel in gaseous, liquid and supercritical $CO_2$ stream. *Int. J. Greenh. Gas. Control* **2019**, *85*, 11–22. [CrossRef]
17. Sun, C.; Sun, J.B.; Luo, J.-L. Unlocking the impurity-induced pipeline corrosion based on phase behavior of impure $CO_2$ streams. *Corros. Sci.* **2020**, *165*, 108367. [CrossRef]
18. Morland, B.; Norby, T.; Tjelta, M.; Sevenningsen, G. Effect of $SO_2$, $O_2$, $NO_2$, and $H_2O$ concentrations on chemical reactions and corrosion of carbon steel in dense phase $CO_2$. *Corrosion* **2019**, *75*, 1327–1338. [CrossRef]
19. Choi, Y.-S.; Hassani, S.; Vu, T.N.; Nešić, S.; Abas, A.Z.B. Effect of $H_2S$ on the corrosion behavior of pipeline steels in supercritical and liquid $CO_2$ environments. *Corrosion* **2016**, *72*, 999–1009. [CrossRef]
20. Sun, J.; Sun, C.; Zhang, G.; Li, X.; Zhao, W.; Jiang, T.; Liu, H.; Cheng, X.; Wang, Y. Effcet of $O_2$ and $H_2S$ impurities on the corrosion behavior of X65 steel in water-saturated supercritical $CO_2$ system. *Corros. Sci.* **2016**, *107*, 31–40. [CrossRef]

21. De Visser, E.; Hendriks, C.; Barrio, M.; Molnvik, M.J.; de Koeijer, G.; Liljemark, S.; Le Gallo, Y. Dynamis $CO_2$ quality recommendations. *Int. J. Greenh. Gas. Control* **2008**, *2*, 478–484. [CrossRef]
22. Xiang, Y.; Wang, Z.; Yang, X.X.; Zheng, L.; Ni, W.D. The upper limit of moisture content for supercritical $CO_2$ pipeline transport. *J. Supercrit. Fluids* **2012**, *67*, 14–21. [CrossRef]
23. Sun, C.; Sun, J.B.; Wang, Y.; Lin, X.Q.; Li, X.D.; Cheng, X.K.; Liu, H.F. Synergistic effect of $O_2$, $H_2S$ and $SO_2$ impurities on the corrosion behavior of X65 steel in water-saturated supercritical $CO_2$ system. *Corros. Sci.* **2016**, *107*, 193–203. [CrossRef]
24. *ISO 27913:2016*; Carbon Dioxide Capture, Transportation and Geological Storage—Pipeline Transportation Systems. International Organization for Standardization: Geneva, Switzerland, 2016.
25. Shirley, P.; Myles, P. *Quality Guidelines for Energy System Studies: $CO_2$ Impurity Design Parameters*; National Energy Technology Laboratory: Pittsburgh, PA, USA, 2019; pp. 12–18.
26. Brown, J.; Graver, B.; Gulbrandsen, E.; Dugstad, A.; Morland, B. Update of DNV recommended practice RP-J202 with focus on $CO_2$ corrosion with impurities. *Energy Procedia* **2014**, *63*, 2432–2441. [CrossRef]
27. *ASTM G1-03:2011*; Standard Practice for Preparing, Cleaning, and Evaluating Corrosion Test Specimens. ASTM International: West Conshohocken, PA, USA, 2011.
28. Liu, J.; Saw, R.E.; Kiang, Y.-H. Calculation of effective penetration depth in X-ray diffraction for pharmaceutical solids. *J. Pharm. Sci.* **2010**, *99*, 3807–3814. [CrossRef]
29. Heuer, J.K.; Stubbins, J.F. An XPS characterization of $FeCO_3$ films from $CO_2$ corrosion. *Corros. Sci.* **1999**, *41*, 1231–1243. [CrossRef]
30. Xiang, Y.; Wang, Z.; Xu, C.; Zhou, C.; Li, Z.; Ni, W. Impact of $SO_2$ concentration on the corrosion rate of X70 steel and iron in water-saturated supercritical $CO_2$ mixed with $SO_2$. *J. Supercrit. Fluids* **2011**, *58*, 286–294. [CrossRef]
31. Dugstad, A.; Halseid, M.; Morland, B. Effect of $SO_2$ and $NO_2$ on corrosion and solid formation in dense phase $CO_2$ pipelines. *Energy Procedia* **2013**, *37*, 2877–2887. [CrossRef]
32. Xu, M.; Zhang, Q.; Yang, X.X.; Wang, Z.; Liu, J.; Li, Z. Impact of surface roughness and humidity on X70 steel corrosion in supercritical $CO_2$ mixture with $SO_2$, $H_2O$, and $O_2$. *J. Supercrit. Fluids* **2016**, *107*, 286–297. [CrossRef]
33. Sim, S.; Cole, I.S.; Choi, Y.-S.; Birbilis, N. A review of the protection strategies against internal corrosion for the safe transport of supercritical $CO_2$ via steel pipelines for CCS purposes. *Int. J. Greenh. Gas. Con.* **2014**, *29*, 185–199. [CrossRef]
34. McIntire, G.; Lippert, J.; Yedelson, J. The effect of dissolved $CO_2$ and $O_2$ on the corrosion of iron. *Corrosion* **1990**, *46*, 91–95. [CrossRef]
35. Choi, Y.S.; Nešić, S.; Young, D. Effect of impurities on the corrosion behavior of $CO_2$ transmission pipeline steel in supercritical $CO_2$-water environments. *Environ. Sci. Technol.* **2010**, *44*, 9233–9238. [CrossRef]

**Disclaimer/Publisher's Note:** The statements, opinions and data contained in all publications are solely those of the individual author(s) and contributor(s) and not of MDPI and/or the editor(s). MDPI and/or the editor(s) disclaim responsibility for any injury to people or property resulting from any ideas, methods, instructions or products referred to in the content.

Article

# Research and Application of Carbon Capture, Utilization, and Storage–Enhanced Oil Recovery Reservoir Screening Criteria and Method for Continental Reservoirs in China

Jinhong Cao [1,2,3], Ming Gao [3,4,*], Zhaoxia Liu [3,4], Hongwei Yu [3,4], Wanlu Liu [3,4] and Hengfei Yin [3,4]

1. University of Chinese Academy of Sciences, Beijing 100049, China; caojh22@petrochina.com.cn
2. Institute of Porous Flow & Fluid Mechanics, Chinese Academy of Sciences, Langfang 065007, China
3. Research Institute of Petroleum Exploration & Development, PetroChina, Beijing 100083, China; liuzhaoxia@petrochina.com.cn (Z.L.); yhongwei@petrochina.com.cn (H.Y.); liuwanlu@petrochina.com.cn (W.L.); yinhf21@petrochina.com.cn (H.Y.)
4. State Key Laboratory of Enhanced Oil and Gas Recovery, Beijing 100083, China
* Correspondence: gaoming010@petrochina.com.cn

**Abstract:** CCUS-EOR is a crucial technology for reducing carbon emissions and enhancing reservoir recovery. It enables the achievement of dual objectives: improving economic efficiency and protecting the environment. To explore a set of CCUS-EOR reservoir screening criteria suitable for continental reservoirs in China, this study investigated and compared the CCUS-EOR reservoir screening criteria outside and in China, sorted out the main reservoir parameters that affect $CO_2$ flooding, and optimized the indices and scope of CCUS-EOR reservoir screening criteria in China. The weights of parameters with respect to their influences on CCUS-EOR were determined through principal component analysis. The results show that there are 14 key parameters affecting $CO_2$ flooding, which can be categorized into four levels. For the first level, the crude oil-$CO_2$ miscibility index holds the greatest weight of 0.479. It encompasses seven parameters: initial formation pressure, current formation pressure, temperature, depth, $C_2$–$C_{15}$ molar content, residual oil saturation, and minimum miscibility pressure. The second level consists of the crude oil mobility index, which has a weight of 0.249. This index includes four parameters: porosity, permeability, density, and viscosity. The third level pertains to the index of reservoir tectonic characteristics, with a weight of 0.141. It comprises two parameters: permeability variation coefficient and average effective thickness. Lastly, the fourth level focuses on the index of reservoir property change, with a weight of 0.131, which solely considers the pressure maintenance level. Based on the CCUS-EOR reservoir screening criteria and index weights established in this study, comprehensive scores for CCUS-EOR were calculated for six blocks in China. Among these, five blocks are deemed suitable for CCUS-EOR. Based on the comprehensive scoring results, a planning for field application of CCUS-EOR is proposed. The study provides a rational method to evaluate the CCUS-EOR reservoir screening and field application in continental reservoirs in China.

**Keywords:** CCUS-EOR; $CO_2$ flooding; continental reservoir; screening criteria; principal component analysis

**Citation:** Cao, J.; Gao, M.; Liu, Z.; Yu, H.; Liu, W.; Yin, H. Research and Application of Carbon Capture, Utilization, and Storage–Enhanced Oil Recovery Reservoir Screening Criteria and Method for Continental Reservoirs in China. *Energies* **2024**, *17*, 1143. https://doi.org/10.3390/en17051143

Academic Editor: Jalel Azaiez

Received: 30 November 2023
Revised: 24 December 2023
Accepted: 28 December 2023
Published: 28 February 2024

Copyright: © 2024 by the authors. Licensee MDPI, Basel, Switzerland. This article is an open access article distributed under the terms and conditions of the Creative Commons Attribution (CC BY) license (https://creativecommons.org/licenses/by/4.0/).

## 1. Introduction

To meet the production and lifestyle needs of the increasing global population, the demand for non-renewable energy sources is growing significantly [1], and greenhouse gas emissions are also rising with the industrial production [2]. The issues of global climate change and energy demand are becoming increasingly prominent, posing new challenges for environmental protection and energy utilization. $CO_2$ is the primary greenhouse gas in the atmosphere, and human activities such as fossil fuel-based power generation, heating, and transportation are the main sources of $CO_2$ emissions [3]. Since the industrial revolution, the increase in $CO_2$ levels in the atmosphere has contributed to over 60% of environmental pollution [4], the sea-level rise has accelerated [5], and the environmental problems induced by climate change such as the melting of ice caps and polar glaciers are irreversible [6]. If the environmental impacts of human activities will not be controlled, these issues are expected to worsen rapidly in the coming decades [7]. Carbon capture and utilization (CCU) technologies, including Power-to-Gas (PtG) [8], Power-to-Liquid (PtL) [9], and $CO_2$-Enhanced Oil Recovery ($CO_2$-EOR) [10], in conjunction with carbon capture and storage (CCS) methods like physical absorption and chemical absorption [11], form a comprehensive carbon capture, utilization, and storage (CCUS) technology, which has been proven to be effective for mitigating climate change and achieving sustainable development [12]. It is a crucial tool for reducing carbon emissions in sectors such as coal-fired power generation, cement, and steel production [13]. CCUS-EOR (Enhanced Oil Recovery) [14], a key extension of CCUS, can inject the captured $CO_2$ into oil reservoirs to enhance oil recovery and reduce carbon emissions by storing the injected $CO_2$ in the reservoirs, thereby achieving the dual goals of increasing production efficiency and protecting the environment [15].

$CO_2$ flooding was initially proposed in laboratory research in 1930 and achieved industrial applications after 1950 [4]. Since 1970, projects utilizing $CO_2$ for displacing oil in (depleted) reservoirs have gradually increased. According to statistics of the International Energy Agency (IEA), as of 2017, there were over 160 CCUS-EOR projects globally. The United States, Canada, and China were among the early developers of such projects, and in recent years, countries like Brazil, Turkey, Norway, and Saudi Arabia have also employed to CCUS-EOR [16]. Research has indicated that the mechanisms of $CO_2$ in reservoirs include interfacial tension reduction by oil and gas mass transfer, as well as crude oil expansion and viscosity reduction through gas dissolution, thereby improving flooding efficiency [17,18]. Additionally, the injected $CO_2$ can be stored within depleted reservoirs through physical trapping and chemical reactions with rock formations, thereby reducing $CO_2$ levels in the atmosphere [19]. Therefore, CCUS-EOR has become the primary technology for improving oil recovery rate, following chemical flooding and thermal recovery methods. Identifying suitable depleted oil reservoirs for CCUS-EOR can help recover more reserves from reservoirs with more economic benefits. Most large-scale CCUS-EOR projects are concentrated in North America and Canada, where they have achieved high oil production rates [20]. In China, many oil reservoirs were initially developed using water flooding and chemical flooding methods, while since 2019, breakthroughs have been made in the research and engineering demonstration of CCUS-EOR [7], indicating that CCUS-EOR can be used in depleted oil reservoirs for further enhanced oil recovery. By screening key parameters such as reservoir characteristics and crude oil properties for developed reservoirs, the potential of CCUS-EOR is assessed to achieve the effective application of $CO_2$ flooding technology.

Since the occurrence of CCUS-EOR technology, numerous highly applicable reservoir screening criteria have emerged for various oil displacement projects. These criteria primarily focus on reservoir parameters and fluid properties that are closely related to the $CO_2$–crude oil miscibility, and they are mainly applicable to reservoirs in the United States and Canada [15,21]. However, with the evolution of $CO_2$ injection technology, these criteria have exhibited a deviation from the actual application scope and have become less representative [13]. In contrast to reservoirs in other countries/regions where $CO_2$ and crude oil are more miscible, most reservoirs in China face harsher conditions—particularly, the highly heterogeneous continental sedimentary reservoirs are estimated with inferior ultimate recovery to marine sedimentary reservoirs [22], and the reservoirs suitable for CCUS-EOR are deeper. Moreover, in China, the reservoirs are widespread with large spans and reflect distinct characteristics from region to region. Although laboratory studies and field tests have been performed on $CO_2$ flooding for some reservoirs in China, there is currently no comprehensive research summarizing CCUS-EOR reservoir screening criteria for all reservoirs across China. Therefore, it is necessary to further update and refine the existing CCUS-EOR reservoir screening criteria to develop a set of criteria that align with the specific characteristics of reservoirs in China. Furthermore, the formulation of a complete set of CCUS-EOR screening criteria should ideally consider both technical and non-technical indices [23]. Technical indices include indicators of $CO_2$ capture capacity, $CO_2$-EOR capability, and $CO_2$ storage capacity. Non-technical indices encompass economic parameters, policy factors, safety risks, and more. This paper focuses on establishing parameters for screening reservoirs based on their $CO_2$-EOR capabilities without considering other technical and non-technical indices.

In this paper, the CCUS-EOR reservoir screening criteria are investigated through a review of the available literature, reports, and materials. The $CO_2$ flooding projects in the world are compared, and the parameters in the CCUS-EOR reservoir screening criteria are analyzed. Then, through principal component analysis, the influences of screening indices on CCUS-EOR for reservoirs in China are identified. Finally, the CCUS-EOR reservoir screening criteria suitable for reservoirs in China are proposed. The study results will provide theoretical guidance and optimization recommendations for the rapid and accurate screening of reservoirs for CCUS-EOR.

## 2. Investigation of CCUS-EOR Reservoir Screening Criteria

### 2.1. CCUS-EOR Reservoir Screening Criteria outside China

Reservoir screening is the first stage for CCUS-EOR project implementation, and the success of the CCUS-EOR project relies heavily on the reservoir screening results. A set of precise screening criteria can expedite the selection of reservoirs suitable for $CO_2$ flooding, thereby enhancing the overall economic efficiency of the CCUS-EOR project. The first commercial application of CCUS-EOR was achieved in the United States in 1972, followed by large-scale field applications. The CCUS-EOR technology in the United States is more mature and corresponds to more advanced reservoir screening criteria than that in other countries [21].

Initially, American scholars have screened reservoirs for CCUS-EOR by mainly depending on parameters such as reservoir depth, temperature, formation pressure, oil density and viscosity, and oil saturation. It was generally believed that reservoir depth, temperature, and pressure required the miscibility of $CO_2$ and crude oil, that oil viscosity and density represented the mobility of crude oil, and that oil saturation reflected the economics of the reservoir [24–28]. Carcoana discussed reservoir recovery methods in Romania and suggested that a reservoir with net thickness of less than 15 m allows for a good volumetric sweep efficiency for $CO_2$ [29]. Taber and Martin included crude oil components into screening parameters and indicated that $CO_2$ is more miscible with crude oil containing high levels of intermediate hydrocarbon components like $C_{5+}$ [30–32]. Rivas argued against the exclusion of a reservoir solely based on a parameter not meeting the screening criteria and indicated that, among the reservoir parameters, oil viscosity, gas-oil ratio, and bubble point pressure are associated with oil gravity [33]; he used a normalization model and weighted ranking to determine the parameters of reservoirs in Eastern Venezuela suitable for CCUS-EOR. Diaz et al. evaluated 197 reservoirs in Louisiana by employing Rivas' screening technique and, through economic assessments, identified 39 reservoirs feasible for CCUS-EOR [34]. Shaw and Bachu emphasized the importance of crude oil components, density, and reservoir temperature in determining the minimum miscibility pressure (MMP) [35,36]. They recommended that reservoir temperature and pressure ideally should meet the conditions for $CO_2$ to become supercritical, and they also indicated that reservoir depth and oil viscosity can be disregarded as they are related to temperature and oil density.

Algharaib made a screening of 107 reservoirs for CCUS-EOR in the Middle East according to the criteria which took into account the influence of gas caps and pointed out the necessity for keeping the minimum miscibility pressure (MMP) below the reservoir fracture pressure [37]. Wo et al. suggested that even non-miscible $CO_2$ flooding could displace residual oil in water-flooded reservoirs [38]. They proposed two sets of screening criteria for miscible and immiscible flooding in reservoirs in Wyoming, with the main differences being in oil gravity, reservoir depth, and oil viscosity—miscible flooding is considered feasible when the oil gravity is greater than 22° API, the reservoir depth exceeds 762 m, and the oil viscosity is less than 10 cP. Aladasani and Bai collected data from over 160 CCUS-EOR projects reported in various publications worldwide during 1998–2010 and updated the screening criteria for $CO_2$ miscible and immiscible flooding [39,40]. Gao et al. identified five key screening parameters, including reservoir depth, temperature, pressure, oil gravity, and oil components [41]. During a reservoir screening for CCUS-EOR in Abu Dhabi, United Arab Emirates, Hajeri et al. believed that reservoirs with a high vertical permeability or a high ratio of vertical permeability to radial permeability have low potential [42]. The U.S. National Energy Technology Laboratory (NETL) recommended that carbonate or sandstone reservoirs are more suitable for $CO_2$ flooding than other reservoirs and defined depth, temperature, pressure, permeability, oil gravity, viscosity, and residual oil saturation as screening parameters [43]. Koottungal collected the $CO_2$ flooding projects conducted in the United States from 1972 to 2014 and counted the parameter ranges of porosity, depth, oil gravity, viscosity, temperature, and oil saturation, with an emphasis on $CO_2$ miscible flooding [44].

Verma classified the screening parameters for CCUS-EOR projects in the United States depending on the lithology of reservoir rocks (limestone and sandstone dominantly in reservoirs with miscible flooding) [45]. Yin analyzed 134 $CO_2$ flooding projects in the United States and established corresponding screening criteria for carbonate and sandstone reservoirs [46]. Bachu developed CCUS-EOR reservoir screening criteria applicable to Alberta, Canada, which incorporated economic indicators such as initial oil reserves and remaining oil content and included the key parameters significantly affecting screening results such as oil gravity, MMP, and reservoir size [21]. Through statistical analyses on parameters used in global $CO_2$-EOR projects, Zhang et al. formulated the screening criteria for $CO_2$ miscible and immiscible flooding separately [47,48]; $CO_2$ miscible flooding imposes higher requirements on net reservoir thickness, oil viscosity, oil gravity, and MMP, whereas $CO_2$ immiscible flooding requires higher oil saturation to ensure effective oil displacement. Hares established new screening criteria applicable to oilfields in Alberta based on existing CCUS-EOR research, which increased the importance of oil gravity, MMP, and oil viscosity in the screening process [49].

The above findings reveal that the current CCUS-EOR reservoir screening criteria are primarily based on the characteristic parameters corresponding to $CO_2$ miscible flooding projects in the United States and Canada. Reservoirs selected for CCUS-EOR are usually composed of carbonate and sandstone, with favorable porosity and permeability, and contain light to medium oils with a low viscosity. Additionally, these reservoirs have sufficient depth and temperature to enable the pressures to exceed the MMP, allowing for the $CO_2$–crude oil miscibility. Reservoirs targeted for $CO_2$ immiscible flooding should have a high porosity, high permeability, and good homogeneity due to their relatively heavy and more viscous oil contents. It is worth noting that the ultimate recovery improvement achieved through immiscible flooding is generally lower than that through miscible flooding [50]. Tables (Tables 1 and 2) lists the reservoir screening parameters and scope defined by scholars/institutions for CCUS-EOR projects outside China since 1972. Generally, the screening parameters for CCUS-EOR projects that have been carried out are categorized as: (1) characteristic parameters of reservoir, including reservoir depth, pressure, temperature, porosity, permeability, net thickness, oil saturation, and reservoir dip angle; and (2) characteristic parameters of crude oil, including oil gravity, oil viscosity, and oil composition. Oil gravity and oil density are interchangeable using formulas. Oil composition is less considered but mainly represented by oil gravity. Reservoir dip angle is sparsely reported. In this study, the reservoir depth, temperature, pressure, porosity, permeability, net thickness, oil saturation, oil density, oil composition, and oil viscosity are used as screening indices.

Table 1. CCUS-EOR reservoir screening criteria outside China.

| Scholar/Institution | Area | Year | Depth (m) | Temperature (°C) | Pressure (MPa) | Porosity (%) | Permeability (mD) |
|---|---|---|---|---|---|---|---|
| Geffen [24] | United States | 1973 | | | >7.6 | | |
| Lewin & Assoc [25] | United States | 1976 | >914 | | >10.4 | | |
| NPC [26] | United States | 1976 | >701 | <121 | | | |
| McRee [27] | United States | 1977 | >610 | | | | >5 |
| Iyoho [28] | United States | 1978 | >762 | | | | >10 |
| Carcoana [29] | Romania | 1982 | <3000 | <90 | >8 | >18 | >0.1 |
| Taber & Martin [30] | United States | 1983 | >610 | | 8.3–32 | | |
| Klins [51] | United States | 1984 | >914 | | >103 | | |
| Rivas [33] | Venezuela | 1994 | | 54–93 | $0.1 \leq P/M \leq 1.3$ | 9–33 | 18–2500 |
| Diaz et al. [34] | United States | 1996 | | 27–136 | $0.1 \leq P/M \leq 1.47$ | 17.6–34 | 17–3485 |
| Taber et al. [32] | World | 1997 | 762–1219 | | >MMP | | >5 |
| Bachu [35] | Canada | 2004 | | 32–121 | $0.95 \leq P/M$ | | |
| Alberta Research Council [52] | Canada | 2009 | >450 | 28–121 | >MMP and <$P_f$ | ≥3 | ≥5 |
| Algharaib [37] | Middle East | 2009 | >600 | >30 | >MMP | | |
| Wo et al. [38] | United States | 2009 | >762 | | | >7 | >10 |
| NETL [43] | United States | 2010 | 610–2987 | <121 | >8.3–10.3 | | >1–5 |
| Aladasani [39] | World | 2010 | 457–4074 | 28–121 | | 3–37 | 1.5–4500 |
| Gao & Pan [41] | World | 2010 | >762 | | | >12 | >10 |
| Koottungal [44] | United States | 2014 | 487–3600 | 28–127 | | 30 | 1–4500 |
| Yin [46] | United States | 2015 | 350–3642 | 28–127 | | 4–23.7 | >2 |
| Bachu [21] | World | 2016 | 500–4100 | 28–127 | ≥MMP | 3–37 | |
| Zhang et al. [47] | World | 2018 | 426–2590 | 28–112 | | 11.5–33 | 1.4–2750 |
| Zhang et al. [48] | World | 2019 | >350 | <127 | ≥MMP | 3–37 | >0.1 |
| Hares [49] | Canada | 2020 | 500–1400 | 27–127 | ≥MMP | | |

Table 2. CCUS-EOR reservoir screening criteria outside China.

| Scholar/Institution | Area | Year | Oil Density (g/cm³) | Oil Viscosity (cP, mPa·s) | Oil Saturation (%) | Net Thickness (m) | Oil Composition | Reservoir Dip Angle (°) |
|---|---|---|---|---|---|---|---|---|
| Geffen [24] | United States | 1973 | | | >7.6 | | | |
| Lewin & Assoc [25] | United States | 1976 | <0.88 | <3 | >25 | | | |
| NPC [26] | United States | 1976 | <0.88 | <12 | >25 | | | |
| McRee [27] | United States | 1977 | <0.89 | <10 | | | | |
| Iyoho [28] | United States | 1978 | <0.85 | <5 | >25 | | | |
| Carcoana [29] | Romania | 1982 | 0.8–0.88 | <10 | >25 | | | |
| Taber & Martin [30] | United States | 1983 | <0.82 | <2 | >30 | <15 | | |
| Klins [51] | United States | 1984 | <0.9 | <15 | >30 | | $C_5$–$C_{20}$ | |
| Rivas [33] | Venezuela | 1994 | <0.88 | <12 | >25 | | | |
| Diaz et al. [34] | United States | 1996 | 0.70–0.93 | | 30–92 | 1.5–55 | | 5–20 |
| Taber et al. [32] | World | 1997 | 0.79–0.91 | | 8–80 | 1.5–53 | | 0.03–64 |
| Bachu [35] | Canada | 2004 | 0.8–0.89 | 0.3–6 | 15–70 | | $C_5$–$C_{12}$ | |
| Alberta Research Council [52] | Canada | 2009 | 0.79–0.89 | | >25 | | | |
| Algharaib [37] | Middle East | 2009 | 0.8–0.89 | ≤6 | ≥30 | | | |
| Wo et al. [38] | United States | 2009 | <0.91 | <10 | >25 | | | |
| NETL [43] | United States | 2010 | <0.92 | <10 | | | | |
| Aladasani [39] | World | 2010 | <0.89 | ≤12 | >25–30 | | | |
| Gao & Pan [41] | World | 2010 | 0.8–0.89 | <35 | 15–89 | | | |
| Koottungal [44] | United States | 2014 | <0.89 | <10 | | | | |
| Yin [46] | United States | 2015 | 0.8–0.89 | 0.4–6 | 5–50 | | | |
| Bachu [21] | World | 2016 | <0.89 | <6 | >20 | 4.5–81 | | |
| Zhang et al. [47] | World | 2018 | 0.8–0.92 | 0.4–6 | ≥20 | | | |
| Zhang et al. [48] | World | 2019 | 0.83–0.98 | 0.2–936 | 30–86 | 1.6–91 | | |
| Hares [49] | Canada | 2020 | 0.79–0.90 | <4 | >15 | 4.5–250 | | |

Table 3 shows the statistical analysis of CCUS-EOR reservoir screening criteria outside China. It can be seen that the parameter of temperature ranges in 28–127 °C, while the critical temperature of $CO_2$ is 31 °C. In this study, the range of temperature is adjusted to 31–127 °C to ensure that the matched reservoirs have certain conditions for $CO_2$–crude oil miscibility.

**Table 3.** Statistical analysis of CCUS-EOR reservoir screening criteria outside China.

| Screening Parameter | Range |
|---|---|
| Depth (m) | 350–4100 |
| Temperature (°C) | 31–127 |
| Pressure (MPa) | 0.9 MMP $\leq$ P < $P_f$ |
| Porosity (%) | 3–37 |
| Permeability (mD) | 0.1–4500 |
| Oil density (g/cm$^3$) | 0.79–0.92 |
| Oil viscosity (cP, mPa·s) | 0.4–12 |
| Oil saturation (%) | $\geq$20 |
| Net thickness (m) | 1.5–250 |
| Oil composition | $C_5$–$C_{20}$ |

Note: P is the initial formation pressure; $P_f$ is the reservoir fracture pressure.

### 2.2. CCUS-EOR Reservoir Screening Criteria in China

Since 2000, China has accelerated its research and application of CCUS-EOR technology. China National Petroleum Corporation (CNPC) established the first national CCUS-EOR demonstration project in Jilin Oilfield and conducted relevant field tests and large-scale applications in oilfields such as Daqing, Changqing, and Xinjiang. Currently, CCUS-EOR is at a crucial stage of transition from field test to industrialization [10,50]. Compared to the favorable reservoir properties found in North America and other countries, the reservoirs in China are predominantly continental, with strong heterogeneity and low permeability. Furthermore, the reservoirs in China are widely distributed and diverse for varying types and sizes across basins [53]. Additionally, the scarcity of original $CO_2$ sources and the immaturity of high-concentration $CO_2$ capture technologies have created a shortage of $CO_2$ supply, further limiting the number of reservoirs suitable for CCUS-EOR [7,54]. As a result, existing CCUS-EOR reservoir screening criteria are limited in application to reservoirs in China. It is necessary to refine the screen parameters depending on the specific characteristics of reservoirs in China.

Scholars have often determined the required evaluation parameters and their ranges through literature reviews and mechanistic analyses [55,56]. Zheng et al. assigned the screening indices for gas-flooded reservoirs into three categories: oil viscosity, oil density, and oil saturation represent oil properties; permeability, porosity, wettability, and heterogeneity represent rock properties; reservoir depth, temperature, dip angle, and pressure represent reservoir characteristics. They quantified and partitioned these indices using a fuzzy preference model [57]. Xiang et al. applied the CCUS-EOR reservoir screening criteria commonly used in the United States and Canada to screen offshore reservoirs in China and suggested a good prospect for $CO_2$ miscible flooding and near-miscible flooding in reservoirs in the South China Sea [58]. Liang et al. evaluated 183 reservoirs in the Shengli Oilfield based on the existing screening criteria and defined 18 reservoirs suitable for CCUS-EOR [59]. They also identified a high oil gravity and high oil viscosity as the main factors limiting $CO_2$ flooding in these reservoirs. On the basis of previous studies, Liao et al. presented the CCUS-EOR reservoir screening criteria for the Changqing Oilfield, which consider heterogeneity and permeability coefficient in reservoir characteristics, and they stated that the reservoirs with a permeability variation coefficient of <0.75 are suitable for CCUS-EOR and the reservoirs with a permeability coefficient ($K_h$) >$10^{-13}$–$10^{-14}$ can be selected for $CO_2$ flooding [60]. Wang et al. applied the U.S. CCUS-EOR reservoir screening criteria to the oilfields in the Ordos Basin, China, and the results showed that the reservoirs in Yanchang Formation are suitable for $CO_2$ miscible flooding due to high oil gravity, oil viscosity, and oil composition, but their low porosity, low permeability, low reservoir pressure, and high heterogeneity are unfavorable for $CO_2$–crude oil miscibility [61]. Jiao et al. believed that injecting a certain volume of $CO_2$ for a certain duration can facilitate the $CO_2$–crude oil miscibility [62]. Wang et al., during CCUS-EOR reservoir screening for the Junggar Basin, proposed screening criteria for miscible and immiscible flooding depending on reservoir characteristics [63]. The main differences between the two criteria lie in oil density, oil viscosity, and reservoir depth. He et al. divided the screening indexes for gas-flooding reservoirs into three categories: reservoir oil properties, reservoir tectonic characteristics, and economic factors [64]. Meng et al. selected depth, pressure, temperature, porosity, permeability, oil viscosity, and oil density as screening indices and established the screening criteria for $CO_2$ miscible flooding and $CO_2$ immiscible flooding (applicable when the reservoir depth is small and the oil viscosity and density are high) [65]. Wang et al. developed an index screening system comprising 13 parameters based on the existing CCUS-EOR reservoir screening criteria [66]. They indicated that reservoir depth, thickness, initial oil saturation, and sedimentary rhythm have a significant impact on oil recovery, and the geological conditions and reservoir fluid properties carry relatively high weights according to a sensitivity analysis. They selected effective reservoir thickness, reservoir depth, average permeability, temperature, oil density, and oil viscosity as the final screening indices. Yang et al. established the CCUS-EOR reservoir screening criteria according to the characteristics of the reservoirs in the Bohai Bay Basin, and the screening results showed that a total of 613 reservoirs were suitable for CCUS-EOR, with a total potential of 68.3 billion tons, including 45.9 billion tons of potential for miscible flooding [67]. He et al. proposed the screening criteria for three types of $CO_2$ flooding (miscible, near-miscible, and immiscible) based on the ratio of reservoir pressure to MMP [68]. Wang identified reservoir depth, temperature, original pressure, oil gravity, and oil viscosity as important screening indices and also considered the impacts of porosity and initial oil saturation [69].

Tables 4–7 provide the CCUS-EOR reservoir screening criteria, proposed by Chinese scholars in recent years, for miscible, near-miscible, and immiscible flooding. In general, the CCUS-EOR reservoir screening criteria in China mainly include reservoir depth, temperature, pressure, porosity, permeability, oil saturation, heterogeneity, reservoir dip angle, oil density, oil viscosity, and oil composition. The ranges of various indices are similar for miscible and near-miscible flooding, but the main difference rests in the ratio of reservoir pressure to MMP. When the reservoir pressure falls within the range of 0.8–1 MMP, it is generally considered that the miscibility of $CO_2$ and crude oil can be improved by increasing the injection pressure and other methods. Table 8 provides a statistical analysis of the parameter ranges in CCUS-EOR reservoir screening criteria. Compared to the screening criteria in other countries, the screening criteria in China additionally consider reservoir heterogeneity and reservoir dip angle, reflecting the impact of reservoir physical properties on the efficiency of $CO_2$ flooding. Moreover, reservoirs in China have a greater range of depth and temperature for screening and are more diverse in types, making the $CO_2$–crude oil miscibility more difficult. Therefore, the development of screening criteria for $CO_2$ immiscible flooding in China has become more urgent.

Table 4. China's CCUS-EOR reservoir screening criteria for miscible and near-miscible flooding.

| Scholar/Institution | Year | Depth (m) | Temperature (°C) | Pressure (MPa) | Porosity (%) | Permeability (mD) |
|---|---|---|---|---|---|---|
| Xiong et al. [55] | 2004 | | | | 5–25 | 1–1000 |
| Zeng et al. [56] | 2005 | 1200–2500 | | $0.75 \leq P/MMP \leq 3$ | >15 | >50 |
| Zheng et al. [57] | 2005 | 800–3500 | 50–120 | 15–50 | 4–30 | 0.1–500 |
| Shen et al. [70] | 2009 | 800–3500 | 50–120 | 8–35 | | |
| Wang et al. [61] | 2013 | 200–2500 | | | 5–17 | 0.1–7 |
| Wang et al. [63] | 2014 | >600 | 32–120 | | | >1 |
| He et al. [64] | 2015 | 900–3000 | <90 | ≥MMP | | <10 |
| Meng et al. [65] | 2016 | 800–3500 | 50–120 | 8–35 | 4–30 | 0.1–500 |
| Yang et al. [67] | 2017 | 488–4074 | 28–127 | ≥MMP | 3–37 | |
| He et al. [68] | 2020 | >1000 | <120 | ≥MMP | | >1 |
| Wang et al. [69] | 2023 | 488–4074 | 28–127 | ≥MMP | 3–37 | |

Table 5. China's CCUS-EOR reservoir screening criteria for miscible and near-miscible flooding.

| Scholar/Institution | Year | Oil Density (g/cm³) | Oil Viscosity (cP, mPa·s) | Oil Saturation (%) | Net Thickness (m) | Permeability Variation Coefficient | Oil Composition | Reservoir Dip Angle (°) |
|---|---|---|---|---|---|---|---|---|
| Xiong et al. [55] | 2004 | | | | 5–25 | 1–1000 | | |
| Zeng et al. [56] | 2005 | | <20 | 30–80 | | | | |
| Zheng et al. [57] | 2005 | <0.88 | <8 | >30 | 3–20 | | $C_1$–$C_6$ | 0–90 |
| Shen et al. [70] | 2009 | <0.88 | <4 | >25 | | <0.65 | | >10 |
| Wang et al. [61] | 2013 | 0.795–0.9 | <10 | >25 | | | | |
| Wang et al. [63] | 2014 | 0.73–0.86 | 1.3–9 | 40–56 | | | $C_5$–$C_{20}$ | |
| He et al. [64] | 2015 | <0.92 | <188 | >20 | | | | |
| Meng et al. [65] | 2016 | <0.90 | <10 | >30 | | <0.75 | $C_2$–$C_{10}$ | |
| Yang et al. [67] | 2017 | 0.79–0.92 | 1.5–12 | | | | | |
| He et al. [68] | 2020 | | 0.4–6 | ≥26.5 | | | | |
| Wang et al. [69] | 2023 | <0.876 | <10 | >30 | | <0.75 | | |

Table 6. China's CCUS-EOR reservoir screening criteria for immiscible flooding.

| Scholar/Institution | Year | Depth (m) | Temperature (°C) | Pressure (MPa) | Porosity (%) |
|---|---|---|---|---|---|
| Shen et al. [70] | 2009 | 600–900 | | | |
| Wang et al. [63] | 2014 | >550 | | | |
| He et al. [64] | 2015 | >900 | | | |
| Meng et al. [65] | 2016 | 600–900 | | | |
| Yang et al. [67] | 2017 | 350–2590 | 28–92 | <MMP | 17–32 |
| He et al. [68] | 2020 | >600 | <120 | <0.8 MMP | |
| Wang et al. [69] | 2023 | 350–2591 | 28–92 | <MMP | 17–32 |

Table 7. China's CCUS-EOR reservoir screening criteria for immiscible flooding.

| Scholar/Institution | Year | Permeability (mD) | Oil Density (g/cm³) | Oil Viscosity (cP, mPa·s) | Oil Saturation (%) | Permeability Variation Coefficient |
|---|---|---|---|---|---|---|
| Shen et al. [70] | 2009 | 600–900 | | | | |
| Wang et al. [63] | 2014 | | >0.9 | 100–1000 | 30–70 | |
| He et al. [64] | 2015 | | 0.92–0.98 | <600 | | |
| Meng et al. [65] | 2016 | | <0.99 | <600 | >30 | <0.75 |
| Yang et al. [67] | 2017 | | 0.92–0.98 | 100–1000 | | |
| He et al. [68] | 2020 | | | 0.6–592 | ≥30 | |
| Wang et al. [69] | 2023 | >1 | <0.98 | <600 | >40 | <0.55 |

Table 8. Statistical analysis of screening indices for CCUS-EOR in China.

| Screening Index | Range | |
|---|---|---|
| | Miscible and Near-Miscible Flooding | Immiscible Flooding |
| Depth (m) | 600–3500 | >350 |
| Temperature (°C) | 28–127 | <120 |
| Pressure (MPa) | ≥0.8 MMP | <0.8 MMP |
| Porosity (%) | 3–37 | 17–32 |
| Permeability (mD) | >0.1 | |
| Oil density (g/cm$^3$) | <0.92 | 0.92–0.99 |
| Oil viscosity (cP, mPa·s) | <20 | <1000 |
| Oil saturation (%) | >25 | >30 |
| Permeability variation coefficient | <0.75 | <0.75 |
| Oil composition | $C_2$–$C_{15}$ | |
| Reservoir dip angle (°) | >10 | |

## 3. Optimization of CCUS-EOR Reservoir Screening Criteria in China

In China, the increasing number of laboratory experiments and studies on $CO_2$-EOR with great attention and support from the government and petroleum companies for large-scale CCUS projects has deepened the understanding of $CO_2$ flooding mechanisms, enhanced the application of $CO_2$-EOR technology, and expanded the range of reservoirs suitable for $CO_2$ flooding. Under this background, it is necessary to update screening indices and their ranges in China's existing CCUS-EOR reservoir screening criteria, coupled with the results of $CO_2$ flooding laboratory experiments and pilot projects. Tables 9 and 10 shows the parameters of some CCUS-EOR reservoirs from pilot tests conducted in China. It has been found that, apart from relatively unique low-temperature, low-permeability, and low-pressure reservoirs, the reservoirs in China exhibit depths of 1000–3100 m, temperatures of 45–120 °C, formation pressures of 9–42 MPa, porosity of 6–28%, permeability of 0.1–1600 mD, oil density of 0.78–0.9 g/cm$^3$, oil viscosity of 0.3–12 cP, oil saturation > 30%, and average effective thickness of 1.5–17.7 m. Moreover, most of the reservoirs have a strong heterogeneity, with the oil composition dominated by a C7+ medium and heavy hydrocarbons, rarely containing light components. The limited records on formation dip angle suggests a range from 1° to 8°. The MMP ranges from 16 MPa to 55 MPa, while most reservoirs have formation pressures below the MMP, primarily favoring $CO_2$ immiscible flooding.

Table 9. Parameters of CCUS-EOR reservoirs in China.

| Study Area | Depth (m) | Temperature (°C) | Pressure (MPa) | Porosity (%) | Permeability (mD) | Oil Density (g/cm$^3$) | Oil Viscosity (cP, mPa·s) |
|---|---|---|---|---|---|---|---|
| Daluhu Oilfield in Shengli Oil Area [71] | 3147 | 116 | 31.56 | | | | |
| Fang 48 fault block [72] | 1699 | 85.9 | 20.4 | 14.5 | 1.4 | 0.815 | 6.6 |
| Taizhou formation in Caoshe Oilfield [73] | 3065 | 110 | 35.9 | | | <0.9 | |
| Zhongnan fault block in Chujialou Oilfield [73] | 2962.9 | | 28.943 | 21.3 | 241 | | |
| Well Shu 101 in Daqing Oilfield [74] | | 108 | 22.05 | | | 0.78 | 2.8 |
| Dagang Oilfield [75] | 2700 | | 27.21 | 19.04 | 300 | 0.88 | 6.59 |

Table 9. Cont.

| Study Area | Depth (m) | Temperature (°C) | Pressure (MPa) | Porosity (%) | Permeability (mD) | Oil Density (g/cm$^3$) | Oil Viscosity (cP, mPa·s) |
|---|---|---|---|---|---|---|---|
| Well Shu 101 in Yushulin Oilfield [76] | 2044 | 108 | 22.05 | 10.8 | 1.16 | | 3.6 |
| M Oilfield [77] | 2880 | 119.2 | 30.2 | 9 | 6.5 | 0.865 | 5.2 |
| Liubei block in Jidong Oilfield [78] | 2625 | 102 | 29.5 | 17.05 | 273 | 0.794 | 0.329 |
| Caoshe [79] | 3020 | 107 | 35.9 | 13.2 | 24.8 | 0.88 | 7 |
| Fumin [79] | 2090 | 76 | 20.9 | 12 | 854 | 0.82 | 2.4 |
| Sa II in Sanan Oilfield [79] | 1072 | 49 | 11.6 | 25.3 | 1165 | 0.86 | 8.6 |
| Sa I in Sanan Oilfield [79] | 1140 | 45 | 12.3 | 27.6 | 1628 | 0.87 | 9.8 |
| Jingbian [79] | 1590 | 47 | 12.3 | 12.8 | 0.9 | 0.86 | 2.5 |
| Huang 3 testing area in Changqing Oilfield [80] | | 84 | 15.78 | | 0.3–1 | 0.73 | 1.81 |
| Chang 3 reservoir in Weibei Oilfield [81] | 550 | 29.2 | 2.06 | 11.2 | 0.76 | | 6.64 |
| Gao 89-1 block in Shengli Oilfield [82] | 2900 | | 42 | 9.18–14.7 | 0.29–4.92 | 0.86 | 11.83 |
| Chang 4 + 5 reservoir in Wuqi Oilfield [83] | | 60 | 15 | 12.8 | 0.78 | 0.78 | 2.38 |
| Chang 6 formation in Yanchang Oilfield [84] | | 46 | 8.9 | 7–12 | 0.94 | 0.79 | 3.4 |
| M reservoir [85] | 2025 | 75 | 21.3 | 16.3 | 15.7 | | 3.64 |
| A block in CQ Oilfield [86] | 2200 | 75 | 18 | 9.8 | 0.07 | 0.825 | 8.73 |
| Fu3 member in Zhangjiaduo Oilfield [87] | | 107 | 38 | 18.2 | 6.5 | | 4.92 |
| Yan 2 block in Benbutu Oilfield [88] | 2550 | 95 | 186.27 | 12.2 | 9.8 | 0.64 | 0.68 |
| Area A in Tahe Oilfield [89] | 4600 | 110.5 | 128.5 | 21 | 733 | | 2.89 |
| North Xinghe block in Ansai Oilfield [90] | 1250 | 48 | 29.9 | 10.39 | 0.61 | 0.766 | 2.26 |
| Triassic Yanchang formation in Wuqi Oilfield [91] | 2000 | 72.8 | 18.5 | 6.1 | 3.44 | 0.78 | 2.03 |

Table 10. Parameters of CCUS-EOR reservoirs in China.

| Study Area | Oil Saturation (%) | Effective Thickness (m) | Permeability Variation Coefficient/Heterogeneity | Oil Composition | Reservoir Dip Angle (°) | P/MMP | Miscible/Immiscible Flooding |
|---|---|---|---|---|---|---|---|
| Daluhu Oilfield in Shengli Oil Area [71] | | | | $C_{7+}$ | | 1.21 | Miscible |
| Fang 48 fault block [72] | | 6.6 | | | | 0.37 | Immiscible |
| Taizhou formation in Caoshe Oilfield [73] | 30–50 | | Relatively heterogeneous | | | >1 | Miscible |
| Zhongnan fault block in Chujialou Oilfield [73] | 35 | | Highly heterogeneous | | | <1 | Immiscible |
| Well Shu 101 in Daqing Oilfield [74] | | | | | | <1 | Immiscible |
| Dagang Oilfield [75] | | 10 | 0.5 | $C_{11+}$ | | 1.18 | Miscible |
| Well Shu 101 in Yushulin Oilfield [76] | | 17.7 | | $C_8$–$C_{25}$ | 2–4 | 0.68 | Immiscible |
| M Oilfield [77] | | | | | | 1.12 | Miscible |
| Liubei block in Jidong Oilfield [78] | | | | $C_{7+}$ | | 0.98 | Near-miscible |
| Caoshe [79] | 31 | 17 | | | | 1.22 | Miscible |
| Fumin [79] | 36 | 6.1 | | | | 0.96 | Near-miscible |
| Sa II in Sanan Oilfield [79] | 51.7 | 8.6 | | | | 0.46 | Immiscible |

Table 10. Cont.

| Study Area | Oil Saturation (%) | Effective Thickness (m) | Permeability Variation Coefficient/Heterogeneity | Oil Composition | Reservoir Dip Angle (°) | P/MMP | Miscible/Immiscible Flooding |
|---|---|---|---|---|---|---|---|
| Sa I in Sanan Oilfield [79] | 45.8 | 9.2 | | | | 0.48 | Immiscible |
| Jingbian [79] | 48 | 12 | | | | 0.52 | Immiscible |
| Huang 3 testing area in Changqing Oilfield [80] | | | Heterogeneous | $C_2$–$C_{10}$ | | 0.98 | Near-miscible |
| Chang 3 reservoir in Weibei Oilfield [81] | | | | | 1.1 | 0.13 | Immiscible |
| Gao 89-1 block in Shengli Oilfield [82] | | 1.5 | Highly heterogeneous | | 5–8 | 1.45 | Miscible |
| Chang 4 + 5 reservoir in Wuqi Oilfield [83] | 55 | 7.69 | | | | 0.84 | Immiscible |
| Chang 6 formation in Yanchang Oilfield [84] | 42.2 | 14.1 | | | | 0.62 | Immiscible |
| M reservoir [85] | | | | Light components | | 0.78 | Immiscible |
| A block in CQ Oilfield [86] | | | | Light components | | 0.75 | Immiscible |
| Fu3 member in Zhangjiaduo Oilfield [87] | | | | | | 1.29 | Miscible |
| Yan 2 block in Benbutu Oilfield [88] | | | | | | 0.76 | Near-miscible |
| Area A in Tahe Oilfield [89] | | 15 | 0.74 | $C_2$–$C_{15}$ | 0.8 | 1.22 | Miscible |
| North Xinghe block in Ansai Oilfield [90] | | | Highly heterogeneous | $C_2$–$C_{15}$ | | 0.46 | Immiscible |
| Triassic Yanchang formation in Wuqi Oilfield [91] | | | | $C_{7+}$ | | 1.00 | Miscible |

By comparison, the average depths and temperatures of reservoirs in China are higher than those of marine reservoirs in other countries. Therefore, the lower limits of screening indices used in China should be adjusted upward accordingly. In China, reservoirs are predominantly tight with a wide range of low permeability and a strong heterogeneity. Moreover, the reservoirs contain crude oil with a higher content of heavy components than reservoirs in other countries, resulting in higher MMP values. So, $CO_2$ immiscible flooding is the primary mode of CCUS-EOR for such reservoirs. It is believed that China's screening criteria should focus more on the impact of reservoir heterogeneity. An investigation on reservoir heterogeneity relying on the parameters that can quantify the reservoir heterogeneity, such as permeability variation coefficient, can allow for a more rational screening result. The formation pressure referred to in the existing studies is the initial formation pressure, which, however, often changes after multiple rounds of exploitation. Therefore, the current pressure maintenance level can better reflect the relationship between reservoir pressure and MMP. Furthermore, as many reservoirs in China have been developed by water flooding, polymer flooding and other techniques, the selection of residual oil saturation, instead of oil saturation, as the screening index agrees more with the actual reservoir conditions.

To further clarify the screening indices for CCUS-EOR reservoirs in China, the current screening criteria are updated and improved with respect to indices and data. Specifically, the reservoir pressure index is categorized into initial formation pressure, current formation pressure, and pressure maintenance level. In view of oil composition, the molar content of $C_2$–$C_{15}$ is analyzed quantitatively. Additional investigation is performed on the characteristic parameters of similar reservoirs or oils. The influence of formation dip on $CO_2$-EOR has been rarely reported in previous studies and materials, and it cannot be examined sufficiently by using the limited sample size; therefore, it is excluded from the analysis below.

The various parameters impacting $CO_2$-EOR capabilities are interrelated. For instance, reservoir temperature tends to increase with the depth of the oil reservoir. Therefore, a correlation analysis is considered to explore the relationships between different parameters.

Furthermore, there are numerous indicators influencing the screening criteria, necessitating the categorization of similar indicators to identify the main factors impacting the criteria. Consequently, principal component analysis (PCA) is employed to classify and evaluate the different parameters of the oil reservoir.

*3.1. Pearson Correlation Analysis*

According to the data of parameters of reservoirs for CCUS-EOR in China, 13 data reflecting relatively complete reservoir information were selected for missing value processing. Based on the assumption of linear correlation, the linear relationships between parameters were analyzed by using the Pearson correlation coefficient. It is defined that two parameters are highly correlated when the absolute value of the Pearson correlation coefficient $|r|$ is >0.8, moderately correlated when $|r|$ is 0.3–0.8, and not correlated when $|r|$ is <0.3.

Among the initially selected screening parameters, the data of reservoir temperature, depth, initial formation pressure, current formation pressure, porosity, permeability, and viscosity are integral. Regarding the missing data, the density is supplemented from its linear correlation with viscosity; the permeability variation coefficient is quantified through the description of reservoir heterogeneity; the molar content of $C_2$–$C_{15}$ is filled according to its scatterplot relationships with density, viscosity, and MMP. The Pearson correlation coefficient matrix for these processed parameters is shown in Figure 1.

It is found that there is a strong positive correlation ($|r| > 0.85$) between the reservoir depth and the initial formation pressure or current formation pressure, a moderate positive correlation ($|r| > 0.75$) between the reservoir depth and the temperature or permeability, and a moderate negative correlation ($|r| > 0.55$) between the reservoir depth and the residual oil saturation or molar content of $C_2$–$C_{15}$. The reservoir temperature exhibits a strong positive correlation with the initial formation pressure and a strong negative correlation with the residual oil saturation. The current formation pressure and initial formation pressure show similar linear correlations with other parameters. The pressure maintenance level exhibits certain correlations with the current formation pressure, porosity, viscosity, molar content of $C_2$–$C_{15}$, and MMP. The porosity has a strong positive correlation with the permeability and shows a moderate positive correlation with depth and pressure. The permeability shows a positive correlation with depth, pressure, and porosity. The density shows a positive correlation with viscosity and a negative correlation with residual oil saturation and permeability variation coefficient. The viscosity shows positive correlations with pressure maintenance level, density, effective thickness, and MMP. The effective thickness has a positive correlation with viscosity and permeability variation coefficient. The molar content of $C_2$–$C_{15}$ exhibits a negative correlation with depth, temperature, and pressure. MMP has a strong positive correlation with current formation pressure and porosity and a positive correlation with temperature and viscosity ($|r| > 0.35$). The Pearson correlation analysis results demonstrate certain correlations among the parameters, allowing for the application of PCA.

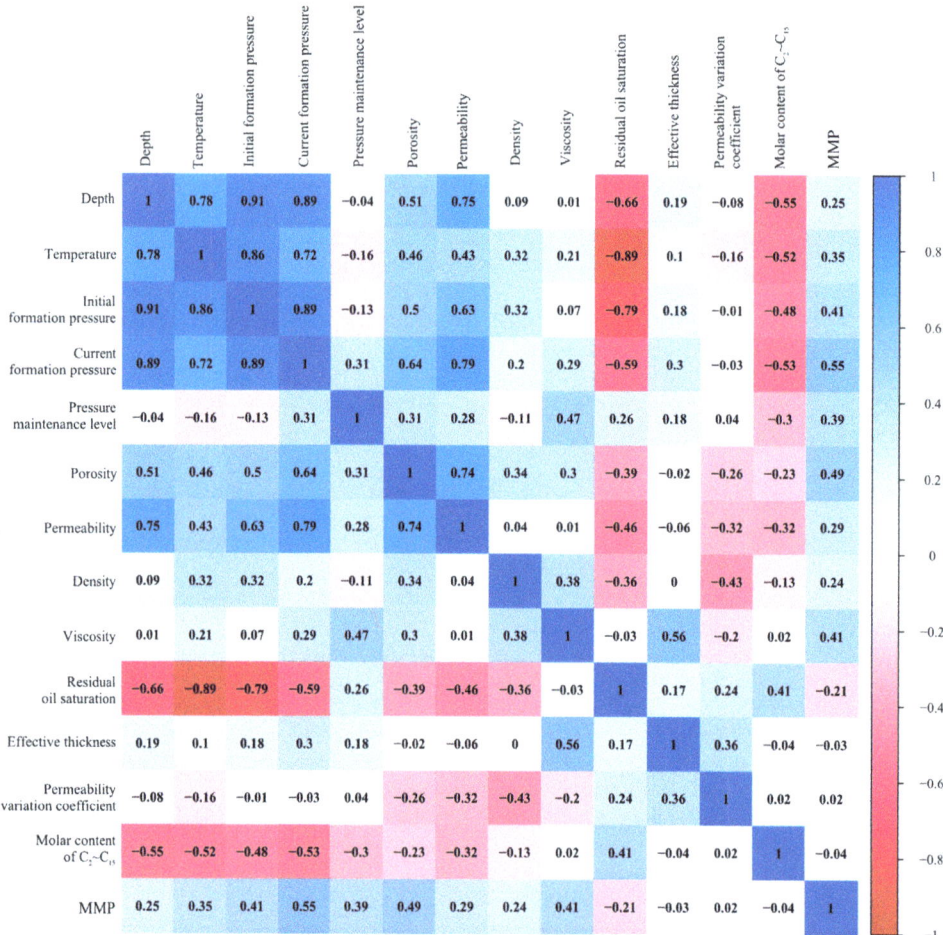

**Figure 1.** Pearson correlation coefficient matrix for reservoir parameters.

## 3.2. PCA of Screening Indices

The preliminary linear analysis reveals that most of the parameters are influenced by several other parameters. Therefore, it is necessary to further reduce the dimensionality of the parameters through principal component analysis (PCA) to simplify the data structure. PCA can transform the original variables into a set of mutually uncorrelated principal components through a linear transformation, thereby reducing the dimensionality of variables while retaining most of the information in the original data. It is suitable for dimensionality reduction and data simplification. Essentially, through an eigenvalue decomposition of the covariance matrix of the data, the principal components that explain the variability in the data are identified, and the weights of eigenvalues are determined. The analysis steps are as follows:

(1) Collect the data of parameters that influence $CO_2$-EOR and assess the feasibility of PCA.
(2) Normalize the data and calculate the covariance matrix to measure the correlation between the parameters.
(3) Perform eigenvalue decomposition of the covariance matrix to obtain eigenvalues and corresponding eigenvectors.

(4) Select the first few eigenvectors as the principal components and calculate their weights based on the magnitude of their eigenvalues.
(5) Calculate the component scores by linearly combining the original variables with the selected principal components and their weights.
(6) Determine the importance and weights of the parameters based on the component scores.

By using the parameters of CCUS-EOR reservoirs in Table 6 as raw data, the characteristic parameters of 25 reservoirs were selected for PCA after excluding the data from Weibei Oilfield with low temperature, low pressure, and low permeability to ensure the stability of the analysis. Before the analysis, missing value processing and supplementation were performed. Based on the Pearson correlation coefficient matrix and the mechanistic relationships between parameters, five indexes with relatively high levels of missing values, i.e., oil saturation, effective thickness, permeability variation coefficient, and molar content of $C_2$–$C_{15}$, were supplemented using methods such as polynomial regression and logarithmic fitting.

According to the data verification results, the Bartlett's sphericity test yields a $p$-value of 0.000, which is less than 0.05, indicating a suitability for PCA. In the analysis results, the cumulative contribution rate of variance for the first four principal components exceeds 75%, suggesting that relatively little parameter information is lost overall. Therefore, it is considered that these principal components can effectively characterize the CCUS-EOR capability of the reservoirs. The PCA results for the first and second principal components are shown in Figure 2, and the corresponding parameter loadings for the first four principal components are given in Table 11. The analysis reveals that in the first principal component (PC1), the loading coefficients for initial formation pressure, current formation pressure, temperature, depth, residual oil saturation, molar content of $C_2$–$C_{15}$, and MMP are relatively large. This suggests that PC1 can be considered as representing the potential for $CO_2$ miscible flooding in the reservoir. In the second principal component (PC2), the loading coefficients for porosity, permeability, viscosity, and density are relatively large, mainly characterizing the reservoir's fluid mobility. In the third principal component (PC3), the loading coefficients for effective thickness and permeability variation coefficient are relatively large, representing the influence of reservoir tectonic characteristics on $CO_2$-EOR. In the fourth principal component (PC4), the loading coefficient for pressure maintenance level is the largest, primarily reflecting the impact of pressure changes on $CO_2$ flooding in the reservoir.

**Table 11.** Parametric component matrix of principal components.

| Screening Parameter | PC1 | PC2 | PC3 | PC4 |
|---|---|---|---|---|
| Depth | 0.906 | −0.105 | −0.018 | 0.015 |
| Temperature | 0.929 | −0.186 | −0.048 | −0.149 |
| Initial formation pressure | 0.944 | 0.012 | −0.030 | −0.059 |
| Current formation pressure | 0.931 | 0.172 | 0.038 | 0.170 |
| Pressure maintenance level | 0.214 | 0.417 | 0.171 | 0.658 |
| Porosity | 0.078 | 0.860 | −0.241 | 0.157 |
| Permeability | −0.177 | 0.795 | −0.283 | 0.116 |
| Density | 0.217 | 0.584 | 0.053 | −0.561 |
| Viscosity | 0.071 | 0.716 | 0.417 | −0.377 |
| Residual oil saturation | −0.832 | 0.161 | 0.150 | 0.297 |
| Effective thickness | 0.277 | 0.111 | 0.857 | −0.148 |
| Permeability variation coefficient | −0.038 | −0.193 | 0.644 | 0.400 |
| Molar content of $C_2$–$C_{15}$ | −0.759 | 0.239 | 0.151 | −0.164 |
| MMP | 0.559 | 0.326 | −0.053 | 0.388 |

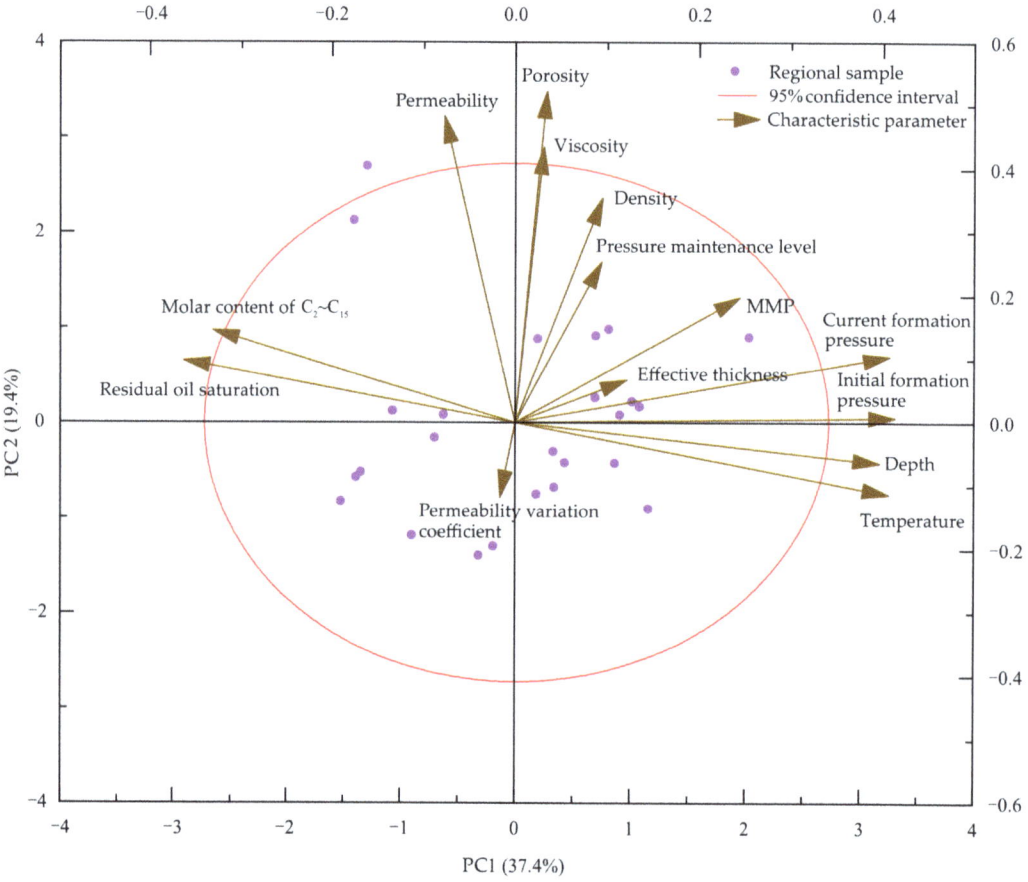

**Figure 2.** PCA results for PC1 and PC2.

Among the four principal components, PC1 contributes 37.41% to the variance, PC2 contributes 19.37%, PC3 contributes 11.04%, and PC4 contributes 10.22%. This indicates that among the factors influencing the screening of reservoirs for CCUS-EOR, the parameters representing the potential of $CO_2$–crude oil miscibility have the most significant impact on the reservoir scores, followed by parameters representing the oil mobility. Parameters characterizing reservoir tectonic characteristics and pressure changes have similar influences on the reservoir screening results. Using the coefficient matrix of the principal components, the linear combinations of parameters for different principal components can be obtained based on the weights of each parameter in different principal components. This allows for the calculation of scores for different principal components. By considering the proportion of variance contributed by each principal component, normalized weights for the four principal components can be calculated. The reservoir's comprehensive score can then be obtained by multiplying the corresponding weights with the principal component matrix. Table 12 provides the normalized weights for the first four principal components. The linear combination for calculating the reservoir's comprehensive score is shown in Equation (1):

$$\begin{cases} Y_1 = 0.396 * A_1 + 0.406 * A_2 + 0.412 * A_3 + 0.407 * A_4 + 0.094 * A_5 + 0.034 * A_6 \\ \quad -0.077 * A_7 + 0.095 * A_8 + 0.031 * A_9 - 0.364 * A_{10} + 0.121 * A_{11} - 0.017 * A_{12} \\ \quad -0.332 * A_{13} + 0.244 * A_{14} \\ Y_2 = -0.064 * A_1 - 0.113 * A_2 + 0.007 * A_3 + 0.104 * A_4 + 0.253 * A_5 + 0.522 * A_6 \\ \quad -0.483 * A_7 + 0.355 * A_8 + 0.435 * A_9 - 0.098 * A_{10} + 0.067 * A_{11} - 0.117 * A_{12} \\ \quad -0.145 * A_{13} + 0.198 * A_{14} \\ Y_3 = -0.014 * A_1 - 0.039 * A_2 - 0.024 * A_3 + 0.031 * A_4 + 0.138 * A_5 - 0.194 * A_6 \\ \quad -0.228 * A_7 + 0.043 * A_8 + 0.335 * A_9 + 0.121 * A_{10} + 0.689 * A_{11} + 0.518 * A_{12} \\ \quad +0.121 * A_{13} - 0.043 * A_{14} \\ Y_4 = 0.013 * A_1 - 0.125 * A_2 - 0.049 * A_3 + 0.142 * A_4 + 0.550 * A_5 + 0.131 * A_6 \\ \quad +0.097 * A_7 - 0.469 * A_8 - 0.315 * A_9 + 0.248 * A_{10} - 0.124 * A_{11} + 0.334 * A_{12} \\ \quad -0.137 * A_{13} + 0.324 * A_{14} \\ Z = 0.479 * Y_1 + 0.249 * Y_2 + 0.141 * Y_3 + 0.131 * Y_4 \\ T = 60 + 10 * Z \end{cases} \quad (1)$$

where $Y_1$ to $Y_4$ represent the scores of PC1 to PC4, respectively; $A_1$ to $A_{14}$ correspond to the normalized values of reservoir parameters from top to bottom as listed in Table 11; $Z$ represents the comprehensive score for an individual reservoir based on its principal components; and $T$ represents the reservoir's comprehensive score after being transformed into T-scores.

**Table 12.** Weights of the first four principal components.

|  | PC1 | PC2 | PC3 | PC4 |
| --- | --- | --- | --- | --- |
| Variance contribution rate (%) | 37.414 | 19.372 | 11.037 | 10.215 |
| Normalized weight | 0.479 | 0.249 | 0.141 | 0.131 |

The calculated comprehensive scores for the reservoirs are shown in Figure 3. It can be seen that in the testing areas and projects for CCUS-EOR, reservoirs with high comprehensive scores are mainly suitable for miscible flooding, while reservoirs with low comprehensive scores are more suitable for immiscible flooding. Reservoirs suitable for near-miscible flooding have comprehensive scores at an intermediate level. The comprehensive scores obtained through PCA can effectively reflect a reservoir's CCUS-EOR capability: the higher the comprehensive score, the stronger the oil mobility and the crude oil–$CO_2$ miscibility; and vice versa. Based on the actual conditions of reservoirs in China and the current range of comprehensive scores in the testing areas, it is considered that a reservoir with a comprehensive score of above 50 points is suitable for CCUS-EOR.

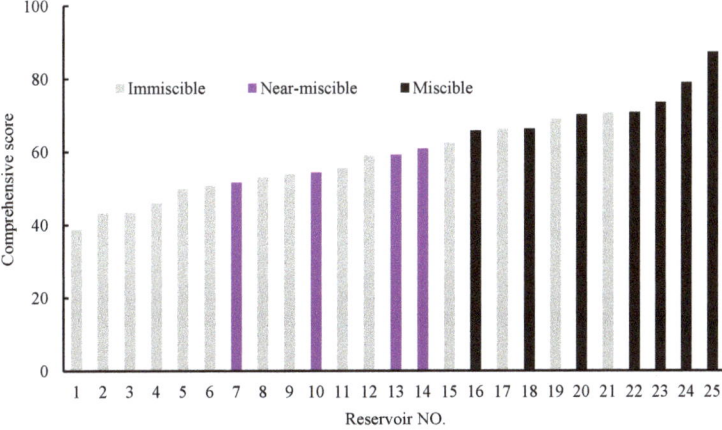

**Figure 3.** Comprehensive score and ranking of tested reservoirs.

### 3.3. Results of CCUS-EOR Reservoir Screening Parameter Range and Weight

Based on the results of the Pearson correlation analysis and PCA, it can be determined that the popular CCUS-EOR reservoir screening parameters in China can be mainly categorized into four groups. The first category of parameters represents the $CO_2$–crude oil miscibility, with a weight of 0.479. These parameters include temperature, depth, initial formation pressure, current formation pressure, residual oil saturation, molar content of $C_2$–$C_{15}$, and MMP. Specifically, temperature, depth, and pressure are highly positively correlated and can reflect the reservoir's ability to reach MMP; residual oil saturation shows a certain negative correlation with these four parameters; the molar content of $C_2$–$C_{15}$ and MMP can reflect the difficulty of miscibility between $CO_2$ and crude oil. The second category of parameters mainly represents the mobility of oil within the reservoirs, with a weight of 0.249. These parameters include porosity, permeability, density, and viscosity. A higher porosity and permeability indicate a better diffusion ability of crude oil and $CO_2$ within the pores. A higher density and viscosity imply stronger interactions at the interface between crude oil and medium. These four parameters also affect the mass transfer and diffusion of $CO_2$. The third category of parameters primarily represents reservoir tectonic characteristics, with a weight of 0.141. These parameters include effective thickness and permeability variation coefficient. The effective thickness can influence the $CO_2$ swept volume and sweep efficiency, while the permeability variation coefficient reflects reservoir heterogeneity and affects the $CO_2$ injection and displacement efficiency, as well as the extent of contact between $CO_2$ and crude oil. The fourth category of parameters mainly represents the impact of reservoir pressure changes, with a weight of 0.131. The pressure maintenance level can, to some extent, reflect changes in crude oil properties within the reservoir and have a certain influence on the potential of reservoir for CCUS-EOR.

Based on the ranges of selected reservoir parameters in the study areas and correlation analysis, some indices of the CCUS-EOR reservoir screening criteria in China are optimized, and some reservoir parameters with missing values are supplemented through a literature review and correlation fitting. Due to the limited data availability and low representativeness of reservoir dip angle, this parameter is removed from the screening parameters. Instead, two parameters, current formation pressure and pressure maintenance level, are added. Furthermore, the oil composition is changed to the molar content of $C_2$–$C_{15}$. The optimized CCUS-EOR reservoir screening criteria in China are shown in Table 13.

Table 13. CCUS-EOR reservoir screening criteria in China.

| Weight | Screening Index | Range | | |
|---|---|---|---|---|
| | | Miscible Flooding | Near-Miscible Flooding | Immiscible Flooding |
| 0.479 | Depth (m) | 2750–4600 | 2550–3050 | 1072–2963 |
| | Temperature (°C) | 107–120 | 84–102.5 | 45–126 |
| | Initial formation pressure (MPa) | 27–50 | 18–30.5 | 7–42 |
| | Current formation pressure (MPa) | 23–48 | 14–22 | 5–26 |
| | Residual oil saturation (%) | 31–40 | 37–43 | 27–55 |
| | $C_2$–$C_{15}$ molar content (mol%) | 40–52 | 34–48 | 35–61 |
| | MMP | 23–40 | 16–30 | 14–55 |
| 0.249 | Porosity (%) | 9–21 | 8–17 | 6–27.6 |
| | Permeability (mD) | 1–735 | 0.5–273 | 0.05–1628 |
| | Oil density (g/cm$^3$) | 0.8–0.88 | 0.64–0.83 | 0.77–0.87 |
| | Oil viscosity (cP, mPa·s) | 2.9–15.1 | 0.3–1.8 | 1.98–9.8 |
| 0.141 | Average effective thickness (m) | 10–60 | 2–12 | 1.5–20 |
| | Permeability variation coefficient | 0.34–0.96 | 0.72–0.8 | 0.7–0.9 |
| 0.131 | Pressure maintenance level (%) | 77–97 | 64–86 | 56–110 |

*3.4. Application Cases*

Six reservoirs with relatively complete screening index data were selected for a comprehensive score assessment to validate the CCUS-EOR evaluation using the linear combination of comprehensive scores [92–97]. The ranking of the reservoir comprehensive scores is shown in Figure 4. It is found that the Y block in Dagang Oilfield has the highest comprehensive score, with a score of 78.9, while the A block in northern Shaanxi has the lowest comprehensive score, with a score of 45.3. Reservoirs feasible for miscible and near-miscible flooding have relatively high comprehensive scores, while reservoirs that require non-miscible flooding have relatively low comprehensive scores. For example, the Y block in Dagang Oilfield has a current formation pressure of 37.92 MPa, MMP of 35.12 MPa, oil density of 0.77 g/cm$^3$, viscosity of 2.27 mPa·s, $C_2$–$C_{15}$ molar content of 43.03%, porosity of 11.75%, and permeability of 17.42 mD, indicating a good $CO_2$–crude oil miscibility. In contrast, the H block in Jilin Oilfield has a current formation pressure of 23 MPa and MMP of 23.2 MPa, making it suitable for near-miscible flooding; however, it has a relatively high permeability variation coefficient (1.11), resulting in a relatively low reservoir comprehensive score. The remaining reservoirs have current formation pressures below MMP, making the $CO_2$–crude oil miscibility impossible. Additionally, their permeability is less than 5 mD, which affects the oil mobility, resulting in relatively low comprehensive scores. Overall, only the A block in northern Shaanxi has a comprehensive score below 50, while the other five blocks are suitable for CCUS-EOR. In conclusion, the comprehensive scores of the reservoirs, calculated using the linear combination formula, can accurately reflect the comprehensive potential of the reservoirs for CCUS-EOR, which provide a basis for screening CCUS-EOR reservoirs.

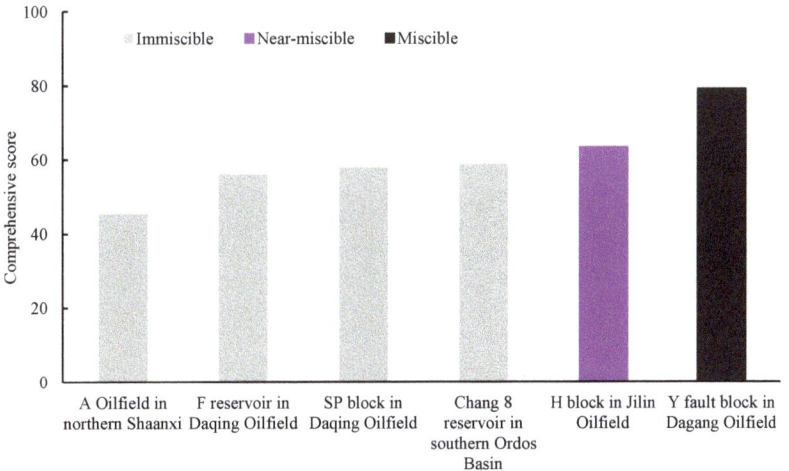

**Figure 4.** Comprehensive score and ranking of predicted reservoirs.

## 4. Conclusions and Prospect

Reservoirs outside China are relatively superior in physical properties and mostly contain light to medium oil; therefore, the applicable CCUS-EOR reservoir screening criteria are mainly constructed depending on the characteristic parameters of $CO_2$ miscible flooding reservoirs. These criteria typically include reservoir depth, temperature, pressure, porosity, permeability, net thickness, oil saturation, oil density, oil composition, and oil viscosity as indices. Specifically, reservoir depth and temperature are mainly used to assist in exploring the relationship between formation pressure and MMP, oil density and oil composition are key factors affecting MMP, and oil saturation reflects the economic potential of reservoirs for

CCUS-EOR. In China, the research and field applications of CCUS-EOR started relatively late, and mainly copied foreign screening criteria initially. However, reservoirs in China are primarily continental with strong heterogeneity, large depth, and high temperature, and they are also diverse in types and characteristics; so, a set of uniform screening criteria is infeasible for these reservoirs. In contrast, the CCUS-EOR reservoir screening criteria in China should incorporate heterogeneity and formation dip angle as indices, since the reservoirs are primarily suitable for $CO_2$ non-miscible flooding.

Based on the results of $CO_2$ flooding laboratory experiments and field applications in oilfields in China, the screening indices and their ranges were optimized. Most reservoirs in China exhibit differences between the current formation pressure and the initial formation pressure after years of exploitation. Therefore, when assessing the potential of reservoirs for CCUS-EOR, it is important to consider the influence of pressure maintenance level. Additionally, the oil saturation is replaced with residual oil saturation, and the molar content of $C_2$–$C_{15}$ in crude oil is also considered. The results of PCA indicate that China's CCUS-EOR reservoir screening indices can be categorized into four groups by their weights. The first category includes depth, temperature, initial formation pressure, current formation pressure, residual oil saturation, molar content of $C_2$–$C_{15}$, and MMP as indices of the $CO_2$–crude oil miscibility, with a weight of 0.419. The second category integrates porosity, permeability, density, and viscosity as indices of the oil mobility, with a weight of 0.249. The third category involves average net thickness and permeability variation coefficient as indices of reservoir tectonic characteristics, with a weight of 0.141. The fourth category considers pressure maintenance level as an index of reservoir property changes, with a weight of 0.131. By calculating the comprehensive scores of reservoirs based on normalized variables and weights, it is possible to quantify the potential of miscibility crude oil and $CO_2$ and make a preliminary ranking and assessment on the potential of reservoir for CCUS-EOR.

Compared to other countries and regions, oil reservoir characteristics in China are more complex, with a wider range of crude oil physical properties. It is not sufficient to simply categorize reservoirs into two types based on the miscibility capabilities between $CO_2$ and crude oil—those suitable for $CO_2$ miscible flooding or $CO_2$ immiscible flooding. It is also necessary to consider a third category of reservoirs that fall between these two types. Therefore, the ranges of screening indices are divided with respect to $CO_2$ miscible flooding, immiscible flooding, and near-miscible flooding. In general, miscible flooding has high requirements for depth, temperature, initial formation pressure, current formation pressure, permeability variation coefficient, and pressure maintenance level. A low MMP of the reservoir allows for miscible flooding. Immiscible flooding has low requirements for depth, temperature, and pressure but requires high residual oil saturation, molar content of $C_2$–$C_{15}$, and permeability. Near-miscible flooding requires parameters between miscible flooding and immiscible flooding. Based on the screening criteria established in this paper, a more efficient determination of the development approach for reservoirs undergoing CCUS-EOR can be made. This provides a more effective technical assessment for preliminarily determining the extraction potential of the reservoirs.

Indeed, when the parameters from $CO_2$ flooding projects in China are used for statistical analyses and PCA, the completeness and representativeness of the data are crucial to the results. For example, reservoir dip angle, a relatively important index, has to be excluded from the analyses because it has been rarely reported in the literature. Moreover, during analyses, the absence of certain reservoir parameters has led to deviations between the results and actual conditions. For instance, the permeability variation coefficients for some reservoirs are estimated based on qualitative descriptions of their heterogeneity. This approach may result in the final range of permeability variation coefficients being narrower than the actual range. Looking forward, it is necessary to complete and update the parameters in China's CCUS-EOR reservoir screening criteria by incorporating more diverse and representative indices, making the criteria more suitable for China's reservoir characteristics. Additionally, the incorporation of parameters such as $CO_2$ storage capacity, safety risks, and economic metrics could be instrumental in further evaluating the comprehensive potential of reservoirs for CCUS-EOR. This expansion would significantly enhance the integrity of China's CCUS-EOR screening criteria.

**Author Contributions:** Conceptualization, J.C. and M.G.; methodology, J.C. and Z.L.; validation, J.C. and W.L.; formal analysis, J.C. and H.Y. (Hongwei Yu); investigation, J.C. and H.Y. (Hongwei Yu); resources, J.C. and W.L.; data curation, J.C. and H.Y. (Hengfei Yin); writing—original draft preparation, J.C.; writing—review and editing, J.C. and M.G.; visualization, J.C. and Z.L.; supervision, M.G. and H.Y. (Hongwei Yu); project administration, M.G.; funding acquisition, M.G. All authors have read and agreed to the published version of the manuscript.

**Funding:** This research was funded by the National Key Research and Development Program of China (grant No. 2023YFF0614100 and No. 2023YFF0614101), and the Major Science and Technology project of the CNPC in China (grant No. 2021ZZ01-03 and No. 2021ZZ01-06).

**Data Availability Statement:** The raw/processed data required to reproduce these findings cannot be shared at this time, as the data also form part of an ongoing study.

**Acknowledgments:** The authors are grateful for the financial support of the CNPC in China.

**Conflicts of Interest:** The authors declare no conflict of interest.

# References

1. Zou, C.; Zhao, Q.; Zhang, G.; Xiong, B. Energy Revolution: From a Fossil Energy Era to a New Energy Era. *Nat. Gas Ind. B* **2016**, *3*, 1–11. [CrossRef]
2. Roy, P.; Mohanty, A.K.; Misra, M. Prospects of Carbon Capture, Utilization and Storage for Mitigating Climate Change. *Environ. Sci. Adv.* **2023**, *2*, 409–423. [CrossRef]
3. Karnauskas, K.B.; Miller, S.L.; Schapiro, A.C. Fossil Fuel Combustion Is Driving Indoor $CO_2$ toward Levels Harmful to Human Cognition. *GeoHealth* **2020**, *4*, e2019GH000237. [CrossRef]
4. Kumar, N.; Augusto Sampaio, M.; Ojha, K.; Hoteit, H.; Mandal, A. Fundamental Aspects, Mechanisms and Emerging Possibilities of $CO_2$ Miscible Flooding in Enhanced Oil Recovery: A Review. *Fuel* **2022**, *330*, 125633. [CrossRef]
5. Kulp, S.A.; Strauss, B.H. New Elevation Data Triple Estimates of Global Vulnerability to Sea-Level Rise and Coastal Flooding. *Nat. Commun.* **2019**, *10*, 4844. [CrossRef]
6. Rignot, E. Sea Level Rise from Melting Glaciers and Ice Sheets Caused by Climate Warming above Pre-Industrial Levels. *Phys.–Uspekhi* **2022**, *65*, 1129–1138.
7. Yao, J.; Han, H.; Yang, Y.; Song, Y.; Li, G. A Review of Recent Progress of Carbon Capture, Utilization, and Storage (CCUS) in China. *Appl. Sci.* **2023**, *13*, 1169. [CrossRef]
8. Di Stasi, C.; Renda, S.; Greco, G.; González, B.; Palma, V.; Manyà, J.J. Wheat-Straw-Derived Activated Biochar as a Renewable Support of Ni-$CeO_2$ Catalysts for $CO_2$ Methanation. *Sustainability* **2021**, *13*, 8939. [CrossRef]
9. Dieterich, V.; Buttler, A.; Hanel, A.; Spliethoff, H.; Fendt, S. Power-to-Liquid via Synthesis of Methanol, DME or Fischer–Tropsch-Fuels: A Review. *Energy Environ. Sci.* **2020**, *13*, 3207–3252. [CrossRef]
10. Yuan, S.; Ma, D.; Li, J.; Zhou, T.; Ji, Z.; Han, H. Progress and Prospects of Carbon Dioxide Capture, EOR-Utilization and Storage Industrialization. *Pet. Explor. Dev.* **2022**, *49*, 955–962. [CrossRef]
11. Ateka, A.; Rodriguez-Vega, P.; Ereña, J.; Aguayo, A.T.; Bilbao, J. A Review on the Valorization of $CO_2$. Focusing on the Thermodynamics and Catalyst Design Studies of the Direct Synthesis of Dimethyl Ether. *Fuel Process Technol.* **2022**, *233*, 107310. [CrossRef]
12. Huang, J.; Chen, Q.; Zhong, P. *National Assesment Report on Development of Carbon Capture Utilization and Storage Technology in China*; Science Press: Beijing, China, 2021.
13. Liu, Z.; Gao, M.; Zhang, X.; Liang, Y.; Guo, Y.; Liu, W.; Bao, J. CCUS and $CO_2$ Injection Field Application in Abroad and China: Status and Progress. *Geoenergy Sci. Eng.* **2023**, *229*, 212011. [CrossRef]
14. Yan, L.; Hu, J.; Fang, Q.; Xia, X.; Lei, B.; Deng, Q. Eco-Development of Oil and Gas Industry: CCUS-EOR Technology. *Front. Earth Sci.* **2023**, *10*, 1063042. [CrossRef]
15. Gozalpour, F.; Ren, S.R.; Tohidi, B. $CO_2$ Eor and Storage in Oil Reservoir. *Oil Gas Sci. Technol.* **2005**, *60*, 537–546. [CrossRef]
16. Hill, L.B.; Li, X.; Wei, N. $CO_2$-EOR in China: A Comparative Review. *Int. J. Greenh. Gas Control* **2020**, *103*, 103173. [CrossRef]
17. Al-Shargabi, M.; Davoodi, S.; Wood, D.A.; Rukavishnikov, V.S.; Minaev, K.M. Carbon Dioxide Applications for Enhanced Oil Recovery Assisted by Nanoparticles: Recent Developments. *ACS Omega* **2022**, *7*, 9984–9994. [CrossRef]
18. Huang, T.; Zhou, X.; Yang, H.; Liao, G.; Zeng, F. $CO_2$ Flooding Strategy to Enhance Heavy Oil Recovery. *Petroleum* **2017**, *3*, 68–78. [CrossRef]
19. Ampomah, W.; Balch, R.; Cather, M.; Rose-Coss, D.; Dai, Z.; Heath, J.; Dewers, T.; Mozley, P. Evaluation of $CO_2$ Storage Mechanisms in $CO_2$ Enhanced Oil Recovery Sites: Application to Morrow Sandstone Reservoir. *Energy Fuels* **2016**, *30*, 8545–8555. [CrossRef]

20. Kuuskraa, V.; Ferguson, R. *Storing $CO_2$ with Enhanced Oil Recovery*; National Energy Technology Laboratory: Washington, DC, USA, 2008.
21. Bachu, S. Identification of Oil Reservoirs Suitable for $CO_2$ -EOR and $CO_2$ Storage (CCUS) Using Reserves Databases, with Application to Alberta, Canada. *Int. J. Greenh. Gas Control* **2016**, *44*, 152–165. [CrossRef]
22. Wang, G.; Zheng, X.; Zhang, Y.; Lü, W.; Wang, F.; Yin, L. A New Screening Method of Low Permeability Reservoirs Suitable for $CO_2$ Flooding. *Pet. Explor. Dev.* **2015**, *42*, 390–396. [CrossRef]
23. Mahjour, S.K.; Faroughi, S.A. Risks and Uncertainties in Carbon Capture, Transport, and Storage Projects: A Comprehensive Review. *Gas Sci. Eng.* **2023**, *119*, 205117. [CrossRef]
24. Geffen, T.M. Improved Oil Recovery Could Help Ease Energy Shortage. *World Oil* **1973**, *177*, 84.
25. Lewin, S. The Potential and Economics of Enhanced Oil Recovery. In *Report US FEA Contract Np; CO-03-50222-000*; FEA: Washington, DC, USA, 1976; p. 274.
26. Council, N.P. *Enhanced Oil Recovery—An Analysis of the Potential for Enhanced Oil Recovery from Known Fields in the United States–1976–2000*; National Petroleum Council: Washington, DC, USA, 1976.
27. McRee, B.C. $CO_2$: How It Works, Where It Works. *Pet. Eng.* **1977**, *52*, 63.
28. Iyoho, A.W. Selecting Enhanced Oil Recovery Processes. *World Oil United States* **1978**, *187*, 6. Available online: https://www.osti.gov/biblio/6293850 (accessed on 22 August 2023).
29. Carcoana, A.N. Enhanced Oil Recovery in Rumania. In Proceedings of the SPE Improved Oil Recovery Conference, Tulsa, OH, USA, 4–7 April 1982; SPE: Richardson, TX, USA, 1982; p. SPE-10699.
30. Taber, J.J. Technical Screening Guides for the Enhanced Recovery of Oil. In Proceedings of the SPE Annual Technical Conference and Exhibition, San Francisco, CA, USA, 5–8 October 1983; OnePetro: Richardson, TX, USA, 1983.
31. Taber, J.J.; Martin, F.D.; Seright, R.S. EOR Screening Criteria Revisited—Part 1: Introduction to Screening Criteria and Enhanced Recovery Field Projects. *SPE Reserv. Eng.* **1997**, *12*, 189–198. [CrossRef]
32. Taber, J.J.; Martin, F.D.; Seright, R.S. EOR Screening Criteria Revisited—Part 2: Applications and Impact of Oil Prices. *SPE Reserv. Eng.* **1997**, *12*, 199–206. [CrossRef]
33. Rivas, O.; Embid, S.; Bolívar, F. Ranking Reservoirs for Carbon Dioxide Flooding Processes. *SPE Adv. Technol. Ser.* **1994**, *2*, 95–103. [CrossRef]
34. Diaz, D.; Bassiouni, Z.; Kimbrell, W.; Wolcott, J. Screening Criteria for Application of Carbon Dioxide Miscible Displacement in Waterflooded Reservoirs Containing Light Oil. In Proceedings of the SPE Improved Oil Recovery Conference, Tulsa, OH, USA, 21–24 April 1996; SPE: Richardson, TX, USA, 1996; p. SPE-35431.
35. Shaw, J.; Bachu, S. Screening, Evaluation, and Ranking of Oil Reservoirs Suitable for $CO_2$-Flood EOR and Carbon Dioxide Sequestration. *J. Can. Pet. Technol.* **2002**, *41*, 9. [CrossRef]
36. Bachu, S.; Shaw, J.C.; Pearson, R.M. Estimation of Oil Recovery and $CO_2$ Storage Capacity in $CO_2$ EOR Incorporating the Effect of Underlying Aquifers. In Proceedings of the SPE/DOE Symposium on Improved Oil Recovery, Tulsa, OH, USA, 17–21 April 2004.
37. Algharaib, M. Potential Applications of $CO_2$-EOR in the Middle East. In Proceedings of the All Days, Manama, Bahrain, 15–18 March 2009; SPE: Richardson, TX, USA, 2009; p. SPE-120231-MS.
38. Wo, S.; Whitman, L.D.; Steidtmann, J.R. Estimates of Potential $CO_2$ Demand for $CO_2$ EOR in Wyoming Basins. In Proceedings of the All Days, Denver, CO, USA, 14–16 April 2009; SPE: Richardson, TX, USA, 2009; p. SPE-122921-MS.
39. Aladasani, A.; Bai, B. Recent Developments and Updated Screening Criteria of Enhanced Oil Recovery Techniques. In Proceedings of the International Oil and Gas Conference and Exhibition in China, Beijing, China, 8–10 June 2010; OnePetro: Richardson, TX, USA, 2010.
40. Al Adasani, A.; Bai, B. Analysis of EOR Projects and Updated Screening Criteria. *J. Pet. Sci. Eng.* **2011**, *79*, 10–24. [CrossRef]
41. Gao, P.; Towler, B.; Pan, G. Strategies for Evaluation of the $CO_2$ Miscible Flooding Process. In Proceedings of the Abu Dhabi International Petroleum Exhibition and Conference, Abu Dhabi, United Arab Emirates, 1–4 November 2010; SPE: Richardson, TX, USA, 2010; p. SPE-138786.
42. Hajeri, S.A.; Negahban, S.; Al-Yafei, G.; Basry, A.A. Design and Implementation of the First CO2-EOR Pilot in Abu Dhabi, UAE. In Proceedings of the SPE EOR Conference at Oil & Gas West Asia, Muscat, Oman, 11–13 April 2010.
43. NETL, N. Carbon Dioxide Enhanced Oil Recovery-Untapped Domestic Energy Supply and Long Term Carbon Storage Solution. *Energy Lab* **2010**. Available online: https://netl.doe.gov/research/coal/energy-systems/gasification/gasifipedia/eor (accessed on 22 August 2023).
44. Koottungal, L. 2014 Worldwide EOR Survey. Available online: https://www.ogj.com/drilling-production/production-operations/ior-eor/article/17210637/2014-worldwide-eor-survey (accessed on 22 August 2023).
45. Verma, M.K. *Fundamentals of Carbon Dioxide-Enhanced Oil Recovery ($CO_2$-EOR): A Supporting Document of the Assessment Methodology for Hydrocarbon Recovery Using $CO_2$-EOR Associated with Carbon Sequestration*; U.S. Geological Survey: Reston, VA, USA, 2015.
46. Yin, M. *$CO_2$ Miscible Flooding Application and Screening Criteria*; Missouri University of Science and Technology: Rolla, MO, USA, 2015.
47. Zhang, N.; Wei, M.; Bai, B. Statistical and Analytical Review of Worldwide $CO_2$ Immiscible Field Applications. *Fuel* **2018**, *220*, 89–100. [CrossRef]
48. Zhang, N.; Yin, M.; Wei, M.; Bai, B. Identification of $CO_2$ Sequestration Opportunities: $CO_2$ Miscible Flooding Guidelines. *Fuel* **2019**, *241*, 459–467. [CrossRef]

49. Hares, R. *Feasibility of CCUS to $CO_2$-EOR in Alberta*; University of Calgary: Calgary, AB, Canada, 2020; Volume 1.
50. Qin, J. Application and Enlightenment of Carbon Dioxide Flooding in the United States of America. *Pet. Explor. Dev.* **2015**, *42*, 232–240. [CrossRef]
51. Klins, M.A. *Carbon Dioxide Flooding*; Basic Mechanism and Project Design; Springer: Berlin/Heidelberg, Germany, 1984.
52. GHG; IEA. $CO_2$ Storage in Depleted Oilfields: Global Application Criteria for Carbon Dioxide Enhanced Oil Recovery. *Chelten. Glos UK Prep. Adv. Resour. Int. Melzer Consult.* **2009**, *12*. Available online: https://ieaghg.org/docs/General_Docs/Reports/2009-12.pdf (accessed on 22 August 2023).
53. Song, Z.; Li, Y.; Song, Y. A Critical Review of CO2 Enhanced Oil Recovery in Tight Oil Reservoirs of North America and China. In Proceedings of the SPE/IATMI Asia Pacific Oil & Gas Conference and Exhibition, Bali, Indonesia, 29–31 October 2019.
54. Guo, H.; Lyu, X.; Meng, E.; Xu, Y.; Zhang, M.; Fu, H.; Zhang, Y.; Song, K. CCUS in China: Challenges and Opportunities. In Proceedings of the SPE Improved Oil Recovery Conference, Virtual, 25–29 April 2022; OnePetro: Richardson, TX, USA, 2022.
55. Xiong, Y. A Screening Candidate Reservoir for Gas in-Jection Way Based on Characteristic Parame-Ter's Comprehensive Weight Optimizationmethod. *J. Southwest Pet. Univ. Sci. Technol. Ed.* **2004**, *26*, 22.
56. Zeng, S.; Yang, Q.; Chen, J. Fuzzy Hierarchy Analysis-Based Selection of Oil Reservoirs for Gas Storage and Gas Injection. *Henan Pet.* **2005**, *19*, 40–46.
57. Zheng, Y. Screening Method Based on Fuzzy Optimum for Gas Injection in Candidate Reservoir. *J. Southwest Pet. Univ. Sci. Technol. Ed.* **2005**, *27*, 44.
58. Xiang, W.; Zhou, W.; Zhang, J.; Yang, G.; Jiang, W.; Sun, L.; Li, J. The Potential of $CO_2$-EOR in China Offshore Oilfield. In Proceedings of the All Days, Perth, Australia, 20–22 October 2008; SPE: Richardson, TX, USA, 2008; p. SPE-115060-MS.
59. Liang, Z.; Shu, W.; Li, Z.; Shaoran, R.; Qing, G. Assessment of $CO_2$ EOR and Its Geo-Storage Potential in Mature Oil Reservoirs, Shengli Oilfield, China. *Pet. Explor. Dev.* **2009**, *36*, 737–742. [CrossRef]
60. Liao, X.; Gao, C.-N.; Wu, P.; Su, K.; Shangguan, Y. Assessment of $CO_2$ EOR and Its Geo-Storage Potential in Mature Oil Reservoirs, Changqing Oil Field, China. In Proceedings of the Carbon Management Technology Conference; Carbon Management Technology Conference, Orlando, FL, USA, 7–9 February 2012.
61. Wang, Y.; Jiao, Z.; Surdam, R.; Zhou, L.; Gao, R.; Chen, Y.; Luo, T.; Wang, H. A Feasibility Study of the Integration of Enhanced Oil Recovery ($CO_2$ Flooding) with $CO_2$ Storage in the Mature Oil Fields of the Ordos Basin, China. *Energy Procedia* **2013**, *37*, 6846–6853. [CrossRef]
62. Jiao, Z.; Zhou, L.; Gao, R.; Luo, T.; Wang, H.; Wang, H.; McLaughlin, F.; Bentley, R.; Quillinan, S. Opportunity and Challenges of Integrated Enhanced Oil Recovery Using $CO_2$ Flooding with Geological $CO_2$ Storage in the Ordos Basin, China. *Energy Procedia* **2014**, *63*, 7761–7771.
63. Wang, H.; Liao, X.; Dou, X.; Shang, B.; Ye, H.; Zhao, D.; Liao, C.; Chen, X. Potential Evaluation of $CO_2$ Sequestration and Enhanced Oil Recovery of Low Permeability Reservoir in the Junggar Basin, China. *Energy Fuels* **2014**, *28*, 3281–3291. [CrossRef]
64. He, L.; Shen, P.; Liao, X.; Gao, Q.; Wang, C.; Li, F. Study on $CO_2$ EOR and Its Geological Sequestration Potential in Oil Field around Yulin City. *J. Pet. Sci. Eng.* **2015**, *134*, 199–204.
65. Meng, X.; Zhou, H.B.; Luo, D.K. Screening, Assessing, and Sorting Chinese Oilfields for $CO_2$-EOR Suitability and Potential Economic and Social Benefits. In Proceedings of the SPE Hydrocarbon Economics and Evaluation Symposium, Houston, TX, USA, 17–18 May 2016; SPE: Richardson, TX, USA, 2016; p. D021S008R005.
66. Wang, X.; Yuan, Q.; Wang, S.; Zeng, F. The First Integrated Approach for CO2 Capture and Enhanced Oil Recovery in China. In Proceedings of the Carbon Management Technology Conference, Houston, TX, USA, 17–20 July 2017; CMTC: Long Beach, CA, USA, 2017; p. CMTC-486571.
67. Yang, W.; Peng, B.; Liu, Q.; Wang, S.; Dong, Y.; Lai, Y. Evaluation of $CO_2$ Enhanced Oil Recovery and $CO_2$ Storage Potential in Oil Reservoirs of Bohai Bay Basin, China. *Int. J. Greenh. Gas Control* **2017**, *65*, 86–98. [CrossRef]
68. He, Y.; Zhao, S.; Ji, B.; Liao, H.; Zhou, Y. Screening Method and Potential Evaluation for EOR by $CO_2$ Flooding in Sandstone Reservoirs. *Editor. Dep. Pet. Geol. Recovery Effic.* **2020**, *27*, 140–145.
69. Wang, P.-T.; Wu, X.; Ge, G.; Wang, X.; Xu, M.; Wang, F.; Zhang, Y.; Wang, H.; Zheng, Y. Evaluation of $CO_2$ Enhanced Oil Recovery and $CO_2$ Storage Potential in Oil Reservoirs of Petroliferous Sedimentary Basin, China. *Sci. Technol. Energy Transit.* **2023**, *78*, 3. [CrossRef]
70. Shen, P.; Liao, X. *The Technology of Carbon Dioxide Stored in Geological Media and Enhanced Oil Recovery*; The Press of the Petroleum Industry: Beijing, China, 2009.
71. Hao, Y.; Bo, Q.; Chen, Y. Laboratory Investigation of $CO_2$ Flooding. *Pet. Explor. Dev.* **2005**, *2*, 110–112.
72. Cheng, J.C.; Lei, Y.; Zhu, W. Pilot Test on $CO_2$ Flooding in Extra-Low Permeability Fuyu Oil Layer in Daqing Placanticline. *Nat. Gas Geosci.* **2008**, *19*, 402–409.
73. Liu, W.; Chen, Z. Pilot Test on $CO_2$ Flooding in Complex and Fault Block Small Reservoirs of North Jiangsu Oilfield. *Nat. Gas Geosci.* **2008**, 147–149+645.
74. Zhao, M.; Li, J.; Wang, Z. The Study on $CO_2$ Immiscible Mechanism in Low Permeability Reservoir. *Sci. Technol. Eng.* **2011**, *11*, 1438–1440.
75. Song, Z.; Li, Z.; Lai, F. Parameter Optimization and Effect Evaluation of $CO_2$ Flooding after Water Flooding. *J. Xian Shiyou Univ. Sci. Ed.* **2012**, *27*, 42–47+4.

76. Yang, T.; Yang, Z. Evaluation of Oil Displacement by $CO_2$ at Fuyang Extralow Permeability Layer in Yushulin Oilfield. *Sci. Technol. Rev.* **2015**, *33*, 52–55.
77. Li, X.; Cao, T. Effect Evaluation of $CO_2$ Flooding in Low Permeability Reservoirs. *Contemp. Chem. Ind.* **2016**, *45*, 2339–2342.
78. Sun, L.; Ji, M.; Zheng, J. EOR Feasibility of $CO_2$ Flooding for Liubei Conglomerate Oil Reservoirs. *Petro-Leum Geol. Oilfield Dev. Daqing* **2016**, *35*, 123–127.
79. Yang, H.; Wang, H.; Nan, Y.; Qu, Y. Suitability Evaluation of Enhanced Oil Recovery by $CO_2$ Flooding. *Lithol. Res.-Ervoirs* **2017**, *29*, 140–146.
80. Wang, S.; Tan, J.; Lei, X. *$CO_2$ Flooding Technology for Low Permeability Sandstone Reservoir of Changqing Oilfield*; Xi'an Shiyou Unversity, Shaanxi Petroleum Society: Xi'an, China, 2018; pp. 2542–2553.
81. Zhang, B.; Zhou, L.; He, X. A Laboratory Study on $CO_2$ Injection of Chang 3 Reservoir of Weibei Oilfield in Ordos Basin. *Pet. Geol. Eng.* **2018**, *32*, 87–90+125.
82. Cao, X.; Lyu, G.; Wang, J. Technology and Application of $CO_2$ Flooding in Ultra-Low Permeability Beach-Bar Sand Reservoir. *Reserv. Eval. Dev.* **2019**, *9*, 41–46.
83. Qi, C.; Li, R.; Zhu, S. Pilot Test on $CO_2$ Flooding of Chang4+ 51 Oil Reservoir in Yougou Region of the Ordos Basin. *Oil Drill. Prod. Technol.* **2019**, *41*, 249–253.
84. Zhao, X.; Yang, H.; Chen, L. Analysis of $CO_2$ Flooding and Storage Potential of Chang6 Reservoir in Huaziping Area of Yanchang Oilfield. *J. Xi'an Shiyou Univ. Nat. Sci.* **2019**, *34*, 62–68.
85. Jin, Z.; Wang, Z.; Mao, C. Dominant Mechanism and Application of $CO_2$ Immiscible Flooding in M Block with Low Permeability. *Reserv. Eval. Dev.* **2020**, *10*, 68–74.
86. Li, S.; Xia, Y.; Lan, J.; Ye, S.; Ma, X.; Zou, J.; Li, M. $CO_2$ Flooding Experiment in the Chang-7 Tight Oil Reservoir of Ordos Basin. *Sci. Technol. Eng.* **2020**, *20*, 2251–2257.
87. Ren, B. Analysis of $CO_2$ Displacement Mechanism in the Third Member of Fujiang Formation in Zhangjiaduo Oilfield. *Petrochem. Ind. Technol.* **2020**, *27*, 97–98.
88. Li, Y.; Zhang, D.; Fan, X.; Zhang, J.; Yang, R.; Ye, H. EOR of $CO_2$ Flooding in Low-Permeability Sandy Conglomerate Reservoirs. *Xinjiang Pet. Geol.* **2022**, *43*, 59.
89. Liu, D. Research on Flood Law and Development Scheme of Sha $NO_3$ Middle Reservoir in Fan 162 Block of Daluhu. Master's Thesis, China University of Petroleum (East China), Qingdao, China, 2020.
90. Li, Y. Laboratory Evaluation on EOR by $CO_2$ Flooding in Ultra-Low Permeability Reservoirs of Changqing Oilfield. Master's Thesis, Xi'an Shiyou University, Xi'an, China, 2019.
91. Ren, D.; Wang, X.; Kou, Z.; Wang, S.; Wang, H.; Wang, X.; Tang, Y.; Jiao, Z.; Zhou, D.; Zhang, R. Feasibility Evaluation of $CO_2$ EOR and Storage in Tight Oil Reservoirs: A Demonstration Project in the Ordos Basin. *Fuel* **2023**, *331*, 125652. [CrossRef]
92. Cai, L. Research on the Mechanism $CO_2$ Flooding on F Reservoir with Ultra-Low Permeability in Daqing. Master's Thesis, Northeast Petroleum University, Daqing, China, 2016.
93. Lyu, J. The Experimental Study on $CO_2$ Flooding for Ultra-Low Permeability Reservoir in Chang 8 Oil-Baring Formation of Southern Erdos. Master's Thesis, China University of Petroleum (East China), Qingdao, China, 2015.
94. Li, M. Evaluation of the Impact of $CO_2$ Injection for Enhanced Oil Recover in the Hei 79 Block of Jilin Oilfield. Master's Thesis, Southwest Petroleum University, Chengdu, China, 2015.
95. Luo, T. Research on Reservoir Characterization and Numerical Simulation of $CO_2$ Enhanced Oil Recovery for Low Permeability Oil Reservoir in a Oilfield in Northern Shaanxi. Doctoral Dissertation, Northwest University, Xi'an, China, 2016.
96. Dong, P. Research on Optimization of Reservoir Engineering Parameters in Deep-Buried Low-Permeability Reservoirs by $CO_2$ Flooding. Master's Thesis, China University of Petroleum (Beijing), Beijing, China, 2020.
97. Zhang, L. Study on the Mechanism and Effect Evaluation of $CO_2$ Gas Flooding Development. Master's Thesis, Northeast Petroleum University, Daqing, China, 2023.

**Disclaimer/Publisher's Note:** The statements, opinions and data contained in all publications are solely those of the individual author(s) and contributor(s) and not of MDPI and/or the editor(s). MDPI and/or the editor(s) disclaim responsibility for any injury to people or property resulting from any ideas, methods, instructions or products referred to in the content.

Article

# Mechanism and Quantitative Characterization of Wettability on Shale Surfaces: An Experimental Study Based on Atomic Force Microscopy (AFM)

Xu Huo [1,2], Linghui Sun [1,2,3,*], Zhengming Yang [1,2,3], Junqian Li [4], Chun Feng [3], Zhirong Zhang [1,2], Xiuxiu Pan [1,2] and Meng Du [1,2]

1. Engineering College, University of Chinese Academy of Sciences, Beijing 100190, China; huoxu21@mails.ucas.ac.cn (X.H.); yzhm69@petrochina.com.cn (Z.Y.); 18392055195@163.com (Z.Z.); panxiuxiu22@mails.ucas.ac.cn (X.P.); dumeng22@mails.ucas.ac.cn (M.D.)
2. Institute of Porous Flow and Fluid Mechanics, Chinese Academy of Sciences, Beijing 100083, China
3. Research Institute of Petroleum Exploration and Development, Petrochina, Beijing 100083, China; fengchun123@petrochina.com.cn
4. School of Geosciences, China University of Petroleum (East China), Qingdao 266580, China; lijunqian@upc.edu.cn
* Correspondence: sunlinghuipetrochina@outlook.com

**Abstract:** Wettability, as a vital tool for analyzing and describing oil flow, plays a significant role in determining oil/water relative permeability, residual oil distribution, and on-site recovery efficiency. Although the contact angle method is widely used for measuring wetting behavior, it is susceptible to the effects of surface roughness, oil–water saturation, and the distribution of mixed wetting within the range of droplet sizes. Additionally, millimeter–scale droplets fail to accurately represent the wetting distribution and the influencing factors at the micro/nano–scale. Therefore, this study presents a comprehensive investigation of the microstructure and wettability of shale samples. The characterization of the samples was performed using scanning electron microscopy (SEM) and atomic force microscopy (AFM) techniques to gain insights into their microscopic features, surface properties, and wettability. Results demonstrate the following: (1) Quartz and clay minerals tended to exhibit rough surface topography, appearing as darker areas (DA) under scanning electron microscopy (SEM). It is worth noting that plagioclase minerals exhibited brighter areas (BA) under SEM. (2) An increase in the content of minerals such as quartz and clay minerals was observed to decrease the surface oil wetting behavior. In contrast, plagioclase feldspar exhibited an opposite trend. (3) Based on the adhesive forces of the samples towards oil or water, a wetting index, I, was established to evaluate the wettability of shale at a microscale. The dimensionless contact angle W, obtained by normalizing the contact angle measurement, also consistently indicated oil wetting behavior. (4) By comparing the differences between I and W, it was observed that surface roughness significantly affected the behavior of water droplets. The presence of roughness impeded the contact between the solid and liquid phases, thus influencing the accuracy of the wetting results. Organic matter also plays a significant role in influencing surface wettability, and its distribution within the shale samples can lead to localized variations in wettability.

**Keywords:** wettability; shale reservoir; surface roughness; mineral composition; AFM; contact angle

## 1. Introduction

Unconventional oil and gas reservoirs have gained significant global attention as conventional reserves have become depleted and the demand for petroleum resources continues to rise. Among these reservoirs, shale oil has emerged as a crucial contributor to maintaining the energy balance [1–3]. However, the exploitation of shale oil resources presents substantial challenges [4,5]. Shale oil production currently faces challenges such as low yields and rapid decline in well productivity. The intricate micro/nano–scale pore–throat network units and the adsorption behavior of organic matter between mineral grains are critical factors influencing the flow of crude oil. The flowability of crude oil is significantly influenced by the complex physicochemical properties derived from the shale surface [6,7]. These properties are directly related to wettability, which characterizes and describes the interactions between the solid and liquid phases [8]. Wettability serves as a direct indicator of the extent and capability of these factors to affect the flow behavior of oil. Indeed, a detailed understanding of wettability is crucial for enhancing shale oil recovery. $CO_2$ sequestration techniques commonly used in tight shale are influenced by various factors, such as fluid properties and surface characteristics. The key lies in understanding the micro– and nanoscale solid–liquid interactions, which include phenomena like changes in pH and the acidification of rocks when $CO_2$ dissolves in reservoir water, altering wettability within the reservoir. Therefore, gaining further insights into solid–liquid interactions is fundamental to enhancing $CO_2$–enhanced oil recovery efficiency, and it can also provide valuable guidance for the storage and utilization of carbon dioxide. Nevertheless, conventional macroscopic evaluation techniques struggle to accurately characterize these intricate patterns. Therefore, it is crucial to develop accurate and comprehensive assessment methods to understand solid–liquid interactions at micro– and nanoscales in shale formations.

Wettability, which describes the interfacial tension relationships between solid, liquid, and gas phases, is influenced by several factors [9–11]. These factors include surface roughness, temperature, pressure, mineral compositions, and organic matter [12]. Among these factors, surface roughness plays a crucial role in determining interfacial tension and, subsequently, wettability. The roughness of the solid surface affects the contact area and the degree of wetting by different phases [13]. Temperature and pressure variations can also alter the interfacial tension, thereby impacting wettability behavior. Furthermore, the mineral composition and organic matter within reservoir formations are recognized as primary factors governing shale wettability [6,7,14–16]. The distribution of minerals within the formation also plays a vital role in determining the overall wettability characteristics of a sample. Reservoirs rich in quartz, feldspar, and clay minerals tend to exhibit hydrophilic behavior, as quartz surfaces possess strong water–attracting capabilities [17,18]. Additionally, expandable clay minerals exhibit significant water absorption and swelling behavior [19], which can modify the roughness and surface forces at the solid–liquid interface. This, in turn, has a profound impact on the overall wettability state of the reservoir. Additionally, organic matter plays a crucial role in modifying the interfacial forces between solids and liquids [20]. One example is asphaltene, which possesses hydrophobic aromatic cores and hydrophilic polar groups. Asphaltene has the ability to spontaneously adsorb onto solid surfaces, thereby influencing the distribution of charged particles and interfacial tension at the solid–liquid interface [21]. This adsorption of asphaltene onto solids has been observed to alter surface wettability. Moreover, when asphaltene is adsorbed at the oil–water interface, it can affect surface potential, interfacial tension, and other related properties [22–24]. Moreover, the non–uniform and stochastic distribution of minerals and organic matter within shale formations leads to a heterogeneous distribution of mixed wettability, with the presence of oil–wet zones being particularly significant. These oil–wet zones have a detrimental effect on the efficiency of oil recovery. It has been observed that even after water flooding, substantial amounts of trapped oil remain on rock surfaces, indicating the limited micro–scale mobility of shale oil [25]. Based on this observation, Xi et al. [5] conducted a study to explore the relationship between mineral types, trapped oil, and pore structure in mixed–wettability shale reservoirs. The results revealed that macroscopic wettability outcomes were directly influenced by different laminae, mineral types, and the distribution of organic matter within the reservoir. Many researchers have extensively studied the interrelationships among the factors influencing wettability [9,10]. Their findings have further demonstrated and elucidated the complexity of shale wettability, which pose higher demands on the accuracy and applicability of wettability characterization methods [26].

In order to clarify and evaluate the wettability of samples, many testing methods have been proposed by predecessors. The measurement of contact angles between the wall surface and liquid droplets has been the most widely used method to quantify surface wettability [27–29]. Contact angle measurement is a valuable and robust theoretical model; however, its initial application was focused on macroscopic wettability and is subject to limitations imposed by various external factors. Challenges arise when the droplets being tested exhibit rolling behavior or when the sample surface is rough, making accurate contact angle determination difficult [30]. Furthermore, these measurements fail to capture the local wettability variations resulting from chemical heterogeneity or mineral distribution. Notably, several researchers [25,26,31] have observed different contact angles within adjacent areas, indicating the presence of trapped oil even within contact angles associated with water–wet conditions. This underscores the direct influence of microscale wettability distribution on contact angle measurements and the limited representativeness of this method. Nevertheless, other methods, such as the Amott wettability index, USBM method, and spontaneous imbibition, face challenges in accurately characterizing neutral wettability, being influenced by pore–throat structures, surface roughness, and the fact that they are complex and time–consuming [20,26,32–36]. Nevertheless, nuclear magnetic resonance only allows for the analysis of wettability variations between pore–throat channels of different diameters [20,36–38]. To gain a more comprehensive understanding of the specific microscopic influences on wettability, it is crucial to integrate these approaches with other techniques. In recent years, molecular dynamic simulations have become increasingly popular [39,40], enabling a detailed exploration of molecular interactions and movements at the microscale. However, these simulations do not provide a quantitative assessment of wettability for different samples. Therefore, there is a pressing need to develop new wettability evaluation methods that offer enhanced accuracy and applicability, particularly in assessing microscale rough surfaces.

AFM serves as a robust tool in the realm of scientific exploration. By leveraging the mechanical interactions generated when a flexible cantilever probe comes into contact with a solid surface, coupled with the knowledge of the cantilever's elasticity, AFM enables the measurement of both surface topography and adhesion forces [41,42]. This technique has found extensive utility across various domains, including medicine, biology, and mineralogy, exemplifying its versatility and significance [43]. In recent years, atomic force microscopy (AFM) has gained popularity for conducting nanoscale investigations. Initially utilized for the surface scanning of rocks to generate 2D and 3D topographic maps [44], AFM has evolved to provide mechanical information by the development of droplet probes, which has enabled the measurement of nanoscale mechanical of interaction between oil droplets and solid surfaces. Shi et al. [21] conducted a study focusing on the surface forces of model oil droplets, including toluene and heptol, and their interactions with shale samples after oil washing. The findings revealed that the hydrophobic interactions on the surface of hydrophobic mica can overcome the steric hindrance caused by asphaltene interfacial adsorption, resulting in strong adhesion and attachment of the oil droplets. Later on, researchers explored surface modifications of gold–coated AFM probes using different chemical reagents to investigate specific interaction mechanisms between functional groups and solids, offering insights into dominant mechanical actions at the experimental level [45–47]. For instance, employing probe–based techniques to delineate adhesion curves between crude oil and calcite/dolomite has facilitated the understanding of the interactions between carbonate reservoir surfaces and crude oil [48]. Notably, AFM allows for the evaluation of mechanical properties between solids and liquids at the nanoscale, utilizing probes with radii ranging from 20 to 40 nm [49–51]. This range encompasses the majority of the smallest mineral sizes. The use of atomic force microscope probes can reduce the errors caused by surface roughness in measurements and enable a more precise and qualitative assessment of rocks using various types of probes. However, it is important to note that, to date, the numerical values of adhesion forces measured by AFM have not been directly linked to rock wettability. Instead, AFM has been utilized as a complementary tool in wettability research. Further investigations are required to integrate AFM measurements with other methods to comprehensively assess the microscopic wettability differences in shale reservoirs.

In this study, a combination of electron microscopy and surface morphology measurements was employed to identify distinct feature areas in the samples. Based on this foundation, an analysis was conducted to examine the correlation between mineral content, surface roughness, and wettability. To assess the adhesion forces between the sample surface and specific oil components, modified gold–sulfur bond probes were utilized to simulate the contact process between oil and rock. Additionally, hydrophilicity measurements of the sample surface were conducted using hydroxyl–functionalized AFM probes. The molecular interactions between the chemical probes and the sample surface provided insights into the variations in adhesion forces at the interface of the sample and water/oil droplets, directly associated with wettability. By coupling the shale surface force–distance curves obtained using two different probes, a dimensionless wettability index was established to evaluate wettability characteristics at microscale point locations. This research using the application of AFM in exploring wettability demonstrated its feasibility and addressed the challenges associated with microscale wetting measurements, laying a solid foundation for further investigations to enhance studies aimed at improving oil displacement efficiency.

## 2. Materials and Methods

### 2.1. Materials

Shale Samples

The shale samples used in this study were sourced from two distinct blocks, namely well JA (JA) and well JB (JB), within the Lucaogou Formation in Xinjiang, China. These samples predominantly consisted of quartz, feldspar, clay minerals (primarily kaolinite, illite, and chlorite), and a minor quantity of carbonate minerals. The mineralogical composition, as depicted in Table 1, represents a significant portion of mixed shale reservoirs and serves as an exemplary model system for terrestrial shale reservoirs.

**Table 1.** Inorganic and organic components of the Lucaogou Formation of JA and JB.

| Sample | Quartz (%) | Potassium Feldspar (%) | Plagioclase Feldspar (%) | Total Clay Content (%) | Calcite (%) | Dolomite (%) | TOC (%) |
|---|---|---|---|---|---|---|---|
| JA–2 | 26.8 | 22.2 | 21.8 | 14.7 | 5.9 | 8.6 | 3.87 |
| JA–3 | 32.8 | 9.2 | 27.7 | 14.8 | 4.3 | 11.2 | 3.64 |
| JA–4 | 34.8 | 6.7 | 24.1 | 21.7 | 0.5 | 12.2 | 5.29 |
| JA–6 | 47.3 | 13.6 | 20.5 | 3.5 | ☐ | 15.1 | 5.18 |
| JA–9 | / | / | / | / | / | / | 4.86 |
| JB–1 | 16.1 | 30.2 | 23.525 | 11.025 | / | / | 3.06 |
| JB–3 | 16.4 | 6.6 | 37.4 | 5.1 | 0.7 | 4.5 | 3.48 |
| JB–4 | 19.9 | 7.2 | 51.8 | 3.4 | 0.5 | 17.6 | 2.71 |
| JB–8 | 19.3 | 7.5 | 33 | 3.7 | / | 35.6 | 2.91 |

These samples were derived from the retrieved 25 mm core samples. Subsequently, a 10 mm drill bit was used to extract material from the center of the samples, and in conjunction with mechanical cutting, they were processed into 2 mm–thick circular thin sections for AFM studies. During this process, the samples were not subjected to oil washing treatment to maintain them in a state as close as possible to their original condition in the geological formation, facilitating the measurement of solid–liquid adhesion forces.

### 2.2. Preparation of Sample Flat Surfaces

The sample surfaces were meticulously prepared using a combination of a Leica mechanical polishing machine and a state–of–the–art triple–beam argon ion milling machine. Firstly, the samples were sliced precisely into 2 mm thick sections to ensure uniformity. Subsequently, a meticulous polishing process was carried out, employing a series of sandpapers with varying surface roughness, namely 9 μm, 2 μm, and 0.5 μm. This step was crucial in eliminating any residual scratches caused by coarser sandpapers, thereby refining the surface quality of the samples. To further enhance the surface smoothness and eliminate any remaining imperfections, the polished samples underwent a meticulous ion milling process. Utilizing an argon ion beam, the samples were subjected to alternating polishing cycles, with voltages set at 5 kV and 2 kV, respectively. A working current of 2.0 mA was employed, and each polishing cycle had a duration of 20 min. This meticulous ion milling process ensured the attainment of an optimal surface condition for subsequent analysis. The surface characteristics of the polished samples were thoroughly examined using atomic force microscopy (AFM) in contact mode. Two–dimensional and three–dimensional morphology analyses were performed to capture the intricate details of the sample surfaces. Additionally, the average roughness parameter was determined to quantitatively assess the surface texture. This comprehensive characterization using AFM allowed for a precise evaluation of the sample surface topography, enabling a deeper understanding of its physical properties and features.

*2.3. Contact Angle Measurements*

In this study, the wettability of three adjacent points on the surface of each shale sample was investigated using the sessile drop method. The sessile drop method is the most widely used technique for measuring contact angles; in this method, a pipette was manually used to drop distilled water onto the marked position (corresponding to the range of surface roughness measurements mentioned later in the text), and the contact moment between the solid–liquid interface and the liquid–gas interface was captured using a high–speed camera. Thus, the surface tension relationships among the solid–gas, gas–liquid, and solid–liquid interfaces were characterized, as shown in Figure 1. The coupling relationship based on the Young's equation can be expressed as follows [10,12,21–23,52]:

$$cos\theta = \frac{\gamma_{Sg} - \gamma_{SL}}{\gamma_{gL}} \quad (1)$$

In the equation, $\gamma_{Sg}$, $\gamma_{SL}$, $\gamma_{gL}$ represent the surface tensions of the solid–gas, solid–liquid, and liquid–gas phases, respectively.

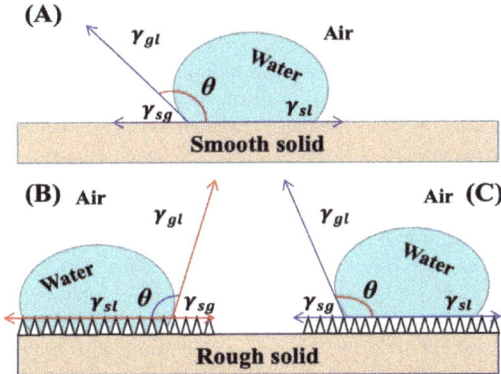

**Figure 1.** The relationship between contact angle and interfacial tension: (**A**) contact angle on a smooth surface, (**B**) Wenzel contact angle on a rough surface, and (**C**) Cassie contact angle on a rough surface.

*2.4. AFM Measurements*

Notably, AFM operates in three primary modes: contact, non–contact, and tapping mode, each offering unique capabilities [44]. In our study, we employed the PeakForce mode, which offers distinct advantages by directly probing the vertical mechanical behavior of shale surfaces. The schematic diagram of the principle is shown in Figure 2. This choice allowed us to delve deeper into the intricacies of surface topography and ascertain the magnitude of adhesion forces. AFM measurements were performed on each polished sample using a tapping mode with a probe attached to the cantilever of the AFM instrument (Bruker, Billerica, MA, USA, model Multimode8). The cantilever probe had an elastic constant (k) of 0.350 N/m, a resonance frequency ($f_0$) of 65 kHz, and a tip radius of 30 nm. All measurements were carried out in ambient air at room temperature (296.15 K) and atmospheric pressure (1 atm). A typical scan rate of 1 Hz was employed during the recording process, utilizing a scan head with a maximum range of 100 μm × 100 μm. The setpoint height was set to 200 nm, and the surface topography information was acquired using the PeakForce Quantitative Nanomechanical (QNM) mode. The scanning area for the experiment was set at 20 μm × 20 μm. Following the characterization of surface topography, chemically modified hydrophobic/hydrophilic probes were employed to perform surface mechanical measurements in force–indentation mode, enabling the determination of adhesion forces' magnitude.

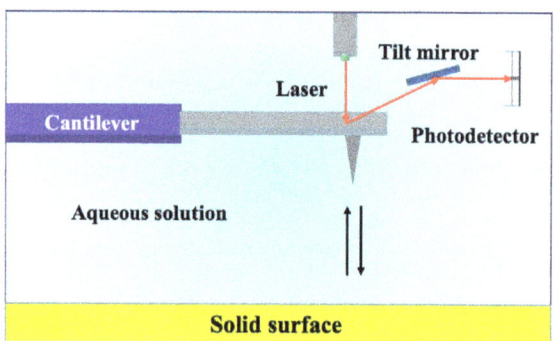

**Figure 2.** Schematic diagram of the atomic force microscope.

### 2.4.1. AFM Probes

The analysis of adhesion forces and surface morphology imaging of the sliced shale samples was conducted using gold–coated cantilevers obtained from Bruker (USA) with a nominal spring constant of 0.350 N/m. These probes were subjected to surface modifications with different chemical moieties prior to usage, enabling the measurement of mechanical interactions at the sample–water interface. Silicon nitride probes were divided into two categories, representing crude oil components and aqueous solutions, respectively. (1) For the hydrophobic modification, the surface was treated with dodecanethiol. The gold–thiol bonds formed between the thiol groups and the gold–coated probes resulted in the self–assembly of a monolayer attached to the AFM tip. (2) To achieve hydrophilicity and represent water–solid interface interactions, the AFM probe surface was modified with hydroxyl groups. Before each experiment, the probes were thoroughly rinsed with ethanol to remove any physically adsorbed substances, followed by drying with nitrogen gas to prevent any potential impact on the experimental data due to adsorbed substances.

### 2.4.2. AFM Surface Roughness Measurement

Roughness plays a crucial role in describing the surface morphology state. To further quantify the variations in surface morphology and understand the changes in wetting behavior under different roughness conditions, several parameters were extracted from the AFM images. Among them, $R_a$ represents the average roughness, which is calculated as the arithmetic mean of the absolute height deviations, and $R_q$ represents the root mean square roughness, which is calculated as the root mean square of the height deviations [44,49,53–55]. When analyzing the roughness of the samples, it is essential to consider a few representative parameters. In this study, we selected $R_a$ as a measure of the average roughness and $R_q$ as an indicator of the root mean square roughness. $R_a$ provides valuable insights into the average roughness of the sample surface and is determined using the following Formula (2) [44]:

$$R_a = \frac{1}{N_x N_y} \sum_{i=1}^{N_x} \sum_{j=1}^{N_y} |z(i,j) - z_{mean}| \qquad (2)$$

The root mean square roughness ($R_q$), which characterizes the degree of surface roughness variation, is calculated using the following Formula (3):

$$R_q = \sqrt{\frac{1}{N_x N_y} \sum_{i=1}^{N_x} \sum_{j=1}^{N_y} (z(i,j) - z_{mean})^2} \qquad (3)$$

where:

$$z_{mean} = \frac{1}{N_x N_y} \sum_{i=1}^{N_x} \sum_{j=1}^{N_y} z_{ij} \qquad (4)$$

$N_x$ and $N_y$ represent the number of points along the $x$-axis and $y$-axis.

2.4.3. AFM Adhesion Measurement

To further elucidate the oil–wet/water–wet characteristics of shale samples, we employed functionalized probes to investigate the intermolecular forces at the probe–rock interface. The measured forces, representing either repulsion or attraction, were analyzed as positive or negative values, respectively. All force–distance curves contributing to the adhesive force map underwent a standardized data processing procedure. Initially, a baseline correction was applied to eliminate any unrelated vertical displacement between the probe tip and the surface, thereby isolating the relevant interaction. The contact point was then identified and set as the reference position. Subsequently, the height signal corresponding to the cantilever deflection was meticulously calibrated to accurately determine the vertical position of the tip. These meticulous steps ensured precise quantification of the adhesive forces. Under distilled water conditions, the adhesive interactions between the modified probes and the shale surface were captured through force–distance curves during the retraction process. To capture the heterogeneity of the shale sample observed under an optical microscope, five representative points were selected from each of the two distinct areas. These selected areas fell within the range of contact angle droplets and were subjected to water/oil adhesion force measurements.

2.5. Other Tests

We conducted a series of conventional tests to characterize the physical properties of the shale surface, including scanning electron microscopy (SEM). By performing secondary electron scans using SEM, we obtained images that revealed the rock features on the sample surface. These SEM images served as a supplementary tool for analyzing surface roughness and adhesion characteristics. Within the designated areas of the marked regions, the sample characteristics were precisely scanned under the fields of view of 10 µm and 5 µm. This detailed scanning enabled the subsequent classification and analysis of the sample features, including the identification of organic matter deposition areas and the understanding of the influence of mineral types on surface roughness.

3. Results

3.1. Contact Angle Measurements on Shale Sample Surfaces

The samples selected for contact angle measurements demonstrated a hydrophobic nature. In the case of JA, the contact angles ranged from 117.1° to 133°, with an average of 124.5° (Table 2). For JB, the contact angles ranged from 112.3° to 131.1°, with an average of 121.5°. These results indicate that JB well exhibited relatively weaker hydrophobicity compared to JA. It is worth noting that the contact angle measurements revealed variations in wettability among different points on the same sample. Previous studies have indicated that the contact angle is influenced by the distribution of mixed wetting within the contact area [6,7,56]. This phenomenon leads to local variations in wetting behavior and challenges the accuracy of wettability descriptions. The macroscopic assessment of wetting properties may overlook the presence of numerous oil droplets adhering to the surface. Therefore, a comprehensive understanding of the wettability characteristics requires a combined analysis of both macroscopic and microscopic aspects, taking into account the intricate wetting distribution at different length scales.

**Table 2.** The results of contact angle measurements.

| Sample | Site 1 | Site 2 | Site 3 | Average |
|---|---|---|---|---|
| JA–2 | 117.1 | 121.6 | 125.3 | 121.3 |
| JA–3 | 129.3 | 124.1 | 133 | 128.8 |
| JA–4 | 123.4 | 121.3 | 128.2 | 124.3 |
| JA–6 | 121.7 | 124 | 122.9 | 122.9 |
| JA–9 | 117.6 | 131.7 | 126.3 | 125.2 |
| JB–1 | 128.6 | 129.9 | 131.1 | 129.7 |
| JB–3 | 124.1 | 116.3 | 119.2 | 119.9 |
| JB–4 | 112.3 | 121.3 | 113.2 | 115.6 |
| JB–8 | 125 | 116.1 | 120.5 | 121.5 |

*3.2. Shale Sample Roughness*

The surface morphology of the samples was characterized using AFM, and the 2D and 3D surface profiles of the shale are illustrated in Figure 3. In the 2D profile, the color depth serves to represent variations in height. Relevant characteristics can also be observed in the 3D topography image. The roughness data, obtained from the deflection of the probe, are summarized in Table 3. For the JA sample, the Ra values ranged from 396 nm to 692 nm, with an average of 535.6 nm. Regarding the JB sample, the $R_a$ values ranged from 45.8 nm to 436 nm, with an average of 207.1 nm. These findings indicate that the JA surface exhibited a rougher texture compared to JB. In terms of $R_q$ values, the JA sample showed a range from 477 nm to 698 nm, with an average of 617.2 nm. On the other hand, the $R_q$ values for JB ranged from 80.7 nm to 539 nm, with an average of 272.8 nm. These results suggest that the surface roughness of the JA sample demonstrated greater variability in comparison to JB. It is noteworthy that the $R_q$ values surpassed the $R_a$ values, indicating that $R_q$ is more sensitive to height variations, which aligns with previous research [57].

**Table 3.** The surface roughness of the sample.

| Smaple | $R_a$ (nm) | $R_q$ (nm) |
|---|---|---|
| JA–2 | 396 | 477 |
| JA–3 | 692 | 605 |
| JA–4 | 529 | 647 |
| JA–6 | 518 | 659 |
| JA–9 | 543 | 698 |
| JB–1 | 436 | 539 |
| JB–3 | 114 | 170 |
| JB–4 | 45.8 | 80.7 |
| JB–8 | 123 | 195 |

It should be noted that there was a significant variation in the average roughness and root–mean–square roughness among the four JB samples, despite using the same rock polishing method (the results are shown in Table 3). This discrepancy can be attributed to the impact of micro–scale pore and crack distributions on the AFM probe tip, which had a radius of approximately 20 nm. The selected areas of the samples in this study primarily exhibited roughness influenced by pores with radii ranging from 10 to 1000 nm.

In Figure 3A, the DA corresponds to smaller height values, indicating the presence of well–developed micro–pores. The utilization of a smaller radius probe enables a more detailed characterization of the influence of heterogeneity on roughness.

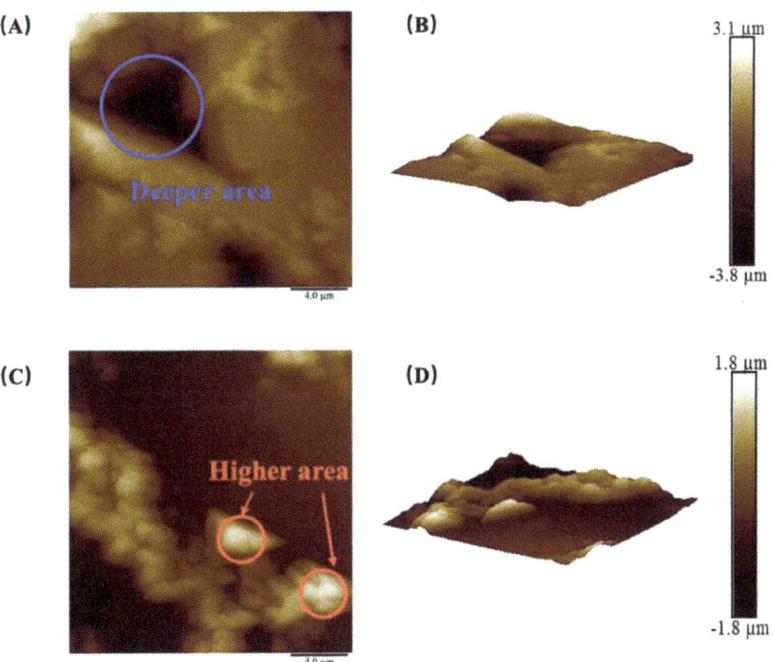

**Figure 3.** (**A**) Two–dimensional surface morphology image and (**B**) 3D surface morphology image of sample JA–3; (**C**) 2D surface morphology image and (**D**) 3D surface morphology image of sample JB–1.

Furthermore, the variations in sample roughness can be attributed to the influence of organic and inorganic content. The experimental samples underwent additional analysis using scanning electron microscopy (SEM), as depicted in Figure 4. Taking sample JA–2 as an example, it was clearly observed that the interior of the dark area (DA) exhibited a more intricate and uneven roughness profile. These areas posed significant obstacles during the scanning process, resulting in higher overall roughness measurements. Notably, the DA emerged as the primary factor contributing to the increased roughness. Moreover, when examining the images at a smaller scale, it was evident that organic matter predominantly occupied the DA, with minimal presence in the bright area (BA). This phenomenon can be attributed to the deposition process of reservoir formation, where organic matter becomes bonded with fine minerals like clay and quartz, effectively filling the intergranular pores between larger mineral particles, such as plagioclase. This process establishes a favorable foundation for shale reservoir formation, contributing to pore development and storage capacity. Additionally, it facilitates the preservation and accumulation of organic matter. Consequently, under the combined influence of these factors, organic matter tends to be concentrated within the interior of the DA. The negative correlation observed between plagioclase feldspar (predominantly found in BA) and total organic carbon (TOC) can be attributed to the effective reduction of organic matter growth space by the development of plagioclase feldspar.

A detailed analysis was conducted to examine the relationships between inorganic and organic content and roughness. The results revealed a positive correlation between the content of minerals such as quartz, clay, and organic matter and surface roughness (from Figure 5A–C). On the other hand, a negative correlation was observed between the presence of plagioclase feldspar and roughness (Figure 5D). By combining the findings from scanning electron microscopy (SEM) studies, the relationship between the mineral composition of the Lucaogou Formation and different characteristic regions was elucidated. The BA in the SEM images correspond to plagioclase feldspar grains that are relatively larger and possess a smoother surface. While quartz grains may also exhibit larger grain sizes, the presence of numerous small surface pores can result in variations in surface roughness. Additionally, the inclusion of clay minerals and the occurrence of authigenic quartz with smaller grain sizes contribute to the formation of DA in association with clay minerals and organic matter (Figure 6).

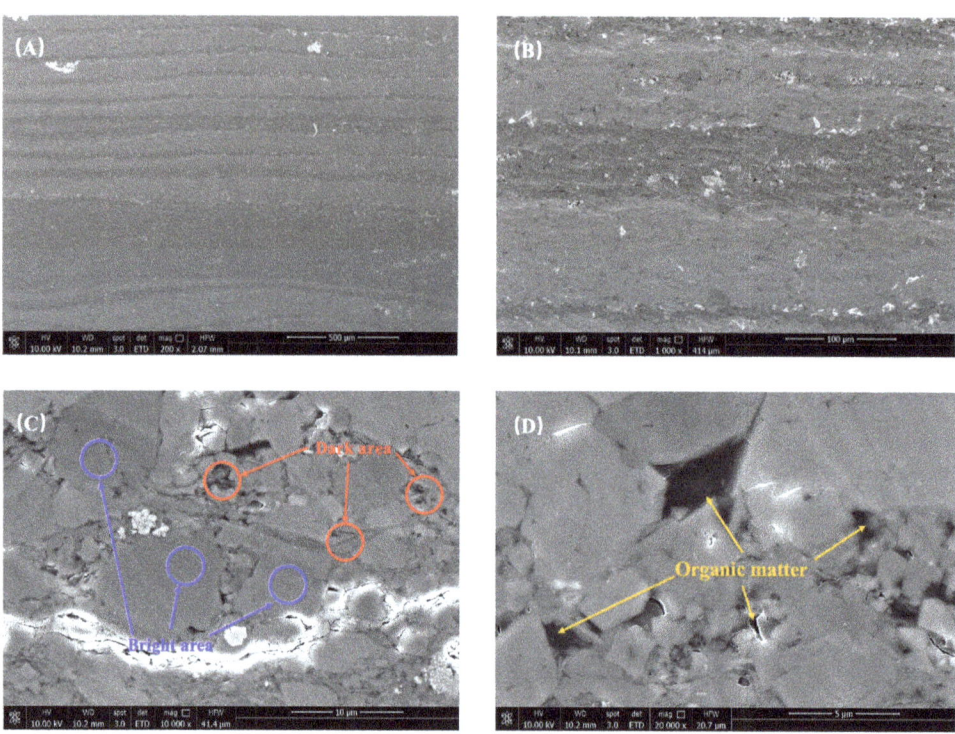

**Figure 4.** The sample surface morphology is illustrated by (**A**,**B**), the distribution of dark and bright areas as shown in (**C**), while (**D**) highlights the occurrence of organic matter.

We employed comprehensive methods, including low–temperature nitrogen adsorption and high–pressure mercury intrusion, to assess the size and distribution of pores and throats within the shale samples. The findings revealed that potassium feldspar and calcite are associated with larger pore throats, while dolomite and plagioclase exhibit medium–sized pore throats. Additionally, clay minerals and quartz, which contain micro– and nanopores, demonstrate a positive correlation with surface roughness. It is worth noting that, despite the development of larger–sized pore throats in plagioclase feldspar, the overall volume of these pore throats is considerably smaller compared to the volume of the mineral itself. As a result, plagioclase feldspar tends to possess a relatively smoother surface. Moreover, the presence of plagioclase feldspar in BA effectively inhibits the formation of DA, as shown in Figure 7. Furthermore, under the scanning electron microscope, it is evident that clay minerals with smaller particle sizes and quartz with higher roughness exhibit more irregular surface features during the agglomeration process (Figure 4C). The presence of associated organic matter, with its plasticity properties, can exacerbate this effect (Figure 4D). These conditions provide the prerequisites for the formation of higher surface roughness. (The correlation between roughness and the content of potassium feldspar, calcite, and dolomite was not evident due to the limited number of samples. Further research is conducted in the next phase to investigate this relationship in more detail.)

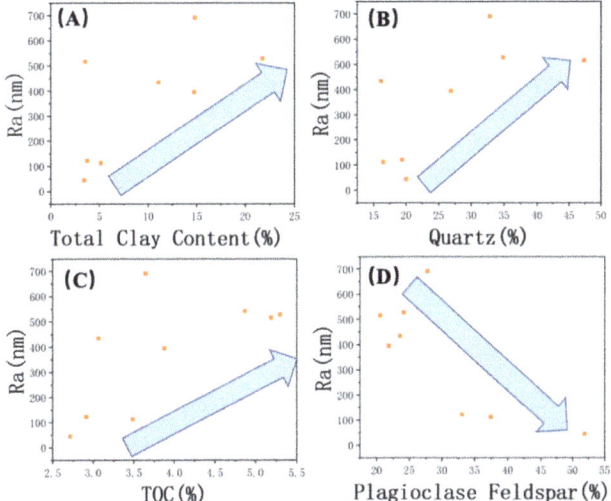

**Figure 5.** Correlation of Ra with (**A**) total clay content, (**B**) quartz, (**C**) TOC, and (**D**) plagioclase. The arrows indicate trends.

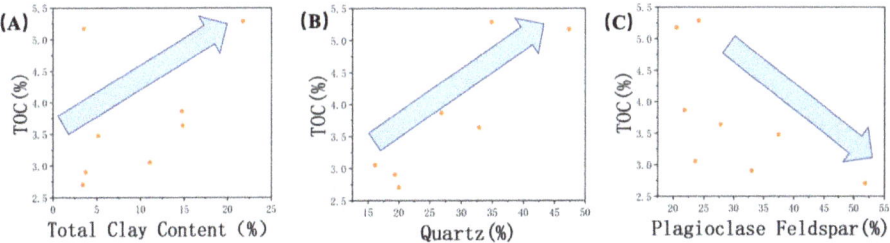

**Figure 6.** Correlation of TOC with (**A**) total clay content, (**B**) quartz, and (**C**) plagioclase. The arrows indicate trends.

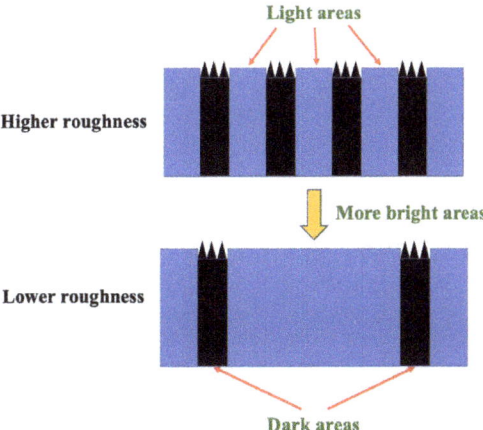

**Figure 7.** Influence of the development of bright areas on roughness.

*3.3. Adhesion Force of Shale Surface*

Figure 8 presents the mechanical variations in the probe–sample system within different characteristic areas. Figure 8A illustrates the interaction between the hydroxyl–functionalized probe and the surface in a distilled water environment. The retraction process occurred at around 2.0 nN, and as the cantilever further deformed, the interaction transitioned gradually from repulsion to attraction. The lowest point of the Y axis is the maximum adhesion. Comparing it with Figure 4D, it is evident that hydrophobic interactions dominated at the same position, as observed in Figure 8C,D. In a horizontal comparison, for probes with the same properties, the DA consistently exhibited stronger oil–wet or water–wet interactions, as deduced from the comparison between Figures 8A and 8D.

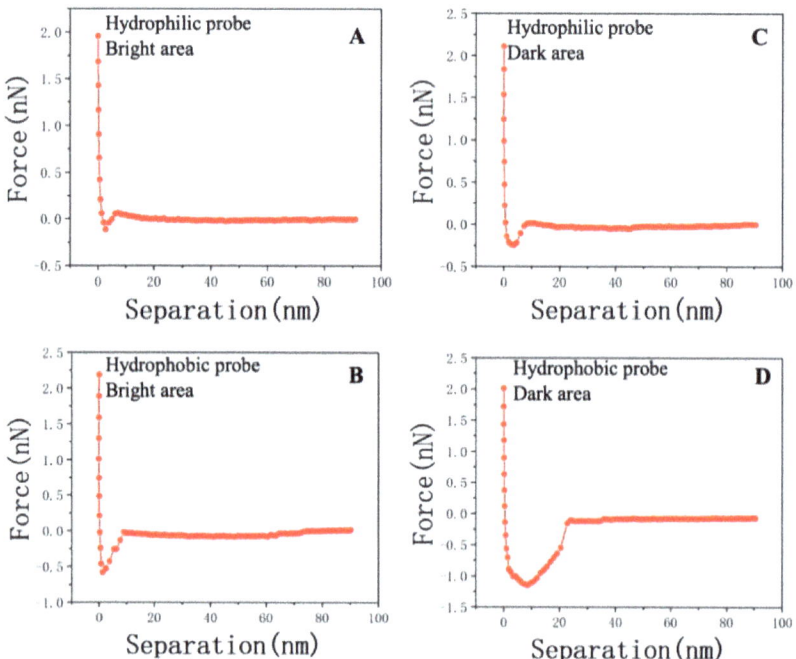

**Figure 8.** (**A**) Adhesion force due to the hydrophilic interaction at point 2 in the BA of sample JA–2. (**B**) Adhesion force due to the hydrophobic interaction at point 2 in the BA of sample JA–2. (**C**) Adhesion force due to the hydrophilic interaction at point 1 in the DA of sample JA–2. (**D**) Adhesion force due to the hydrophobic interaction at point 1 in the DA of sample JA–2.

In order to further elucidate the adhesion characteristics of other locations within JA–2, an analysis was conducted on the adhesion forces, as presented in Table 4. The water adhesion forces of the five distinct positions predominantly originated from the BA, ranging from 0.25 to 0.31 nN, with an average of 0.3 nN. Conversely, the DA exhibited water adhesion forces ranging from 0.11 to 0.16 nN, with an average of 0.12 nN. Notably, a pronounced disparity was observed, indicating that the water adhesion forces in the DA were notably higher than those in the BA. Moreover, employing functionalized probes enabled the characterization of the force–distance curves of the oil–adhesive probes within the BA and DA, at the same selected points. The results mirrored the trend observed for hydrophilic probes, with the oil adhesion forces being comparatively lower in the BA when contrasted with the DA.

It is worth noting that scanning electron microscopy (SEM) imaging revealed a higher accumulation of organic matter within the rough areas. The presence of organic matter in these DA indicates a greater degree of modification. By comparing the differences in the magnitudes of two distinct forces at various locations, it can be inferred that the areas enriched with organic matter exhibited a higher hydrophobic effect compared to hydrophilicity. Consequently, the existence of this DA contributed significantly to the overall hydrophobic characteristics of the sample, thereby enhancing its oil–wetting properties.

Table 4. Hydrophobic/hydrophilic adhesion force magnitudes at different positions of sample JA–2.

| Site | Site 1 | Site 2 | Site 3 | Site 4 | Site 5 | Site 6 | Site 7 | Site 8 | Site 9 | Site 10 |
|---|---|---|---|---|---|---|---|---|---|---|
| Hydrophilic adhesion force | 0.11 | 0.11 | 0.16 | 0.12 | 0.12 | 0.25 | 0.37 | 0.28 | 0.31 | 0.29 |
| Hydrophobic adhesion force | 0.58 | 0.58 | 0.7 | 0.71 | 0.49 | 1.14 | 1.13 | 1 | 1.21 | 0.96 |

The same phenomenon was observed in other samples (with a measurement process similar to that of sample JA–2), as shown in Table 5 (all data in the table are average values).

Table 5. The difference in adhesion force between oil and water.

| Sample | Water Adhesion Force in the BA (nN) | Water Adhesion Force in the DA (nN) | Oil Adhesion Force in the BA (nN) | Oil Adhesion Force in the DA (nN) | The Difference in Adhesion Force between Oil and Water in the BA (nN) | The Difference in Adhesion Force between Oil and Water in the DA (nN) |
|---|---|---|---|---|---|---|
| JA–2 | 0.2 | 0.3 | 0.87 | 1.8 | 0.67 | 1.5 |
| JA–3 | 0.29 | 0.58 | 0.52 | 1.8 | 0.23 | 1.22 |
| JA–4 | 0.49 | 1.2 | 1.25 | 4.75 | 0.76 | 3.55 |
| JA–6 | 0.53 | 1.31 | 1.71 | 4.88 | 1.18 | 3.57 |
| JA–9 | 0.25 | 0.53 | 0.49 | 1.42 | 0.24 | 0.89 |
| JB–1 | 0.27 | 0.52 | 0.57 | 1.55 | 0.3 | 1.03 |
| JB–3 | 0.21 | 0.52 | 0.48 | 1.1 | 0.27 | 0.58 |
| JB–4 | 0.2 | 0.53 | 0.5 | 1.1 | 0.3 | 0.57 |
| JB–8 | 0.2 | 0.42 | 0.58 | 2.49 | 0.38 | 2.07 |

## 4. Discussion

### 4.1. The Influence of Surface Roughness on Contact Angle

In this study, we conducted a comprehensive analysis of the relationship between contact angle and average roughness based on the data presented in Tables 2 and 3. The results are graphically depicted in Figure 9, providing valuable insights into the observed trends. Our analysis revealed a clear and consistent linear increase in the contact angle as the average roughness of the samples increased. The correlation coefficients ($R^2$) obtained were 0.93 and 0.97 for the two experiments, indicating a strong positive relationship between these two variables. This observation suggests that surface roughness plays a crucial role in determining the contact angle. Observed phenomenon can be attributed to the differences in the effective contact area caused by surface roughness. As the roughness of the surface increased, it created variations in the actual contact area between the sample and the liquid phase. Consequently, this disparity in contact area influenced the wetting behavior of the liquid on the surface, resulting in larger contact angles. Moreover, the presence of surface irregularities induced by increased roughness led to a higher proportion of the surface being occupied by air, which further reduced the wetting ability of the liquid and contributed to the observed increase in contact angle.

Young's equation has been widely used to evaluate wetting behavior and describes the wetting properties of materials based on interfacial tension. However, its applicability is limited to ideal flat surfaces that are uniform and smooth (as illustrated in Figure 1). Recognizing the influence of surface roughness, Wenzel introduced the concept of relative roughness to modify Young's equation, aiming to represent the wetting behavior of natural samples [57,58]. In the Wenzel model, the liquid droplet fully penetrates and fills the surface asperities and pores. However, the Wenzel model is suitable for contact angles less than 90° and does not adequately describe the behavior of "liquids standing on the surface of a sample". Therefore, it is necessary to use the Cassie model to discuss and analyze issues with contact angles greater than 90° [59]. In the Cassie model, the hydrophobic surface causes the liquid to be repelled, resulting in air being trapped between the liquid and solid interface. This phenomenon obstructs the contact between the liquid and the surface being measured, leading to measurement errors. The Cassie model can be described by the following equation:

$$cos\theta^* = f_1 cos\theta_1 + f_2 cos\theta_1 \quad (5)$$

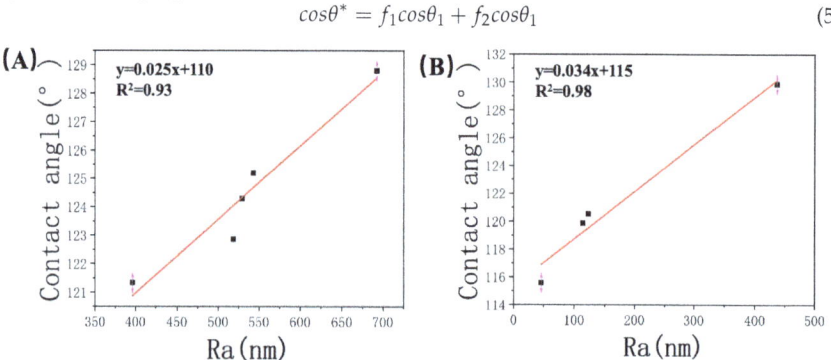

**Figure 9.** Relationship between the roughness of (**A**) JA, (**B**) JB, and the contact angle.

In the equation, $\theta^*$ represents the apparent contact angle of the composite surface; $f_1$, $f_2$ represent the area ratios occupied by the gas and liquid phases on the solid surface (where $f_1 + f_2 = 1$); $\theta_1$, $\theta_1$ represent the intrinsic contact angles at the solid–liquid and gas–liquid interfaces, respectively.

It is important to note that both the Wenzel and Cassie models have their respective limitations and assumptions. The Wenzel model assumes complete wetting and uniform surface roughness, while the Cassie model assumes air trapping and non–wetting behavior. Real–world surfaces often exhibit complex characteristics, including a combination of roughness, heterogeneity, and surface chemistry, which may require more sophisticated models or experimental approaches to accurately describe their wetting behavior.

Based on previous research, it has been established that increased surface roughness on hydrophobic surfaces exacerbates their hydrophobicity. This finding is consistent with the experimental results of our study, where the contact angles of samples from two different shale areas showed a positive correlation with roughness. When comparing JB with JA, the latter exhibited much higher roughness with its uneven surface, providing more favorable conditions for gas entrapment and fewer sites available for water molecules to reside. This further hinders the contact between the liquid and solid surface, enhancing the hydrophobic behavior. As a result, an increase in observed contact angles with increasing roughness can be expected.

*4.2. The Effect of Mineral Content and Organic Matter on Contact Angle*

By comparing the average contact angles, it was observed that JA exhibited stronger oil–wetting behavior compared to JB. To infer the mineralogical control on wetting properties, an assessment of the surface mineralogy of the studied samples was conducted. In our analysis of typical minerals found in the Xinjiang region, we found that the content of quartz and clay minerals exhibited a positive correlation with oil–wettability, while the presence of plagioclase feldspar exhibited an inverse relationship (Figure 10). Indeed, this observation aligns with the discussion in Section 3.2 regarding the relationship between different minerals and surface roughness. The observed trend can be attributed to changes in surface roughness, thus confirming the influence of surface roughness on contact angle. The rougher surface texture of minerals like quartz and clay minerals effectively hinders the contact between water droplets and the rock surface. This results in the hydrophobic parts of the surface being unable to make complete contact with the water droplets, enhancing the expression of hydrophobicity and promoting a higher contact angle. On the other hand, plagioclase feldspar, as a representative mineral with a smoother surface, does not impede the instant contact between the liquid and solid phases.

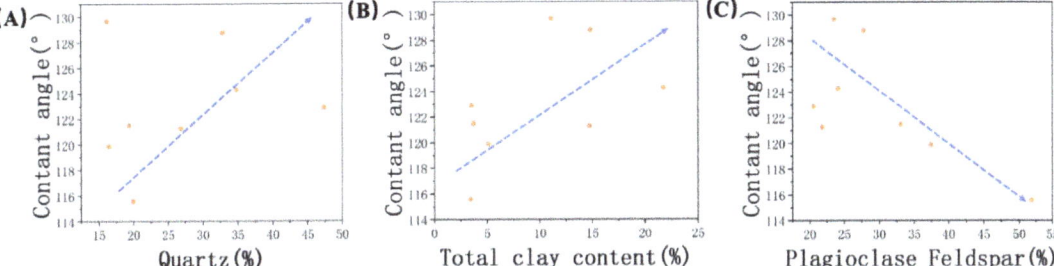

**Figure 10.** Relationship between contact angle and (**A**) quartz, (**B**) total clay content, and (**C**) plagioclase feldspar. The arrows indicate trends.

Based on the above research, we also observed that the wettability of the samples is influenced by the degree of mineral modification by the crude oil. Although, previous researchers [20] classified clay minerals in the Lucaogou Formation samples as hydrophilic minerals. However, it is important not to overlook the interrelationship between clay minerals and organic matter, as they are commonly co–developed within shale formations (as shown in Figure 6A). These two components can have a significant influence on each other.

The wetting behavior of clay is predominantly controlled by residual organic matter, and the process of mineral modification by organic matter on shale wetting cannot be completely ruled out. Otherwise, this may lead to a biased representation of water wetting behavior in areas with relatively low organic matter content. Organic matter is considered to be the primary contributor to oil–wetting behavior, as highlighted by Passey et al. [52]. They proposed that organic matter adsorbed on hydrophilic rocks can cause a transition from hydrophilic to oil–wet surfaces, ultimately exhibiting oil–wet characteristics in the rock's pore structure. Therefore, the presence of organic matter adds complexity to the wetting behavior of shale, as the wetting properties of organic–rich shale are influenced by both mineral composition and organic matter. Su et al. [28] found that shale rocks exhibiting mixed wetting behavior have a higher total organic carbon (TOC) content compared to water–wet rocks. This is attributed to the wetting behavior on the oil–wet rock surface primarily being influenced by the characteristics of organic matter. Consistent with the TOC results in this study (as shown in Table 1), the oil–wetting behavior is found to be enhanced with increasing TOC, with JA exhibiting higher TOC content compared to JB. Therefore, the occurrence and modification of organic matter on the rock surface are considered to be the fundamental factors influencing the formation of shale pore oil–wetting behavior. Areas with a longer exposure to organic matter attachment are more likely to exhibit higher oil–wetting characteristics. This is one of the reasons why sample JA in this study showed stronger oil–wetting behavior compared to JB. Further discussion and analysis on this topic can be found in the following two subsections.

*4.3. The Influence of Wall Oil/Water Adhesion Forces on Wettability*

Regarding the measurement results of the contact angle, we employed atomic force microscopy (AFM) to investigate the adhesive forces of water and oil in different feature areas. Subsequently, we compared the obtained values using a Gaussian fitting and found a strong correlation between the fitted curves and the data, with correlation coefficients of 0.96 and 0.92, respectively. Figure 11 displays the Gaussian distributions (solid lines) fitted to the data. Notably, the water adhesive force on the sample surface was predominantly distributed around 0.35 ± 0.2 nN/m, which was significantly lower than the oil adhesive force, exhibiting a prominent peak at 0.94 ± 0.5 nN/m. Then, we examined the differences in adhesive forces between oil and water during the retraction process and their corresponding discrepancies (Table 5), which shed light on the governing factors of wettability. The results indicated that the oil adhesion forces in both types of characteristic regions were higher than the water adhesion forces (Figure 12).

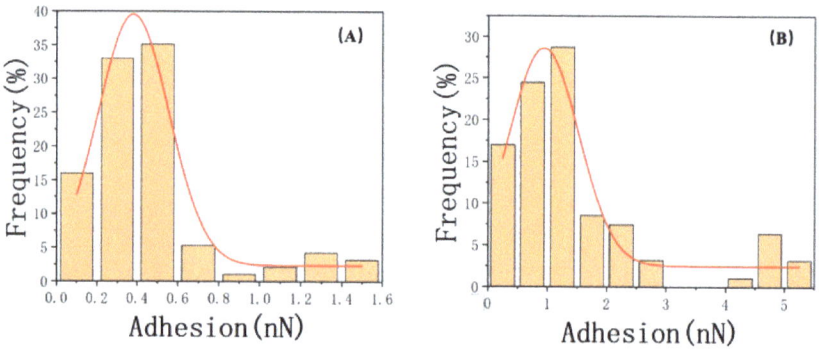

**Figure 11.** Gaussian normal–distribution fitting of adhesion for (**A**) water and (**B**) oil.

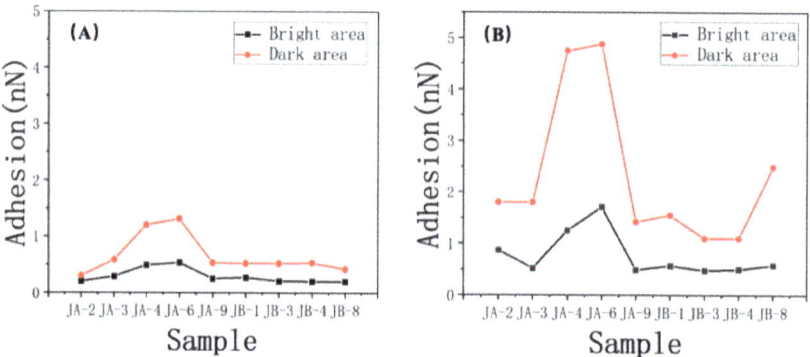

**Figure 12.** Oil and water adhesion in the (**A**) bright area and (**B**) dark area.

However, there is a larger difference between oil and water adhesion forces in DA. In contrast, BA exhibited lower discrepancies (Figure 13). This observation implies that the oil droplet contributed significantly more to the wetting performance in DA compared to BA. Above, we discussed how the presence of organic matter can simultaneously enhance oil–wetting and water–wetting properties. However, it is worth noting that the wetting performance achieved by the oil droplet is notably stronger, underscoring its dominant role in determining wettability. This finding aligns with previous studies that highlighted the influence of organic matter on the surface properties, particularly its propensity to transform the surface from hydrophilic to hydrophobic. Furthermore, the higher interaction of gold–sulfur bonds observed on the DA surface compared to the bright surface corroborated the contribution of DA to the sample's oil–wetting characteristics. To validate the universality of this pattern, we applied the same analysis to the data from other samples, as presented in Table 5. The consistent findings across multiple samples provided further support for the observed trend.

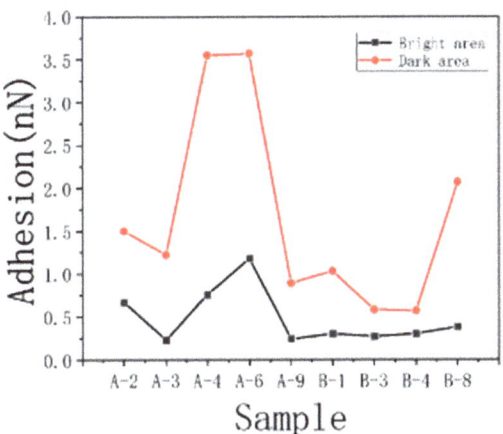

**Figure 13.** Comparison of the adhesion difference between oil and water in different regions.

In order to investigate the relationship between shale adhesion forces and wettability, we conducted a fitting analysis of the oil/water adhesion force difference and the sum of the forces and proposed an AFM–based method to assess reservoir wettability. By measuring the adhesion forces at the oil–solid and water–solid interfaces, we established a dimensionless wettability index, denoted as $I$, to characterize the strength of wettability at the nanoscale. The specific formula is as follows:

$$I = \frac{F_{os} - F_{ws}}{F_{os} + F_{ws}} \qquad (6)$$

In the equation, $F_{os}$ represents the magnitude of the adhesion force between oil and solid, and $F_{ws}$ represents the magnitude of the adhesion force between water and solid. When $0 \leq I \leq 1$, the sample is considered oil–wet, and when $-1 \leq I < 0$, the sample is considered water–wet. Since the wettability index is normalized, its maximum value is 1. This evaluation method is still applicable to water–wet rock cores with a wettability index less than 0.

The wettability between the two distinct areas was assessed using the newly developed AFM evaluation method. The results, presented in Table 6, compare the average wettability indices of the BA and DA in each rock core. The adhesion behavior exhibited by hydrophobic and hydrophilic surfaces differed significantly under the two probes, highlighting the contrasting oil–rock/water–rock interactions that serve as the foundation for the nanomechanical wettability evaluation method. The findings revealed that the average wettability index of BA in each rock core sample was consistently lower than that of DA, indicative of a stronger hydrophobic nature in DA. These outcomes imply that the presence of DA contributes to enhanced hydrophobic behavior.

**Table 6.** Wetting index of different characteristic regions.

| Sample | JA–2 | JA–3 | JA–4 | JA–6 | JA–9 | JB–1 | JB–3 | JB–4 | JB–8 |
|---|---|---|---|---|---|---|---|---|---|
| I in the DA | 0.57 | 0.50 | 0.59 | 0.58 | 0.44 | 0.49 | 0.36 | 0.35 | 0.71 |
| I in the BA | 0.61 | 0.41 | 0.52 | 0.55 | 0.39 | 0.43 | 0.38 | 0.38 | 0.57 |

A comparative analysis was conducted on the wetting index (I) for different regions. Significant differences were observed in the wetting index (I) for samples JA–2, JB–3, and JB–4 compared to the other samples (Figure 14). In comparison to DA, BA contributed to a stronger oil–wetting behavior. Through comparison, it was observed that these three samples exhibited smaller roughness values within their respective regions (as shown in the Table 3), with the BA occupying a larger area. In the case of similar oil–wetting characteristics, the presence of more extensive pore throats associated with the development of larger pores in plagioclase feldspar was observed. This provides more space for the retention of organic matter, resulting in a higher concentration of organic material within the pore throats developed in BA. This process also laid the foundation for the surface modification of plagioclase feldspar in the corresponding regions, leading to a stronger surface modification effect and higher oil–wetting behavior in those areas.

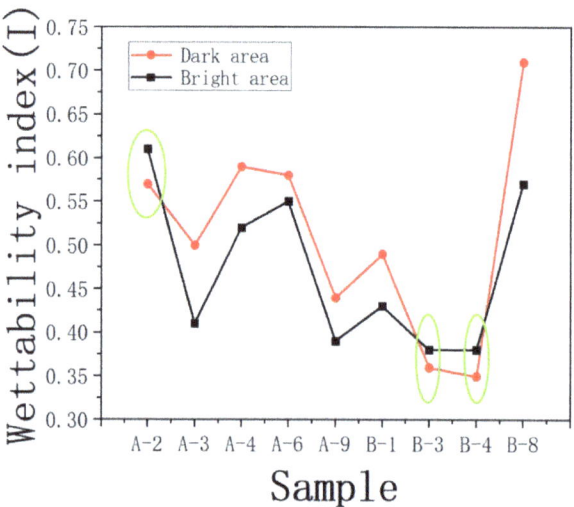

**Figure 14.** Wettability index of different characteristic regions I.

It is also noteworthy that samples JB-3 and JB-4 exhibited a lower development of clay minerals compared to other samples in the JB region. The lower content of clay minerals in these samples is one of the factors contributing to the weaker oil–wetting behavior observed in DA. In addition, a discussion was conducted on the oil–wetting behavior of the surface of sample JB-8, which exhibited similar roughness to sample JB-3. It was found that the oil–wetting behavior in the DA of sample JB-8 was significantly higher than that in the BA. This is different from the observations in samples JA-2, JB-3, and JB-4, where the wetting indices in both areas were very similar. The discrepancy in oil–wetting behavior between the DA and BA of sample JB-8 can be attributed to the differences in the development of large–radius pores. The bright region of sample JB-8 exhibited fewer large–radius pores, which led to a greater retention of organic matter in situ. This facilitated stronger chemical modifications in DA. Indeed, the discussions highlighted the significant influence of organic matter distribution on the surface modification and wetting behavior of shale rocks. The location and concentration of organic matter play a crucial role in determining the wetting characteristics of shale formations.

This observation provides valuable insights into the wettability characteristics of the examined shale reservoirs and underscores the significance of comprehending the spatial distribution and variability of wettability within the reservoir.

The AFM wettability evaluation method was concurrently analyzed with the contact angle method to verify the accuracy of the AFM approach. To facilitate a more intuitive representation of oil wettability, the contact angle measurements were transformed into dimensionless values for direct comparison, as shown in Equation (7):

$$W = \frac{\theta - 90°}{90°} \qquad (7)$$

In the equation, $\theta$ represents the contact angle between gas, water, and the solid surface measured using the sessile drop method. When $0 \leq W \leq 1$, the sample is considered oil–wet, and when $-1 \leq W < 0$, the sample is considered water–wet. Ensuring the cleanliness of the sample surface and the interior of the pipette during the measurement process is crucial. Therefore, in the preparation process, distilled water and nitrogen gas were used to clean the surface residues of the shale, and the pipette and dropper were also cleaned to ensure the accuracy of the experimental results.

The established AFM wettability index (I) was compared to the wettability index (W) obtained through the contact angle method, facilitating a comprehensive analysis of the differences in solid–liquid interface interactions. This comparison allowed for a deeper understanding of the factors influencing wettability on shale surfaces. The comparison between the two methods revealed that the AFM wettability evaluation method provided consistent and complementary results to the contact angle measurements. By utilizing dimensionless values obtained from the contact angle method, a quantitative comparison was made between the two techniques, effectively highlighting both their similarities and differences in assessing oil wettability.

The obtained results from both the AFM wettability index (I) and the contact angle–based wettability index (W) indicated a preference for oil–wet behavior. However, there are notable differences in the degree of oil–wettability, which can be attributed to the influence of surface roughness. The use of AFM probes with a specific radius offered an advantage by minimizing the impact of surface inhomogeneity during the measurement process. As described by the data in Table 7. The wettability index (W) revealed that the JB sample exhibited a wettability range of 0.27 to 0.4 under different roughness conditions. Conversely, the JA sample demonstrated comparable differences in the wettability index, indicating a stronger hydrophobic nature attributed to the presence of surface roughness. Furthermore, by analyzing the discrepancies between I and W, the samples JA–2, JA–4, JA–6, and JB–8 exhibited significant differences in their wetting results, which can be attributed to the incomplete representation of oil–wettability by the contact angle. The rough areas of these four samples demonstrated the highest oil–wetting characteristics among the experimental samples. The strong heterogeneity in these regions prevented the water droplet from fully contacting the sample surface, thereby inhibiting the complete expression of higher oil–wetting characteristics. As a result, this discrepancy between the contact angle (W) and the wetting index (I) values was observed, indicating the limitations of the contact angle measurement in fully capturing the oil–wetting behavior.

**Table 7.** Index W and I.

| Sample | JA–2 | JA–3 | JA–4 | JA–6 | JA–9 | JB–1 | JB–3 | JB–4 | JB–8 |
|---|---|---|---|---|---|---|---|---|---|
| W | 0.31 | 0.38 | 0.34 | 0.33 | 0.35 | 0.40 | 0.31 | 0.27 | 0.31 |
| I | 0.61 | 0.4 | 0.51 | 0.55 | 0.38 | 0.42 | 0.39 | 0.38 | 0.57 |
| I–W | 0.3 | 0.02 | 0.17 | 0.22 | 0.03 | 0.02 | 0.08 | 0.11 | 0.26 |

In summary, the AFM method offers an enhanced level of precision in evaluating surface wettability properties by effectively addressing the impact of roughness and providing valuable insights into the intrinsic wettability of the rock core. As a result, this method demonstrated applicability in assessing wettability for reservoirs.

*4.4. The Impact of Mixed Wetting Distribution on Overall Reservoir Wettability*

The non–uniqueness of wettability magnitudes is apparent from the varied values of the wettability index (I) obtained at different points on the samples. Deglint et al. [7] proposed that microscale contact angles indicate the wetting behavior ranging from hydrophilic to mixed–wetting, while macroscopic contact angle values reflect the phenomenon of mixed wetting. This suggests that the wetting distribution at the microscale has a significant influence on the overall wetting behavior. The heterogeneous distribution of minerals and organic matter contributes to the diverse wettability properties observed on the surface of the rock core, which is the fundamental cause of mixed wettability. However, it is important to note that all measured locations on the samples in this study consistently exhibited oil–wet behavior, including both DA and BA. The presence of hydrophilic minerals did not result in any water–wet sites due to the extended immersion of the samples in oil–bearing formations. This prolonged exposure led to the formation of an oil film and surface modifications, promoting the development of oil–wettability. Meanwhile, the development of oil–wetting characteristics was related to the difference in pore sizes between adjacent regions. When large pores are extensively developed, more organic matter tends to enter these larger pore throats and undergo wetting modifications on a larger scale and over a larger range of the mineral surface. These factors contribute to the overall sample, exhibiting a more pronounced oil–wet state.

## 5. Conclusions

Wettability plays a critical role in assessing reservoir characteristics, such as movable oil content and recovery potential. The traditional contact angle measurement method has several limitations and challenges in the measurement process. To overcome the limitations of traditional wettability evaluation methods, this study employed atomic force microscopy (AFM) and contact angle measurements to assess the surface morphology, roughness, and mechanical properties of mixed shale reservoir samples from Xinjiang, China.

1. Correlation analysis was conducted regarding the mineral content and surface roughness. It was found that an increase in the content of minerals such as quartz, clay, and organic matter contributed to enhanced surface roughness. On the other hand, the presence of plagioclase feldspar showed an opposite trend. Building upon this, an analysis of the wetting behavior based on different mineral contents was conducted, revealing that the influence of mineral content on the contact angle originated from variations in surface roughness.
2. By investigating the adhesive interactions between hydrophilic/hydrophobic probes and the shale surface at the nanoscale, significant disparities in energy and force were observed between oil–repellent and water–repellent pore walls. The analysis demonstrated that, as the hydrophobicity of the rock core increased, the interaction forces between oil and rock became significantly stronger, as evident from both contact angle measurements and AFM force curves.
3. Based on the measurement results of adhesive forces, the impact of organic matter on the overall wettability of the samples was thoroughly examined. The findings highlighted that, in contrast to mineral components, the presence of organic matter emerged as the key factor influencing in situ reservoir wettability. The development of intergranular and intragranular pore networks provided space for the occurrence of crude oil, while also laying the foundation for its own oil–wetting characteristics. Notably, in the DA containing organic matter, the oil–solid adhesion force exhibited a significant increase, a phenomenon that conventional methods fail to assess.
4. Leveraging force mechanical findings, a novel method based on AFM, was developed to evaluate mixed wettability in reservoirs. The calculated wettability indices, W and I, exhibited discrepancies. It was determined that these differences originated from surface roughness, which was effectively mitigated by the nanoscale radius of the probe, thereby providing reliable support. This approach enabled the characterization of wettability at the nano/micrometer scale, while incorporating the evaluation of crude oil detachment difficulty through mechanical curves.

In conclusion, this study contributes valuable insights into the interplay between organic matter, mineral distribution, roughness, and wetting behavior.

**Author Contributions:** Conceptualization, L.S.; Methodology, X.H.; Formal analysis, X.H. and J.L.; Investigation, C.F., Z.Z., X.P. and M.D.; Writing—original draft, X.H.; Supervision, Z.Y.; Project administration, L.S.; Funding acquisition, Z.Y. and J.L. All authors have read and agreed to the published version of the manuscript.

**Funding:** This research was supported by the National Natural Science Foundation of China (No. 41972123) "Microscopic occurrence of pore water in shale matrix and its control mechanism on shale gas output"; The Major Project of CNPC's "CCUS oil displacement geological body fine description and reservoir en-gineering key technology research" (No. 2021ZZ01-03); Petrochina Science and Technology Research Project "shale oil development mechanism and development technology research" (No. 2022KT-10-01).

**Data Availability Statement:** Data are contained within the article.

**Conflicts of Interest:** The authors declare no conflict of interest.

## References

1. Wang, M.; Li, M.; Li, J.B.; Liang, X.; Zhang, J.X. The key parameter of shale oil resource evaluation: Oil content. *Pet. Sci.* **2022**, *19*, 1443–1459. [CrossRef]
2. Hu, T.; Pang, X.; Jiang, F.; Wang, Q.; Liu, X.; Wang, Z.; Jiang, S.; Wu, G.; Li, C.; Xu, T.; et al. Movable oil content evaluation of lacustrine organic–rich shales: Methods and a novel quantitative evaluation model. *Earth-Sci. Rev.* **2021**, *214*, 103545. [CrossRef]
3. Zou, C.; Ma, F.; Pan, S.; Zhang, X.; Wu, S.; Fu, G.; Wang, H.; Yang, Z. Formation and distribution potential of global shale oil and the developments of continental shale oil theory and technology in China. *Earth Sci. Front.* **2023**, *30*, 128. [CrossRef]
4. Wang, M.; Ma, R.; Li, J.; Lu, S.; Li, C.; Guo, Z.; Li, Z. Occurrence mechanism of lacustrine shale oil in the Paleogene Shahejie Formation of Jiyang depression, Bohai Bay Basin, China. *Pet. Explor. Dev. Online* **2019**, *46*, 833–846. [CrossRef]
5. Xi, K.; Zhang, Y.; Cao, Y.; Gong, J.; Li, K.; Lin, M. Control of micro–wettability of pore–throat on shale oil occurrence: A case study of laminated shale of Permian Lucaogou Formation in Jimusar Sag, Junggar Basin, NW China. *Pet. Explor. Dev.* **2023**, *50*, 334–345. [CrossRef]
6. AlRatrout, A.; Blunt, M.J.; Bijeljic, B. Wettability in complex porous materials, the mixed–wet state, and its relationship to surface roughness. *Proc. Natl. Acad. Sci. USA* **2018**, *115*, 8901–8906. [CrossRef]
7. Deglint, H.J.; Clarkson, C.R.; Ghanizadeh, A.; Debuhr, C.; Wood, J. Comparison of micro–and macro–wettability measurements and evaluation of micro–scale imbibition rates for unconventional reservoirs: Implications for modeling multi-phase flow at the micro–scale. *J. Nat. Gas Sci. Eng.* **2019**, *62*, 38–67. [CrossRef]
8. Liu, J.; Sheng, J.J. Experimental investigation of surfactant enhanced spontaneous imbibition in Chinese shale oil reservoirs using NMR tests. *J. Ind. Eng. Chem.* **2019**, *72*, 414–422. [CrossRef]
9. Ding, F.; Gao, M. Pore wettability for enhanced oil recovery, contaminant adsorption and oil/water separation: A review. *Adv. Colloid Interface Sci.* **2021**, *289*, 102377. [CrossRef]
10. Deng, X.; Kamal, M.S.; Patil, S.; Hussain, S.; Zhou, X. A review on wettability alteration in carbonate rocks: Wettability modifiers. *Energy Fuels* **2019**, *34*, 31–54. [CrossRef]
11. Liu, F.; Wang, M. Review of low salinity waterflooding mechanisms: Wettability alteration and its impact on oil recovery. *Fuel* **2020**, *267*, 117112. [CrossRef]
12. Arif, M.; Abu–Khamsin, S.A.; Iglauer, S. Wettability of rock/$CO_2$/brine and rock/oil/$CO_2$–enriched–brine systems: Critical parametric analysis and future outlook. *Adv. Colloid Interface Sci.* **2019**, *268*, 91–113. [CrossRef]
13. Amin, J.S.; Nikooee, E.; Ayatollahi, S.; Alamdari, A. Investigating wettability alteration due to asphaltene precipitation: Imprints in surface multifractal characteristics. *Appl. Surf. Sci.* **2010**, *256*, 6466–6472. [CrossRef]
14. Pan, B.; Clarkson, C.R.; Younis, A.; Song, C.; Debuhr, C.; Ghanizadeh, A.; Birss, V. New methods to evaluate impacts of osmotic pressure and surfactant on fracturing fluid loss and effect of contact angle on spontaneous imbibition data scaling in unconventional reservoirs. *Fuel* **2022**, *328*, 125328. [CrossRef]
15. Jagadisan, A.; Heidari, Z. Impacts of competitive water adsorption of kerogen and clay minerals on wettability of organic–rich mudrocks. *SPE Reserv. Eval. Eng.* **2020**, *23*, 1180–1189. [CrossRef]
16. Ferrari, J.V.; de Oliveira Silveira, B.M.; Arismendi–Florez, J.J.; Fagundes, T.; Silva, M.; Skinner, R.; Ulsen, C.; Carneiro, C. Influence of carbonate reservoir mineral heterogeneities on contact angle measurements. *J. Pet. Sci. Eng.* **2021**, *199*, 108313. [CrossRef]
17. Zdziennicka, A.; Szymczyk, K.; Jańczuk, B. Correlation between surface free energy of quartz and its wettability by aqueous solutions of nonionic, anionic and cationic surfactants. *J. Colloid Interface Sci.* **2009**, *340*, 243–248. [CrossRef] [PubMed]
18. Laird, D.A. Layer charge influences on the hydration of expandable 2:1 phyllosilicates. *Clays Clay Miner.* **1999**, *47*, 630–636. [CrossRef]
19. Shi, K.Y.; Chen, J.Q.; Pang, X.Q.; Jiang, F.; Hui, S.; Zhao, Z.; Chen, D.; Cong, Q.; Wang, T.; Xiao, H.; et al. Wettability of different clay mineral surfaces in shale: Implications from molecular dynamics simulations. *Pet. Sci.* **2023**, *20*, 689–704. [CrossRef]
20. Su, S.; Jiang, Z.; Shan, X.; Zhu, Y.; Wang, P.; Luo, X.; Li, Z.; Zhu, R.; Wang, X. The wettability of shale by NMR measurements and its controlling factors. *J. Pet. Sci. Eng.* **2018**, *169*, 309–316. [CrossRef]
21. Shi, C.; Xie, L.; Zhang, L.; Lu, X.; Zeng, H. Probing the interaction mechanism between oil droplets with asphaltenes and solid surfaces using AFM. *J. Colloid Interface Sci.* **2020**, *558*, 173–181. [CrossRef] [PubMed]
22. Taqvi, S.T.; Almansoori, A.; Bassioni, G. Understanding the role of asphaltene in wettability alteration using ζ potential measurements. *Energy Fuels* **2016**, *30*, 1927–1932. [CrossRef]

23. Rocha, J.A.; Baydak, E.N.; Yarranton, H.W.; Sztukowski, D.M.; Ali–Marcano, V.; Gong, L.; Shi, C.; Zeng, H. Role of aqueous phase chemistry, interfacial film properties, and surface coverage in stabilizing water–in–bitumen emulsions. *Energy Fuels* **2016**, *30*, 5240–5252. [CrossRef]
24. Salou, M.; Siffert, B.; Jada, A. Interfacial characteristics of petroleum bitumens in contact with acid water. *Fuel* **1998**, *77*, 343–346. [CrossRef]
25. Jin, X.; Li, G.; Meng, S.; Wang, X.; Liu, C.; Tao, J.; Liu, H. Microscale comprehensive evaluation of continental shale oil recoverability. *Pet. Explor. Dev.* **2021**, *48*, 256–268. [CrossRef]
26. Yao, Y.; Wei, M.; Kang, W. A review of wettability alteration using surfactants in carbonate reservoirs. *Adv. Colloid Interface Sci.* **2021**, *294*, 102477. [CrossRef]
27. Roshan, H.; Al–Yaseri, A.Z.; Sarmadivaleh, M.; Iglauer, S. On wettability of shale rocks. *J. Colloid Interface Sci.* **2016**, *475*, 104–111. [CrossRef]
28. Mirchi, V.; Dejam, M.; Alvarado, V. Interfacial tension and contact angle measurements for hydrogen–methane mixtures/brine/oil–wet rocks at reservoir conditions. *Int. J. Hydrogen Energy* **2022**, *47*, 34963–34975. [CrossRef]
29. de Oliveira Silveira, B.M.; dos Santos Gioria, R.; Florez, J.J.A.; Fagundes, T.; Silva, M.; Skinner, R.; Ulsen, C.; Carneiro, C.; Ferrari, J. Influence of oil aging time, pressure and temperature on contact angle measurements of reservoir mineral surfaces. *Fuel* **2022**, *310*, 122414. [CrossRef]
30. Cwickel, D.; Paz, Y.; Marmur, A. Contact angle measurement on rough surfaces: The missing link. *Surf. Innov.* **2017**, *5*, 190–193. [CrossRef]
31. Sharifigaliuk, H.; Mahmood, S.M.; Al–Bazzaz, W.; Khosravl, V. Complexities driving wettability evaluation of shales toward unconventional approaches: A comprehensive review. *Energy Fuels* **2021**, *35*, 1011–1023. [CrossRef]
32. Gao, Z.; Fan, Y.; Hu, Q.; Jiang, Z.; Chen, Y. The effects of pore structure on wettability and methane adsorption capability of Longmaxi Formation shale from the southern Sichuan Basin in China. *AAPG Bull.* **2020**, *104*, 1375–1399. [CrossRef]
33. Wu, Y.; Luo, Q.; Chu, Z.; Li, Y.; Sun, L.; Shi, X.; Li, Z.; Ren, W.; Yuan, B. Microscale Reservoir Wettability Evaluation Based on Oil–Rock Nanomechanics. *Energy Fuels* **2023**, *37*, 6697–6704. [CrossRef]
34. Peng, S.; Reed, R.M.; Xiao, X.; Yang, Y.; Liu, Y. Tracer–guided characterization of dominant pore networks and implications for permeability and wettability in shale. *J. Geophys. Res. Solid Earth* **2019**, *124*, 1459–1479. [CrossRef]
35. Alinejad, A.; Dehghanpour, H. Evaluating porous media wettability from changes in Helmholtz free energy using spontaneous imbibition profiles. *Adv. Water Resour.* **2021**, *157*, 104038. [CrossRef]
36. Gupta, I.; Rai, C.; Sondergeld, C. Study impact of sample treatment and insitu fluids on shale wettability measurement using NMR. *J. Pet. Sci. Eng.* **2019**, *176*, 352–361. [CrossRef]
37. Liang, C.; Xiao, L.; Jia, Z.; Guo, L.; Luo, S.; Wang, Z. Mixed Wettability Modeling and Nuclear Magnetic Resonance Characterization in Tight Sandstone. *Energy Fuels* **2023**, *37*, 1962–1974. [CrossRef]
38. Newgord, C.; Tandon, S.; Heidari, Z. Simultaneous assessment of wettability and water saturation using 2D NMR measurements. *Fuel* **2020**, *270*, 117431. [CrossRef]
39. Yong, W.; Zhou, Y. A molecular dynamics investigation on methane flow and water droplets sliding in organic shale pores with nano–structured roughness. *Transp. Porous Media* **2022**, *144*, 69–87. [CrossRef]
40. Tian, H.; Liu, F.; Jin, X.; Wang, M. Competitive effects of interfacial interactions on ion–tuned wettability by atomic simulations. *J. Colloid Interface Sci.* **2019**, *540*, 495–500. [CrossRef]
41. Koetniyom, W.; Suhatcho, T.; Treetong, A.; Thiwawong, T. AFM force distance curve measurement for surface investigation of polymers compound blend with metal nanoparticles. *Mater. Today Proc.* **2017**, *4*, 6205–6211. [CrossRef]
42. Luo, Y.; Andersson, S.B. A continuous sampling pattern design algorithm for atomic force microscopy images. *Ultramicroscopy* **2019**, *196*, 167–179. [CrossRef] [PubMed]
43. Kulkarni, T.; Tam, A.; Mukhopadhyay, D.; Bhattacharya, S. AFM study: Cell cycle and probe geometry influences nanomechanical characterization of Panc1 cells. *Biochim. Biophys. Acta (BBA)-Gen. Subj.* **2019**, *1863*, 802–812. [CrossRef]
44. Lu, Y.; Liu, D.; Cai, Y.; Gao, C.; Jia, Q.; Zhou, Y. AFM measurement of roughness, adhesive force and wettability in various rank coal samples from Qinshui and Junggar basin, China. *Fuel* **2022**, *317*, 123556. [CrossRef]
45. Liu, F.; Yang, H.; Wang, J.; Zhang, M.; Chen, T.; Hu, G.; Zhang, W.; Wu, J.; Xu, S.; Wu, X.; et al. Salinity–dependent adhesion of model molecules of crude oil at quartz surface with different wettability. *Fuel* **2018**, *223*, 401–407. [CrossRef]
46. Liu, S.; Xie, L.; Liu, J.; Zhong, H.; Wang, Y.; Zeng, H. Probing the interactions of hydroxamic acid and mineral surfaces: Molecular mechanism underlying the selective separation. *Chem. Eng. J.* **2019**, *374*, 123–132. [CrossRef]
47. Wang, J.; Li, J.; Xie, L.; Shi, C.; Liu, Q.; Zeng, H. Interactions between elemental selenium and hydrophilic/hydrophobic surfaces: Direct force measurements using AFM. *Chem. Eng. J.* **2016**, *303*, 646–654. [CrossRef]
48. Dickinson, L.R.; Suijkerbuijk, B.M.J.M.; Berg, S.; Marcelis, F.; Schniepp, H. Atomic force spectroscopy using colloidal tips functionalized with dried crude oil: A versatile tool to investigate oil–mineral interactions. *Energy Fuels* **2016**, *30*, 9193–9202. [CrossRef]
49. Lu, Z.; Liu, Q.; Xu, Z.; Zeng, H. Probing anisotropic surface properties of molybdenite by direct force measurements. *Langmuir* **2015**, *31*, 11409–11418. [CrossRef] [PubMed]
50. Stevens, R.M. New carbon nanotube AFM probe technology. *Mater. Today* **2009**, *12*, 42–45. [CrossRef]

51. Gao, Z.; Xie, L.; Cui, X.; Hu, Y.; Sun, W.; Zeng, H. Probing anisotropic surface properties and surface forces of fluorite crystals. *Langmuir* **2018**, *34*, 2511–2521. [CrossRef] [PubMed]
52. Siddiqui, M.A.Q.; Ali, S.; Fei, H.; Roshan, H. Current understanding of shale wettability: A review on contact angle measurements. *Earth-Sci. Rev.* **2018**, *181*, 1–11. [CrossRef]
53. Li, Y.; Yang, J.; Pan, Z.; Tong, W. Nanoscale pore structure and mechanical property analysis of coal: An insight combining AFM and SEM images. *Fuel* **2020**, *260*, 116352. [CrossRef]
54. Wang, S.; Liu, S.; Sun, Y.; Jiang, D.; Zhang, X. Investigation of coal components of Late Permian different ranks bark coal using AFM and Micro–FTIR. *Fuel* **2017**, *187*, 51–57. [CrossRef]
55. Liu, X.; Nie, B.; Wang, W.; Wang, Z.; Zhang, L. The use of AFM in quantitative analysis of pore characteristics in coal and coal–bearing shale. *Mar. Pet. Geol.* **2019**, *105*, 331–337. [CrossRef]
56. Alhammadi, A.M.; AlRatrout, A.; Singh, K.; Bijeljic, B.; Blunt, M. In situ characterization of mixed–wettability in a reservoir rock at subsurface conditions. *Sci. Rep.* **2017**, *7*, 10753. [CrossRef]
57. Zhu, M.; Liu, Y.; Chen, M.; Xu, Z.; Li, L.; Zhou, Y. Metal mesh–based special wettability materials for oil–water separation: A review of the recent development. *J. Pet. Sci. Eng.* **2021**, *205*, 108889. [CrossRef]
58. Wenzel, R.N. Resistance of solid surfaces to wetting by water. *Ind. Eng. Chem.* **1936**, *28*, 988–994. [CrossRef]
59. Watanabe, T.; Yoshida, N. Wettability control of a solid surface by utilizing photocatalysis. *Chem. Rec.* **2008**, *8*, 279–290. [CrossRef]

**Disclaimer/Publisher's Note:** The statements, opinions and data contained in all publications are solely those of the individual author(s) and contributor(s) and not of MDPI and/or the editor(s). MDPI and/or the editor(s) disclaim responsibility for any injury to people or property resulting from any ideas, methods, instructions or products referred to in the content.

Article

# Machine-Learning-Based Approach to Optimize $CO_2$-WAG Flooding in Low Permeability Oil Reservoirs

Ming Gao [1,2,*], Zhaoxia Liu [1,2], Shihao Qian [3], Wanlu Liu [1,2], Weirong Li [3], Hengfei Yin [1,2] and Jinhong Cao [1,2]

[1] PetroChina Research Institute of Petroleum Exploration & Development, Beijing 100083, China; liuzhaoxia@petrochina.com.cn (Z.L.)
[2] State Key Laboratory of Enhanced Oil and Gas Recovery, Beijing 100083, China
[3] Department of Petroleum Engineering, Xi'an Shiyou University, Xi'an 710065, China; ltxh990111@163.com (S.Q.)
* Correspondence: gaoming010@petrochina.com.cn

**Abstract:** One of the main applications of carbon capture, utilization, and storage (CCUS) technology in the industry is carbon-dioxide-enhanced oil recovery ($CO_2$-EOR). However, accurately and rapidly assessing their application potential remains a major challenge. In this study, a numerical model of the $CO_2$-WAG technique was developed using the reservoir numerical simulation software CMG (Version 2021), which is widely used in the field of reservoir engineering. Then, 10,000 different reservoir models were randomly generated using the Monte Carlo method for numerical simulations, with each having different formation physical parameters, fluid parameters, initial conditions, and injection and production parameters. Among them, 70% were used as the training set and 30% as the test set. A comprehensive analysis was conducted using eight different machine learning regression methods to train and evaluate the dataset. After evaluation, the XGBoost algorithm emerged as the top-performing method and was selected as the optimal approach for the prediction and optimization. By integrating the production prediction model with a particle swarm optimizer (PSO), a workflow for optimizing the $CO_2$-EOR parameters was developed. This process enables the rapid optimization of the $CO_2$-EOR parameters and the prediction of the production for each period based on cumulative production under different geological conditions. The proposed XGBoost-PSO proxy model accurately, reliably, and efficiently predicts production, thereby making it an important tool for optimizing $CO_2$-EOR design.

**Keywords:** $CO_2$-EOR; XGBoost regression; PSO; parameter optimization

## 1. Introduction

Oil resources have always been the primary source of fossil energy for global energy demand. However, extracting the remaining oil from complex reservoir formations remains a challenge, thereby making it increasingly important to enhance extraction efficiency [1]. In addition, the use of fossil fuels leads to the emission of a large amount of $CO_2$, which is a greenhouse gas, thereby resulting in global climate change, which has become a major challenge facing the world today [2]. Carbon capture, utilization, and storage (CCUS) technology is considered to be an important approach to mitigating global climate change [3]. Based on the challenges mentioned above, $CO_2$-enhanced oil recovery ($CO_2$-EOR) technology is a method within CCUS that can enhance oil recovery by using the $CO_2$ in a miscible displacement process and effectively sequestering $CO_2$ in the lower portion of the reservoir. This technology has relatively low requirements for the purity of the $CO_2$ and allows for the recycling of $CO_2$, thereby reducing process costs [4]. $CO_2$-EOR, with its potential to significantly enhance oil recovery while achieving carbon capture, utilization, and storage, represents an EOR method that offers both societal and economic benefits.

The injection of $CO_2$ is applied to enhance the oil recovery (EOR) due to its superior capabilities in improving the fluid properties under reservoir conditions. The fundamental mechanisms of $CO_2$-EOR include reducing interfacial tension (IFT), lowering oil viscosity, oil swelling, and light hydrocarbon extraction. These mechanisms contribute to the enhanced recovery of the oil in reservoirs [5–7]. Compared to other gases such as natural gas, air, and nitrogen ($N_2$), carbon dioxide ($CO_2$) has a lower minimum miscibility pressure (MMP) than oil. Therefore, selecting $CO_2$ as the injection gas for displacing oil can achieve better miscibility and more effectively recover the oil [8]. In addition to continuous $CO_2$ injection for miscible displacement, the $CO_2$-WAG (water alternating gas) technique has been proposed to improve the flowability of $CO_2$ in the reservoir and to prevent $CO_2$ fingering. This technique involves alternating injections of $CO_2$ and water, which enhance the efficiency of the $CO_2$ propagation and oil displacement [9,10]. The $CO_2$-WAG technique was first used by Mobil Corporation in 1957 in a sandstone reservoir in Alberta, Canada. In addition, the $CO_2$-WAG technique alleviates the issue of rapid $CO_2$ breakthrough and increases the resistance to gas phase flow. It also reduces the resistance to the water phase flow and increases the mobility ratio [11]. According to surveys, the WAG technique has achieved significant success and has been employed in 80% of the oilfield projects in the United States, thereby demonstrating its superiority in enhancing oilfield development and improving oil recovery [12]. Christensen et al. [13] conducted a study on 59 WAG fields and found that, in all WAG cases, the average oil recovery rate increased by 10%. This demonstrates the positive impact and effectiveness of WAG technology in enhancing oil field production. Al-Bayati et al. [14] investigated the impact of core-scale heterogeneity on the oil recovery efficiency of $CO_2$-WAG injection. The research findings indicated that $CO_2$-WAG injection exhibited better performance in homogeneous, layered, and composite samples. Sun et al. [15] investigated the feasibility of the $CO_2$ phase through porous media in WAG injection scenarios and successfully increased the oil recovery factor (RF) by approximately 46%. The gas-to-water injection ratio was identified as a crucial parameter affecting the efficiency of water-gas alternating injection [16,17]. Khather et al. [18] investigated the impact of $CO_2$–carbonate interaction on the oil and gas recovery in three heterogeneous carbonate rock core samples with different initial oil saturations (low and moderate permeability). Overall, $CO_2$-WAG injection after water flooding resulted in an increase in the recovery factor of over 30% for the three rock cores. Ren et al. [19] conducted experiments on $CO_2$-EOR and storage in oilfields in the Ordos Basin in China using two $CO_2$ injection schemes: continuous injection (CI) and water alternating gas (WAG) injection. The results showed that the equal injection of $CO_2$ and WAG significantly increased the crude oil production.

Currently, the optimization of $CO_2$-EOR technology is a focal point of attention for many oilfield and reservoir engineers. Rodrigues et al. [20] utilized the CMG reservoir numerical simulation software (Version 2021) to optimize the application of WAG in a sub-salt offshore oilfield in Brazil. They proposed a $CO_2$-WAG operational design method suitable for carbonate reservoirs, with a focus on economic viability, $CO_2$ recycling efficiency, and project risks. However, traditional parameter optimization methods are time-consuming and labor-intensive. They tend to overlook complex nonlinear relationships and the underlying influencing factors. Moreover, these methods are often based on specific models and algorithms, thus lacking the flexibility to adapt to different oilfield situations and variations. They have certain limitations and lack adaptability. Therefore, the introduction of more advanced optimization techniques, such as machine learning and metaheuristic algorithms, can better address the complexity and uncertainty of oilfields, thereby enhancing the optimization efficiency and accuracy.

At the current stage, rapidly evolving intelligent algorithms, such as machine learning, have found significant applications in the field of petroleum exploration and development. Sen et al. [21] employed a Specialized RNN Unit (SRU) model, which is a type of recurrent neural network (RNN), to optimize the parameters and predict the production in actual $CO_2$-EOR projects. The injection rate, injection pressure, cumulative injection volume of the injection wells, and bottom hole flowing pressure of the production wells were used as inputs for the SRU model, while the fluid production of the production wells served as the output. Li et al. [22] utilized the random forest (RF) regression algorithm to predict the performance of the $CO_2$-WAG technique, including oil well production, $CO_2$ storage volume, and $CO_2$ storage efficiency. The $CO_2$-WAG cycle, $CO_2$ injection rate, and water-gas ratio were identified as the main injection parameters. The prediction results showed a close approximation between the predicted values and the actual values in the test set. The average absolute prediction deviations for cumulative oil production, $CO_2$ storage volume, and $CO_2$ storage efficiency were 1.10%, 3.04%, and 2.24%, respectively. He et al. [23] proposed an optimization workflow for $CO_2$-EOR operations based on machine learning methods and heuristic optimization algorithms. Their workflow included a power consumption prediction using a Gaussian process regression (GPR) model, which combines a nonlinear autoregressive neural network with external inputs (NARX) model for oil production prediction and an operational optimization model. The optimization results were significant; the optimization parameters used included the duration of the water/gas alternating injection cycles, the bottom hole pressure of the production wells, and the injection rate of water.

Some researchers, in order to swiftly explore the solution space and find the global optimal solution, have combined metaheuristic algorithms with machine learning. By harnessing the predictive capability of machine learning to guide the search process of metaheuristic algorithms, they can quickly identify the optimal solution and achieve better results in parameter optimization. In 2018, Mohagheghia et al. [24] utilized a robust evolutionary algorithm to automatically optimize the performance of the hydrocarbon WAG technique used in the E segment of the Norne oilfield. They employed the net present value (NPV) as the objective function and two global semi-random search strategies, namely, the genetic algorithm (GA) and particle swarm optimization (PSO). Parameters such as the water injection volume, gas injection volume, bottom hole pressure of producing wells, cycle ratio, cycle duration, injected hydrocarbon gas fraction, and total WAG cycle were optimized. You et al. [25] combined Gaussian-SVR (support vector regression) with a Gaussian kernel to construct a surrogate model, and the hyperparameters of the surrogate model were optimized using Bayesian optimization. The trained surrogate model was then coupled with a multi-objective particle swarm optimization (MOPSO) protocol. This approach was used to optimize the complex $CO_2$-WAG process, which involves many control parameters. The optimization parameters included operational variables for controlling the $CO_2$-WAG process, such as the duration of the water/gas alternating injection cycle, the bottom hole pressure control, and the injection rates for each well. Jaber [26] utilized the genetic algorithm (GA) technique based on the surrogate model to optimize the most influential parameters in the $CO_2$-WAG process in the Subba-Nahr Umr reservoir. Four operational variables were considered for optimizing the $CO_2$-WAG displacement: the $CO_2$-to-water slug size ratio (WAG), cyclic length (CL), bottom hole pressure (BHP), and $CO_2$ slug size (SZ). The results demonstrated that the highest incremental oil recovery ($\Delta$FOE) of 9.7% in the Subba-Nahr Umr reservoir could be achieved with a WAG ratio of 1.5, a cyclic length of 3 months, a bottom hole pressure of 2221 psi, and a $CO_2$ slug size of 0.91. Based on the above, it can be observed that, in most cases of $CO_2$-EOR parameter optimization, the dataset is relatively small, and the optimization objective functions often only include specific time points of production, which cannot form a complete production curve. As a result, there are limitations and particularities. Due to the lack of complete production curves and large-scale time series data, machine learning and other prediction methods may not fully leverage their advantages and may struggle to achieve global optimization results.

This study proposes a comprehensive workflow for optimizing $CO_2$-EOR (WAG) parameters by combining reservoir numerical simulation with machine learning. In Section 2, the machine learning methods used in this study are described, along with the workflow. Section 3 focuses on establishing the geological and numerical models of the reservoir. In Section 4, the study conducted a correlation analysis of geological and operational parameters. The performance of production prediction models based on different machine learning models was evaluated, and the best machine learning model was selected. In Section 5, the selected machine learning model, combined with particle swarm optimization (PSO), was used for capacity prediction and parameter optimization. Discussions and conclusions are presented in Section 6.

## 2. Methods

This section describes the methodological principles and workflow of the main algorithms used in this study. Eight machine learning methods were employed to build the prediction models, including linear regression [27,28], ridge regression [29], decision tree (DT) [30], random forest (RF) [31,32], gradient boosting decision tree (GBDT) [33], extreme gradient boosting (XGBoost) [34], K-nearest neighbors (KNN) [35], and neural network (NN) [36]. This study proposes a coupled model of the machine learning algorithm XGBoost and particle swarm optimization (PSO) [37] to address the optimization problem. Therefore, the focus is on introducing the XGBoost algorithm and the particle swarm optimization algorithm (PSO).

### 2.1. XGBoost Algorithm

XGBoost is an expandable tree boosting system proposed by Chen et al. [34]. It is an improved version of the gradient boosting decision tree (GBDT) algorithm [38] and is widely used in classification and regression tasks. The basic idea of XGBoost is similar to GBDT, but it incorporates several optimizations, which include the following:

1. Optimizing the loss function by employing a second-order Taylor expansion to enhance computational accuracy.
2. Simplifying the model using regularization terms to avoid overfitting [39].
3. Utilizing a block storage structure to enable parallel computing and improve efficiency.

The structure of the XGBoost algorithm is illustrated in Figure 1, and the model details are described below.

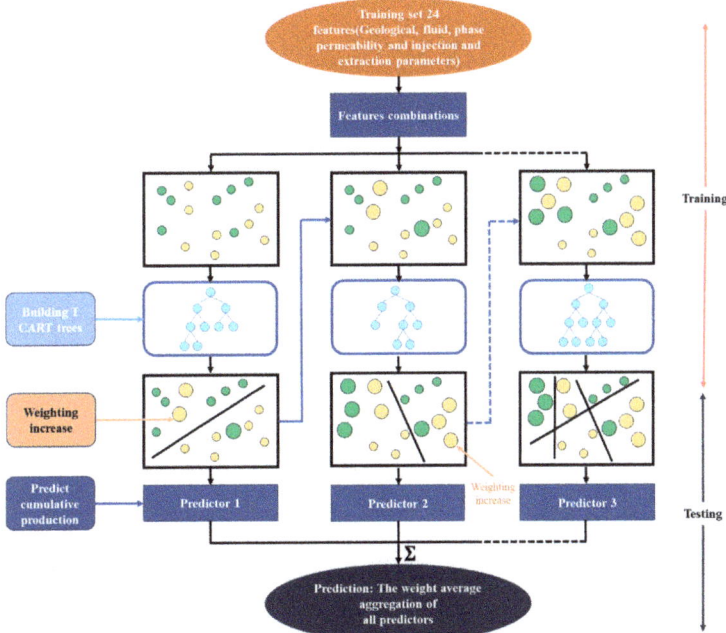

**Figure 1.** The structure of the XGBoost algorithm [40].

Given a training dataset $T = \{(x_1, y_1), (x_2, y_2), \ldots, (x_n, y_n)\}$, a loss function $l(y_i, \hat{y}_i)$, and a regularization term $\Omega(f_k)$, the objective function can be expressed as follows:

$$\mathcal{L}(\phi) = \sum_i l(y_i, \hat{y}_i) + \sum_k \Omega(f_k), \tag{1}$$

where $\mathcal{L}(\phi)$ is the representation in the linear space, $i$ denotes the $i$-th sample, $k$ represents the $k$-th tree, $\hat{y}_i$ hat $i$ is the predicted value of the $i$-th sample $x_i$, and $\sum_k \Omega(f_k)$ represents the complexity of the trees.

Due to the expression of the objective function in GBDT, we can rewrite it as follows:

$$\hat{y}_i = \sum_{k=1}^{K} f_k(x_i) = \hat{y}_i^{(t-1)} + f_t(x_i). \tag{2}$$

In this case, the expression of $\mathcal{L}(\phi)$ can be transformed into the following form:

$$\mathcal{L}^{(t)} = \sum_{i=1}^{n} l\left(y_i, \hat{y}_i^{(t-1)} + f_t(x_i)\right) + \sum_k \Omega(f_k). \tag{3}$$

### 2.2. Particle Swarm Optimization (PSO)

Particle swarm optimization (PSO) is an evolutionary computation technique that was first introduced by Eberhart and Kennedy in 1995 [37]. The basic concept of PSO originates from the study of the foraging behavior in bird flocks and is a simplified model of swarm intelligence algorithms. The algorithm was initially inspired by the regular patterns observed in the movements of prey bird flocks, which led to the development of a simplified model using collective intelligence. PSO utilizes collaboration and information sharing among individuals within a swarm to search for the optimal solution [41].

Figure 2 shows the flow of the PSO algorithm, where each particle individually searches for the optimal solution in the search space. The optimal solution is recorded as the current individual extremum and shared with the other particles in the entire particle population. The particles move at a certain speed in the search space, wherein they dynamically adjust their respective speed and position according to their own flight experience and the flight experience of other particles [42].

The equation to update particle velocity in the PSO algorithm is as follows:

$$V_{new} = \omega V_{id} + C_1 random(0,1)(P_{id} - X_{id}) + C_2 random(0,1)(P_{gd} - X_{id}), \tag{4}$$

where $V_{id}$ is the current velocity of the particle; $\omega$ is the inertia factor (with velocity there is motion inertia); $random(0,1)$ is the random number generation function that generates random numbers between 0 and 1; $P_{id}$ is the current position of the particle; $X_{id}$ is the global best position of this particle; $P_{gd}$ represents the current best position among all particles in the population; and $C_1$ and $C_2$ denote the learning factors, which learn from the best position in the history of this particle and the best position in the population, respectively.

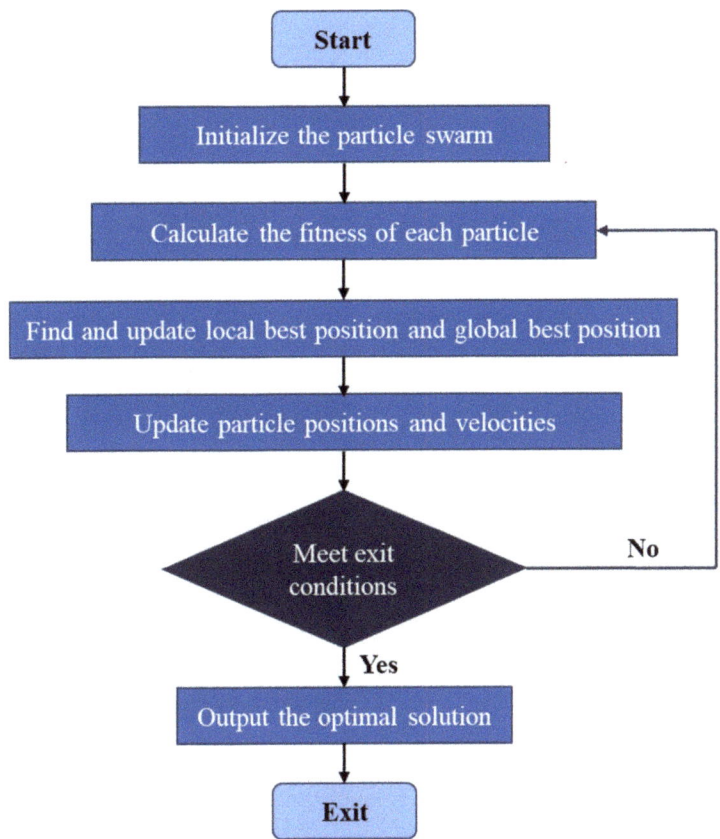

**Figure 2.** Particle swarm optimization Process.

*2.3. Workflow*

As shown in Figure 3, the process of prediction and the parameters optimization of $CO_2$-EOR can be divided into three steps:

**Step 1: Numerical Model and Database Establishment**. Extensive literature research is conducted to gather knowledge on optimizing $CO_2$-EOR parameters and production profiles. The reservoir numerical simulation software CMG (Version 2021) was utilized to build the $CO_2$-EOR numerical model. By employing the Monte Carlo method, 10,000 sets of different reservoir models were randomly generated and simulated to obtain corresponding production curves for various geological parameters, fluid parameters, relative permeability parameters, and injection/production parameters.

**Step 2: Machine Learning Model Selection.** Firstly, a correlation analysis was conducted to assess the relationships between different $CO_2$-EOR parameters. Then, using the dataset generated in the first step, which consisted of 10,000 sets of diverse parameters and corresponding production curves, the machine learning models were trained and evaluated. Eight different machine learning models were employed and trained with the dataset to determine their performance in predicting $CO_2$-EOR production. Through thorough evaluation and comparison, the XGBoost algorithm was selected as the best-performing machine learning method for this study.

**Step 3: $CO_2$-EOR Production Prediction and Parameter Optimization.** The XGBoost-PSO proxy model was employed to predict $CO_2$-EOR production and optimize the $CO_2$-EOR parameters.

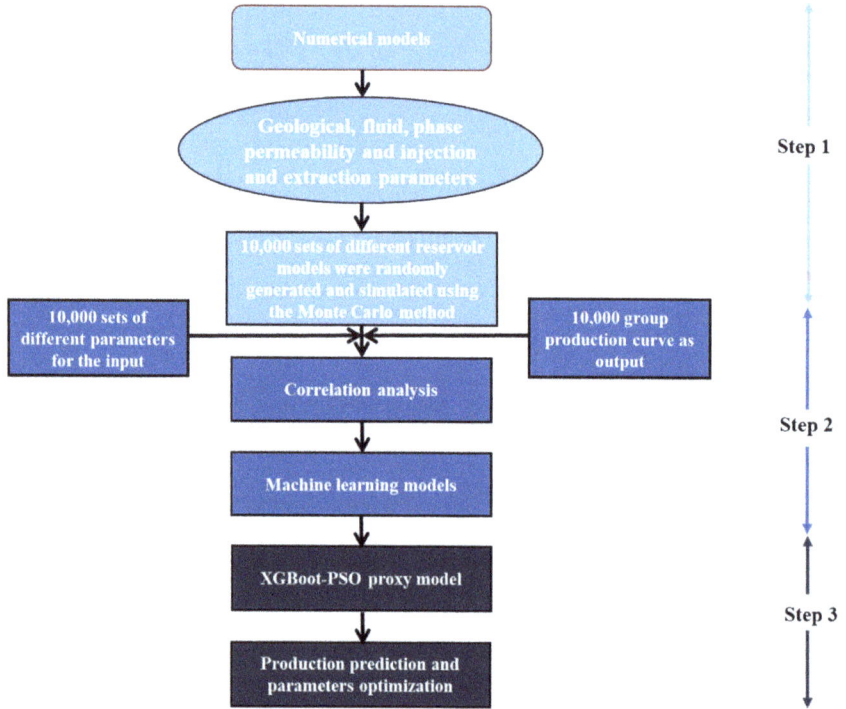

**Figure 3.** Workflow diagram.

## 3. Establishment of Numerical Model and Database

### 3.1. Establishment of $CO_2$-EOR Numerical Model

First, based on the actual geological parameters of the oilfield, a characterization model was established, which took into account factors such as well spacing, fluid properties, and heterogeneity. The numerical model consisted of a grid with dimensions of 21 × 21 × 5, with a grid spacing of 10 m in the I direction, 10 m in the J direction, and 5 m in the K direction. Therefore, the feature model had dimensions of 210 m in length, 210 m in width, and 25 m in depth. The well pattern was deployed as a 1/4 five-spot pattern, with one injector well and one producer well per pattern, as shown in Figure 4. The basic parameters of the feature model are described in Table 1.

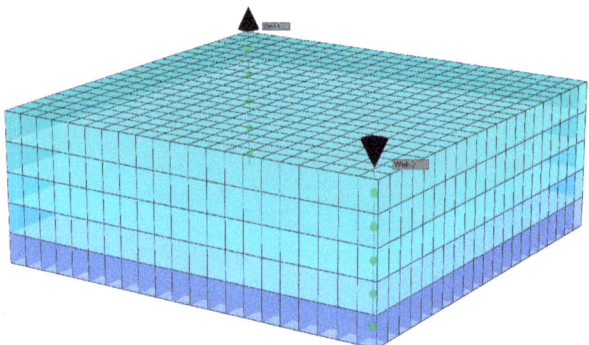

**Figure 4.** Schematic of the reservoir model in the feature model.

Table 1. Basic parameters of the feature model.

| Parameter Type | Parameters | Value | Unit |
|---|---|---|---|
| Geological parameters | Initial pressure | 20 | MPa |
| | Temperature | 70 | °C |
| | Porosity | 0.2 | / |
| | Permeability | 30 | mD |
| | Spacing in the I direction | 10 | m |
| | Spacing in the J direction | 10 | m |
| | Spacing in the K direction | 4 | m |
| Fluid parameters | Oil density | 799.2 | kg/m$^3$ |
| | Gas specific gravity | 0.70 | / |
| | Residual oil saturation index | 86.10 | / |
| | Oil viscosity | 7.67 | mPa·s |
| | Water saturation | 0.30 | / |
| | Oil saturation | 0.70 | / |
| | Phase mixing parameter | 0.70 | / |
| Phase saturation parameters | Residual water saturation | 0.3 | / |
| | Residual oil saturation in oil-water system | 0.2 | / |
| | Residual oil saturation in gas-liquid system | 0.15 | / |
| | Residual gas saturation | 0.15 | / |
| Injection/production parameters | Gas injection well bottom flow pressure | 30 | MPa |
| | Water injection well bottom flow pressure | 21 | MPa |
| | Production well bottom flow pressure | 5 | MPa |
| | WF ending time | 3650 | Day |
| | WAG gas injection | 60 | m$^3$/day |
| | WAG water injection | 60 | m$^3$/day |

*3.2. Establishing the Database*

After building the geologic model, a large dataset needs to be generated to train the predictive model built using machine learning. In this study, a numerical model was used to randomly generate cumulative production data for 10,000 sets of geological and completion parameters. This study investigated a total of 24 parameters, including geological parameters, fluid parameters, initial conditions, and injection/production parameters. The parameters included in this study are as follows. The geological parameters included the following: initial pressure, porosity, permeability, temperature, and spacing in the I, J, and K directions. The fluid parameters included the following: oil density, gas specific gravity, residual oil saturation index, water saturation, oil saturation, oil viscosity, and phase mixing parameter. The phase saturation parameters included the following: residual water saturation, residual oil saturation in the oil-water system, residual oil saturation in the gas-liquid system, and residual gas saturation. The injection/production parameters included the following: gas injection well bottom flow pressure, water injection well bottom flow pressure, production well bottom flow pressure, WF ending time, WAG gas injection rate, and WAG water injection rate. The range of the values for each parameter is shown in Table 2, and the distribution of each parameter is illustrated in Figure 5. The applicable range for each parameter in the table was primarily based on the actual conditions of $CO_2$-driven oil reservoirs in China.

Table 2. Range of values for each parameter.

| Parameters [16,17,22] | Minimum Value | Maximum Value | Unit | Symbol in Figure 5 |
| --- | --- | --- | --- | --- |
| Initial pressure | 15 | 25 | MPa | Initial pressure |
| Temperature | 45.00 | 120.00 | °C | Temperature |
| Porosity | 0.15 | 0.25 | / | Por |
| Permeability | 3.5 | 240.0 | mD | PERMI |
| Spacing in the I direction | 5 | 20 | m | Di |
| Spacing in the J direction | 5 | 20 | m | Dj |
| Spacing in the K direction | 2 | 5 | m | Dk |
| Oil density | 700 | 900 | $kg/m^3$ | Oil density |
| Gas specific gravity | 0.53 | 0.87 | / | Gas gravity |
| Residual oil saturation index | 10.00 | 200.00 | / | Rsi |
| Oil viscosity | 0.15 | 15.00 | mPa·s | Viso |
| Water saturation | 0.20 | 0.40 | / | Sw |
| Oil saturation | 0.60 | 0.80 | / | So |
| Residual gas saturation | 0.50 | 0.85 | / | Omegas |
| Residual water saturation | 0.20 | 0.40 | / | SWCON |
| Residual oil saturation in oil-water system | 0.15 | 0.25 | / | SOIRW |
| Residual oil saturation in gas-liquid system | 0.10 | 0.20 | / | SORG |
| Residual gas saturation | 0.10 | 0.20 | / | SGCON |
| Gas injection well bottom flow pressure | 22 | 38 | MPa | INJG BHP |
| Water injection well bottom flow pressure | 20 | 40 | MPa | INJW BHP |
| Production well bottom flow pressure | 3 | 5 | MPa | Prod BHP |
| WF ending time | 2700 | 4600 | Day | WF Ending time |
| WAG gas injection | 45.00 | 180.00 | $m^3/day$ | WAG INJG |
| WAG water injection | 45.00 | 180.00 | $m^3/day$ | WAG INJW |

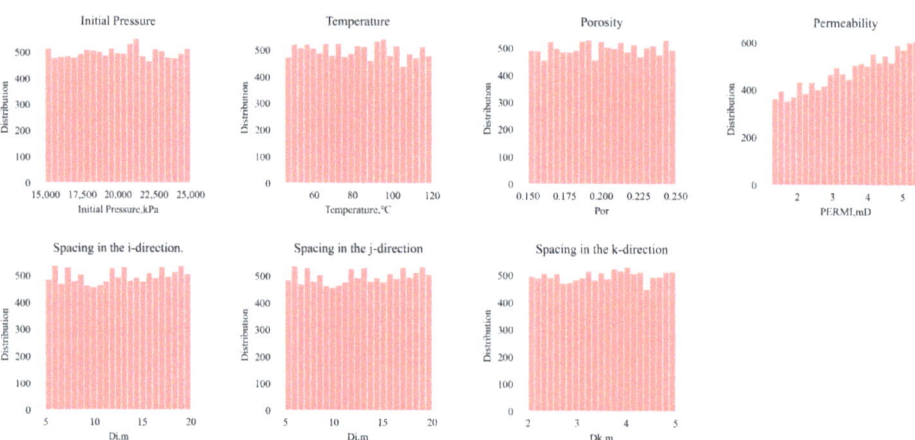

Figure 5. Cont.

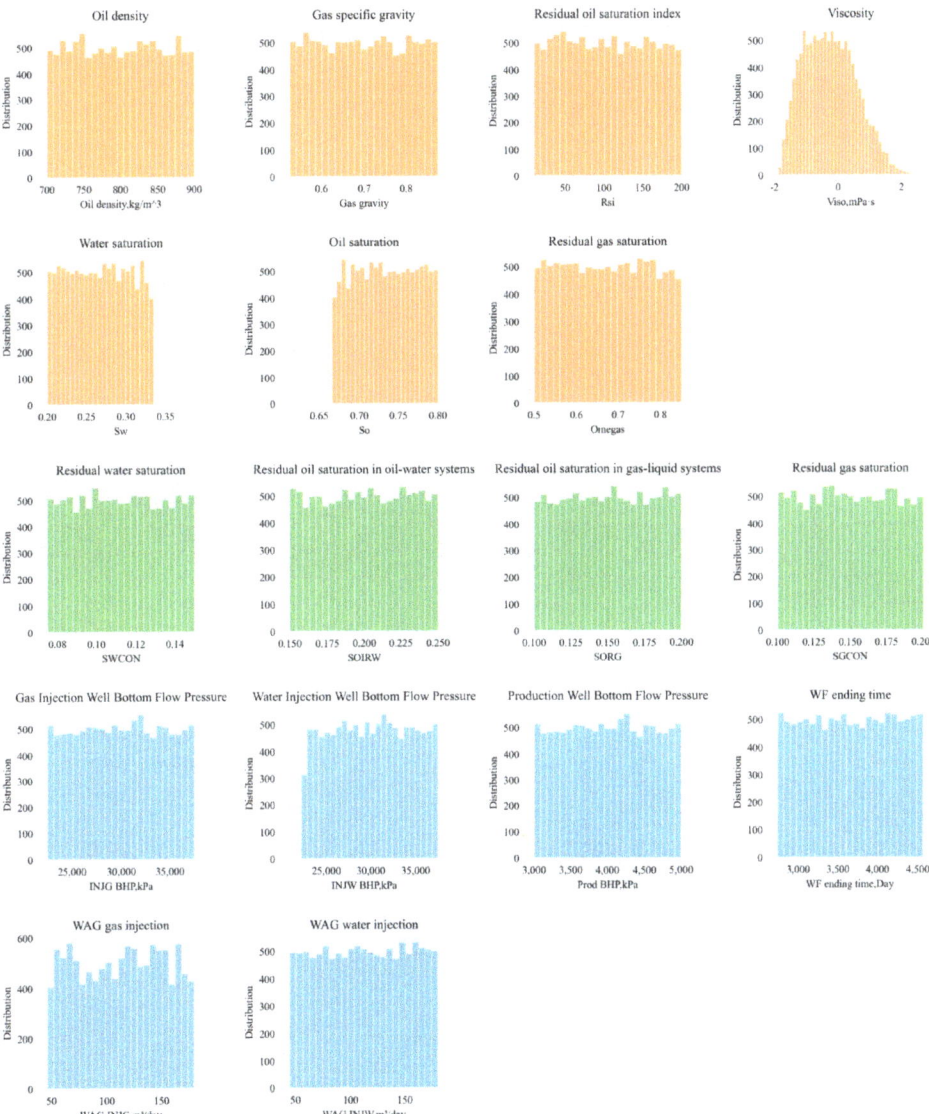

**Figure 5.** Distribution of each parameter. The red color represents geological parameters, the orange color represents fluid parameters, the green color represents phase saturation parameters, and the blue color represents injection/production parameters.

The objective function used in this study was the cumulative oil production, which is the output obtained by simulating the monthly production for each combination using the numerical simulation model. Figures 6 and 7 respectively illustrate the cumulative oil production curve and the distribution of the cumulative oil production. Based on the data and the accompanying figures, it can be observed that the minimum cumulative oil production was $10^4$ m$^3$, while the maximum cumulative oil production was $7.2 \times 10^5$ m$^3$. The majority of the distribution fell within the range of 0–$10^5$ m$^3$ of oil production.

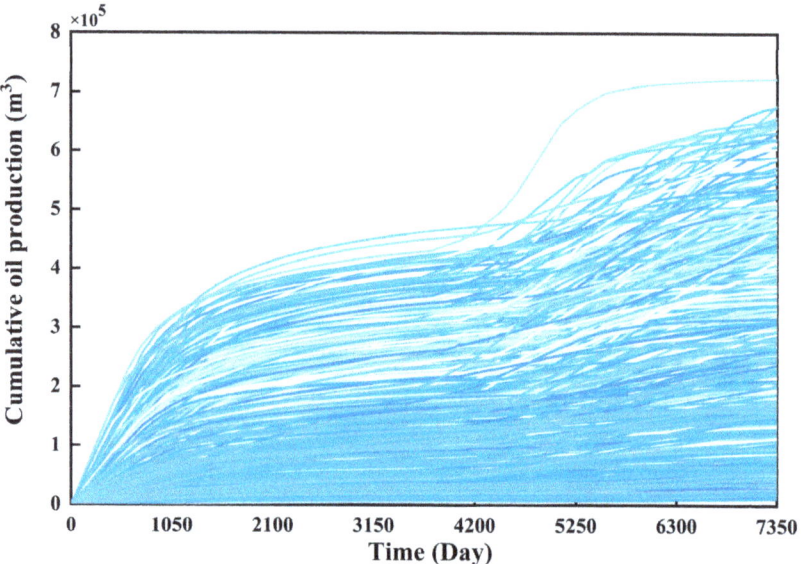

**Figure 6.** Cumulative oil production curve.

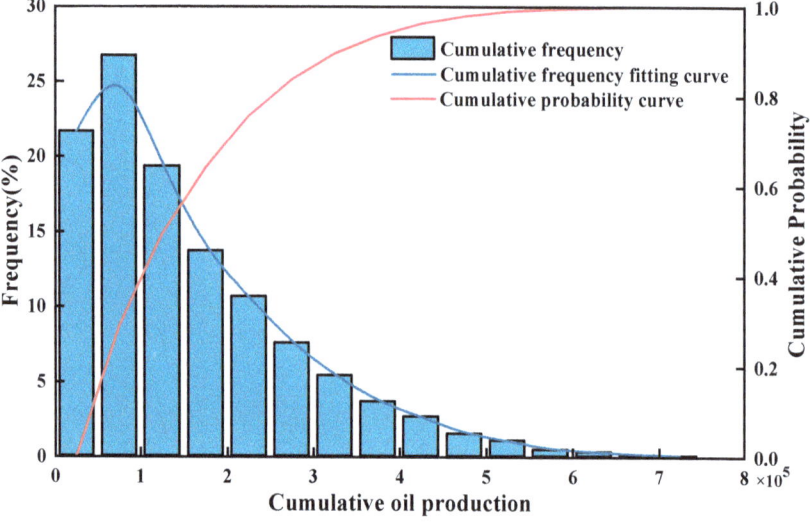

**Figure 7.** Cumulative oil production distribution curve.

## 4. Machine Learning Model Preference

### 4.1. Correlation Analysis

By observing the results of the correlation analysis in Figure 8, it can be concluded that there are strong linear correlations between cumulative oil production in $CO_2$-EOR and geological parameters, fluid parameters, phase saturation parameters, and injection/production parameters. The discovery of these correlations is significant for gaining a deeper understanding of reservoir characteristics and for optimizing the $CO_2$-EOR process.

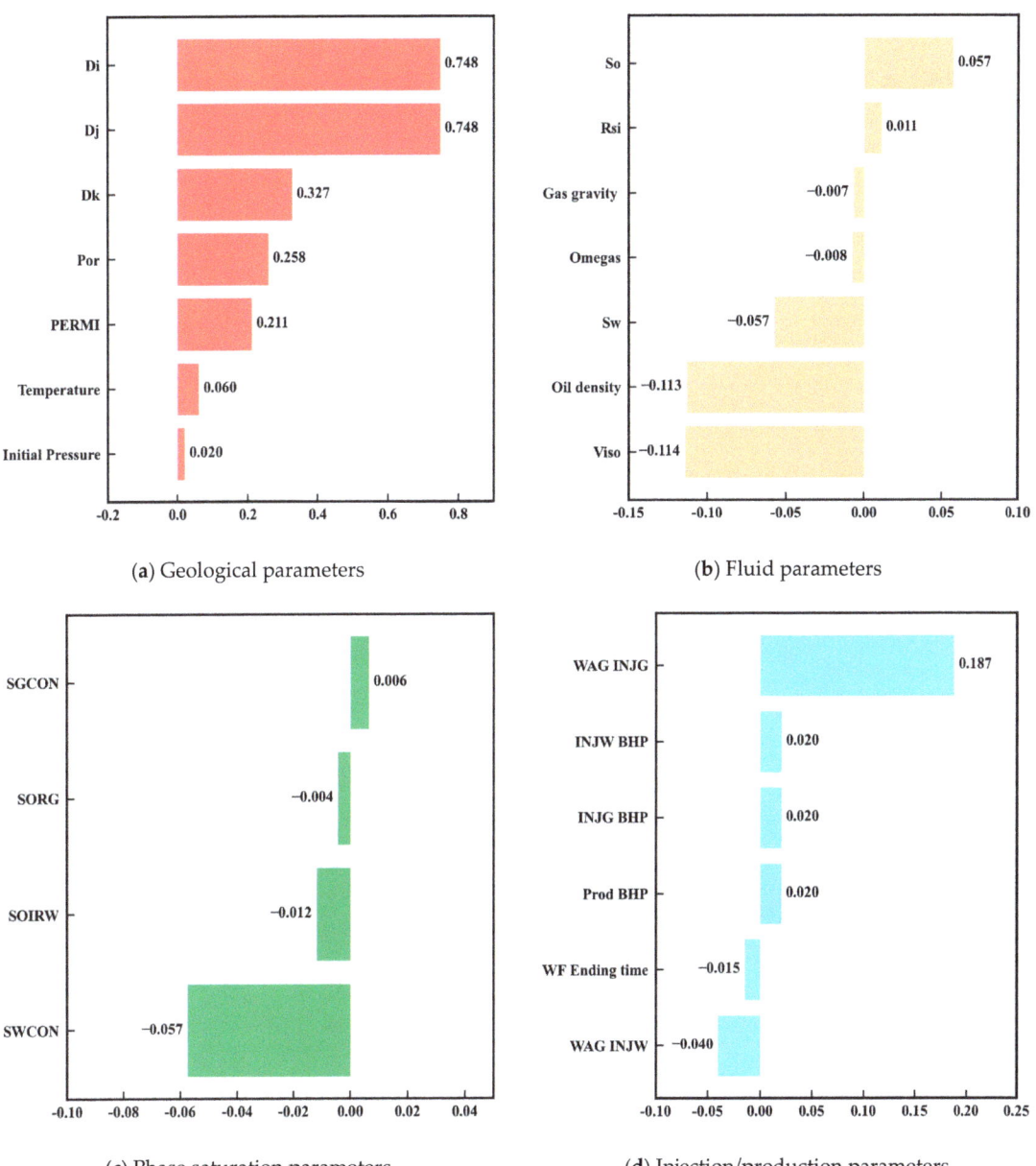

**Figure 8.** Correlation analysis of parameters with cumulative oil production.

From the perspective of the geological parameters (Figure 8a), there was a positive correlation between cumulative oil production in $CO_2$-EOR and certain factors. Notably, there were strong linear correlations between the cumulative oil production and the spacing in the I, J, and K directions, which had correlation coefficients of 0.748, 0.748, and 0.327, respectively. This indicates that the spacing in these directions significantly influences the oil production during the $CO_2$-EOR process. Additionally, the porosity and permeability showed correlations with the cumulative oil production in $CO_2$-EOR, which yielded correlation coefficients of 0.258 and 0.211, respectively. This suggests that, as porosity and permeability increase, the cumulative oil production in $CO_2$-EOR also increases. Porosity represents the void space in the reservoir, while permeability reflects the capacity for fluid flow within the reservoir. Higher porosity and permeability values indicate larger effective storage capacity and better fluid migration capability, thereby enabling $CO_2$ to react more fully with crude oil, which increases the cumulative oil production.

From the perspective of the fluid parameters and phase saturation parameters (Figure 8a,b), there was a weak correlation with the cumulative oil production in $CO_2$-EOR. For instance, in the fluid parameters, the correlation coefficients for the residual oil saturation index, gas specific gravity, and phase mixing parameter were 0.011, −0.007, and −0.008, respectively. Similarly, in the phase permeability parameters, the correlation coefficients for the residual gas saturation, residual oil saturation in the oil–water system, and residual oil saturation in the gas–liquid system were 0.006, −0.004, and −0.012, respectively. These correlation coefficients being close to zero indicate that there is a weak linear relationship between the phase permeability parameters and the cumulative oil production in $CO_2$-EOR.

Furthermore, from the perspective of the injection–production parameters (Figure 8d), there was a strong linear correlation between the $CO_2$-EOR cumulative oil production and $CO_2$-WAG injection volume. The correlation coefficient for the $CO_2$-WAG injection volume was 0.187. This indicates that increasing the $CO_2$-WAG injection volume can effectively enhance the displacement efficiency of the $CO_2$ and increase oil production in the reservoir. Optimizing these parameters can lead to more efficient oil recovery in the $CO_2$-EOR process.

*4.2. Machine Learning Model Building*

The dataset was split into 70% for training and 30% for testing. Eight machine learning models, including linear regression, ridge regression, decision tree, random forest, gradient boosting decision tree, extreme gradient boosting, K-nearest neighbors, and the neural network, were established. By comparing their accuracies, the model with the highest accuracy was selected as the optimal model.

To evaluate the prediction accuracy of the machine learning models, the coefficient of determination ($R^2$) was selected as the metric [43]. The $R^2$ value ranges from 0 to 1, with a higher value indicating a better fit of the model. The specific formula to calculate $R^2$ is as follows:

$$R^2 = 1 - \frac{\sum_{i=1}^{K}(\hat{y}_i - y_i)^2}{\sum_{i=1}^{K}(\overline{y}_i - y_i)^2} \quad (5)$$

where $\overline{y}_i$ is the mean value of $y_i$.

Figure 9 presents the scatter plots of the predicted results versus the true results for the eight predictive models investigated in this study. The corresponding coefficient of determination ($R^2$) values for the selected predictive models are shown in Table 3. Among them, the linear regression, ridge regression, K-nearest neighbors, and decision tree models exhibited scattered predicted points and true points around the 45-degree line, thereby indicating poor predictive performance, with test $R^2$ values of 0.95, 0.81, 0.79, and 0.91, respectively. In contrast, the extreme gradient boosting (XGBoost) model showed a concentration of predicted points and true points along the 45-degree line, with a high test $R^2$ value of 0.98, thereby indicating a low error and good predictive performance.

Table 3. Comparison of predictive performance of the models.

| Machine Learning Algorithms | Train $R^2$ | Test $R^2$ |
|---|---|---|
| Linear Regression | 0.82 | 0.75 |
| Ridge Regression | 0.82 | 0.81 |
| Decision Tree | 1.00 | 0.91 |
| Random Forest | 0.97 | 0.96 |

Table 3. Cont.

| Machine Learning Algorithms | Train $R^2$ | Test $R^2$ |
|---|---|---|
| K Nearest Neighbors | 0.80 | 0.79 |
| Neural Network | 0.96 | 0.96 |
| Gradient Boosting Decision Tree | 0.97 | 0.96 |
| XGBoost | 0.99 | 0.98 |

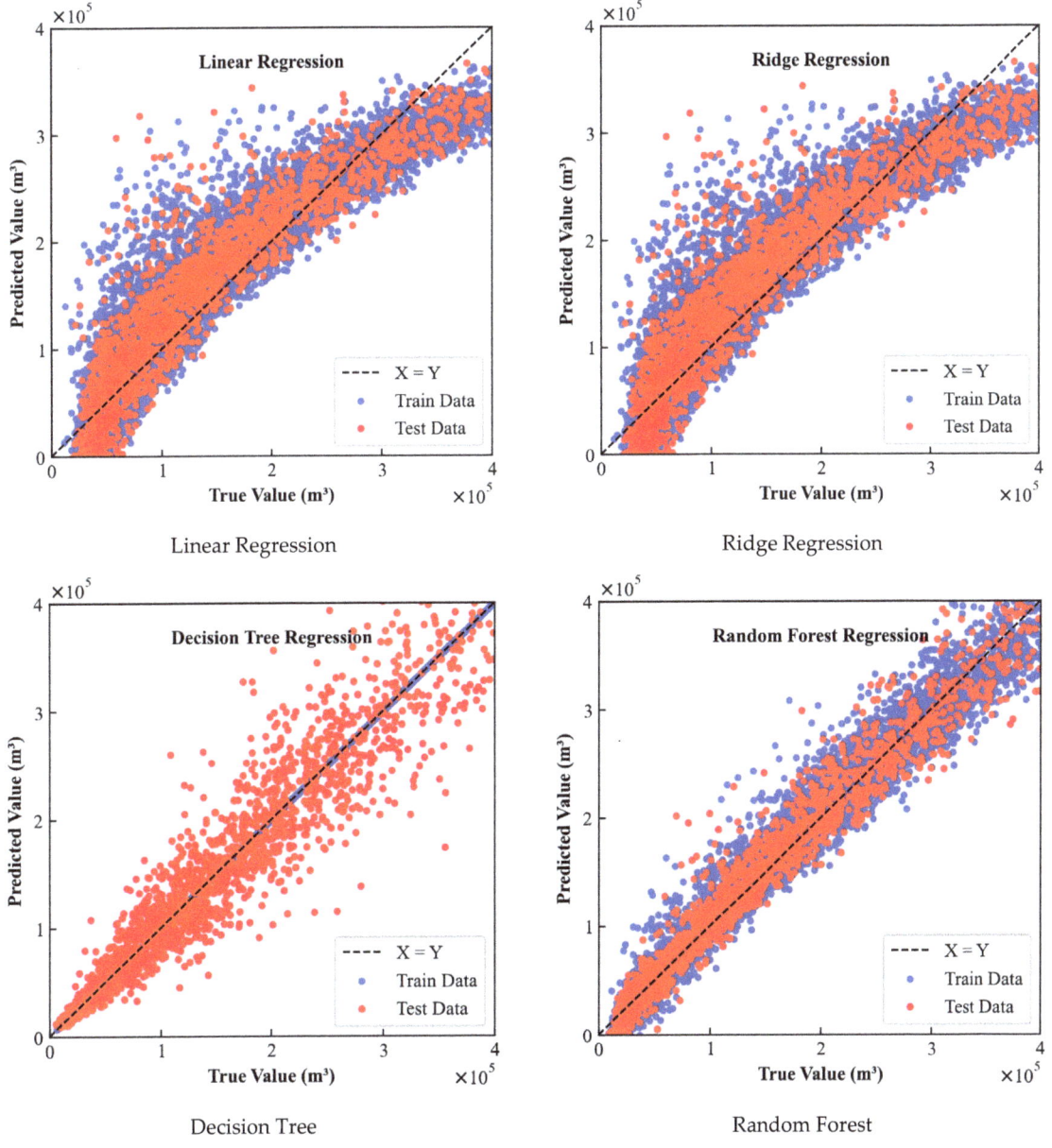

Figure 9. Cont.

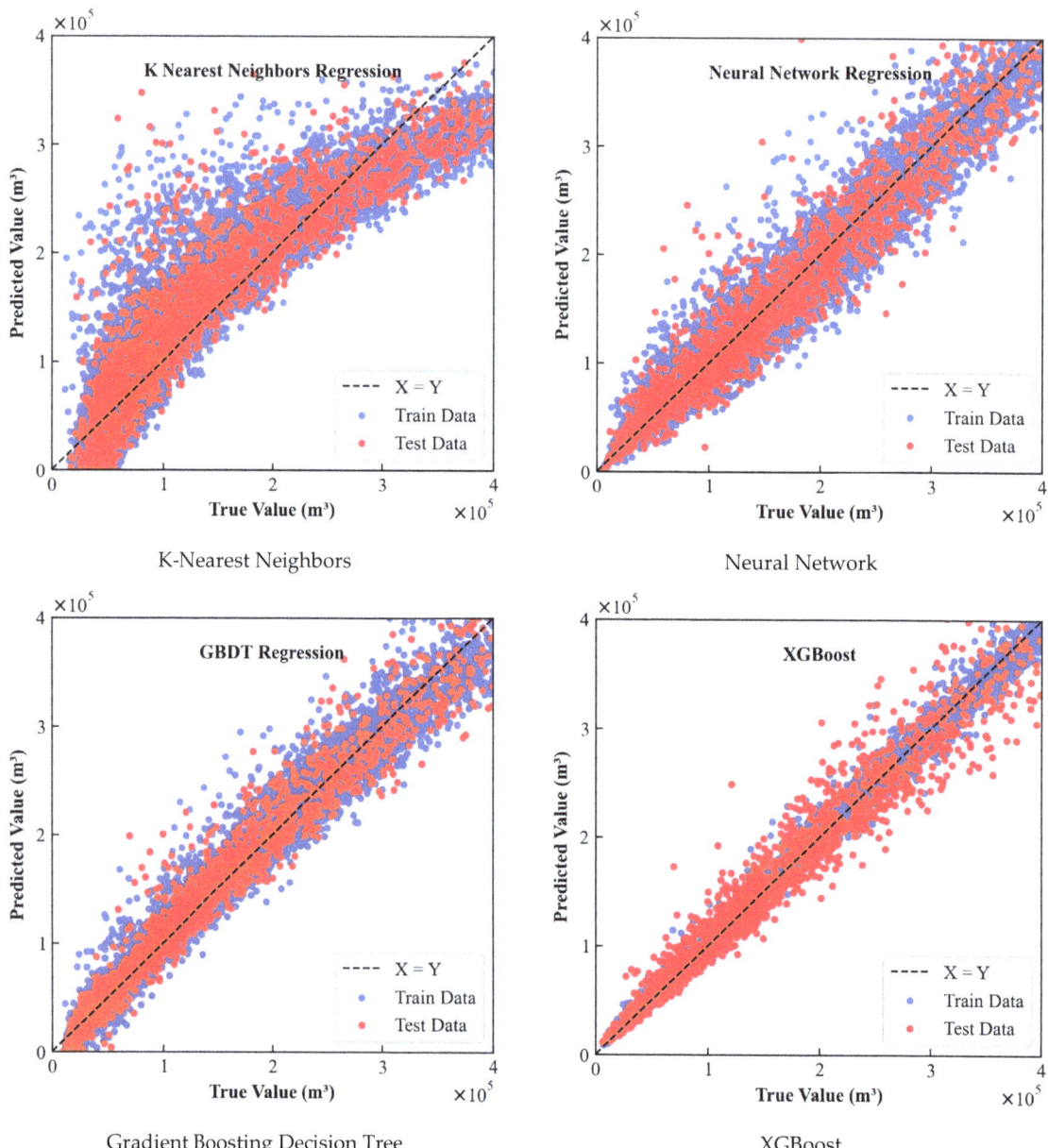

**Figure 9.** Model performance for each model using training and test sets.

Based on the results, it can be observed that, among the eight machine learning methods, the extreme gradient boosting (XGBoost) model exhibited the best predictive performance. It achieved a training $R^2$ of 0.99 and a test $R^2$ of 0.98. This model is suitable for use as a predictive optimization model to optimize the injection and production parameters, as well as to predict cumulative oil production. Table 4 provides the hyperparameters of the XGBoost predictive model.

**Table 4.** Machine learning XGBoot model hyperparameters.

| Model Hyperparameters | Minimum Value | Maximum Values |
|---|---|---|
| Number of boosting stages | 10 | 500 |
| Learning rate | 0.01 | 0.2 |
| Maximum depth of tree | 3 | 10 |
| Gamma | 0 | 0.4 |
| Minimum child weight | 1 | 5 |
| Subsample | 0.5 | 1 |
| Colsample by tree | 0.5 | 1 |

## 5. $CO_2$-EOR Parameter Optimization

### 5.1. Coupling of PSO Optimization and XGBoost Model

Particle swarm optimization (PSO) is a widely recognized metaheuristic algorithm that is known for its ability to effectively explore the solution space and find global optima by simulating the collective behavior of a swarm of particles. In this study, PSO was used to search for the optimal combination of $CO_2$-WAG parameters that maximizes the cumulative oil production. When applying the PSO algorithm, a trained XGBoost model was used to evaluate the suitability of a large number of project design parameters. With the aid of the surrogate model, the computational burden of the optimization procedure was significantly reduced, thereby allowing for more iterations of the PSO algorithm. The parameters of the final PSO model, along with the XGBoost parameters, are provided in Table 5, and the optimization process is depicted in Figure 10.

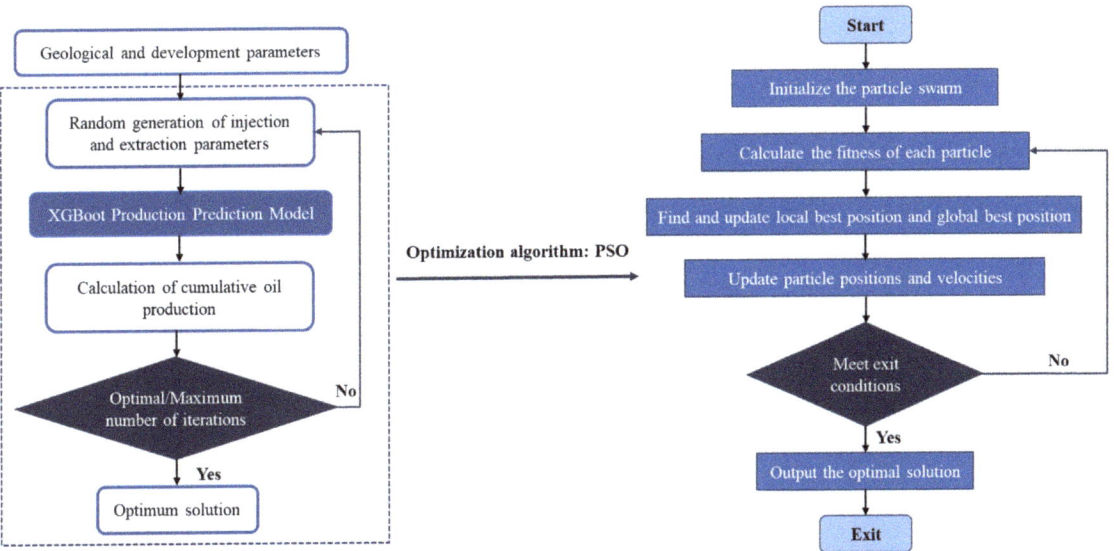

**Figure 10.** Optimization workflow.

**Table 5.** Hyperparameters of PSO algorithm and XGBoost model.

| Parameters in PSO Algorithm | Value |
|---:|:---:|
| Population number group size | 15 |
| Maximum number of iterations maximum | 50 |
| Inertia weight ($\omega$) | 0.8 |
| Learning factor (c1) | 2 |
| Learning factor (c2) | 2 |

*5.2. Production Prediction and Parameter Optimization*

In the process of exploiting reservoirs using $CO_2$-EOR technology, various operational and injection/production parameters have an impact on the cumulative oil production. Therefore, the XGBoost-PSO optimization model was employed to optimize the operational parameters, thereby aiming to enhance the cumulative oil production and recovery factor of the reservoir. During the optimization process, a set of key parameters was considered, including water injection well bottom flow pressure, gas injection well bottom flow pressure, production well bottom flow pressure, WAG gas injection, and WAG water injection.

By optimizing these parameters, the final optimization results were obtained, and they are shown in Table 6. Figures 11 and 12 illustrate the optimized cumulative oil production and daily oil production, respectively. From the figures, it is evident that the cumulative oil production and daily oil production under the $CO_2$-WAG method were significantly higher than under the WF method. This finding indicates the immense potential of $CO_2$-EOR technology in improving oil recovery. The optimized cumulative oil production successfully increased from 425,916 $m^3$ to 475,047 $m^3$. This implies that, by optimizing the operational parameters, the oil production potential of the reservoir can be further enhanced.

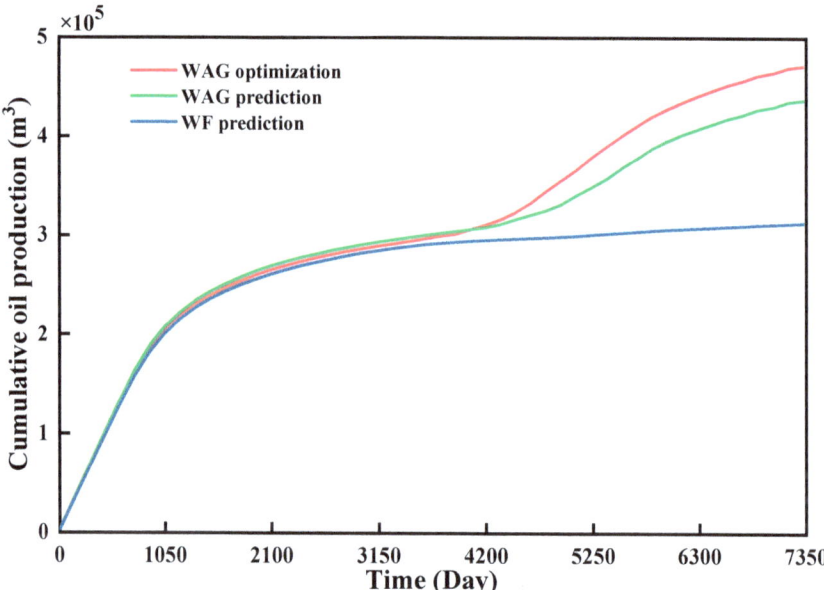

**Figure 11.** Comparison of cumulative oil production before and after optimization.

Table 6. Optimization of basic parameters of $CO_2$-EOR.

| Optimization Parameters | WF | WAG | Optimization of WAG | Unit |
|---|---|---|---|---|
| Gas injection well bottom flow pressure | 28.50 | 28.50 | 25.33 | MPa |
| Water injection well bottom flow pressure | 21 | 21 | 33.76 | MPa |
| Production well bottom flow pressure | 4.00 | 4.00 | 4.29 | MPa |
| WF ending time | 3777 | 3777 | 3533 | day |
| WAG gas injection | 0 | 144 | 120 | m$^3$/day |
| WAG water injection | 174 | 174 | 66.71 | m$^3$/day |
| Cumulative oil production | 319,234 | 425,916 | 475,047 | m$^3$ |

By establishing the XGBoost-PSO optimization model and optimizing the operational parameters, this process can provide reliable guidance and decision support for reservoir development. This optimization model not only improved the cumulative oil production and recovery factor, but also provides crucial support for the long-term sustainable development of the oilfield. Therefore, further research and optimization of this model are necessary to further enhance the efficiency and benefits of reservoir development.

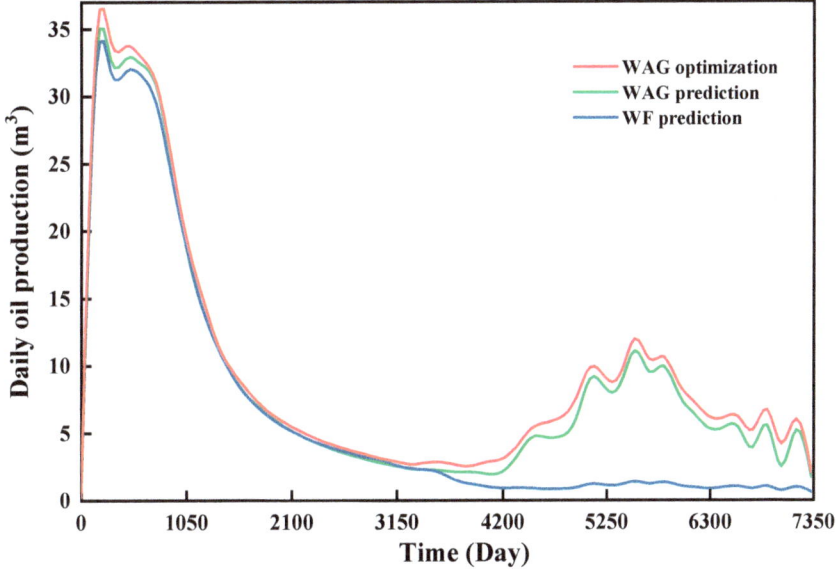

Figure 12. Comparison of daily oil production before and after optimization.

### 6. Discussion and Conclusions

Machine learning is a popular research method used in data processing, and this study utilized a predictive model that has significant room for improvement. In this study, we used the XGBoost model, which allows for the customization of parameters such as the number of hidden layers, the number of neurons, and the learning rate to suit specific needs. Additionally, these hyperparameters can be adjusted to enhance the model's predictive accuracy. Optimization techniques such as PSO or GA can also be introduced to further strengthen the model's hyperparameters. Furthermore, the evaluation and optimization in this study only considered the cumulative production, daily production, and recovery rate as the objective functions. Future research can consider incorporating other metrics such as net present value (NPV) as objective functions.

This study utilized the XGBoost machine learning algorithm to establish a workflow for evaluating the cumulative gas production in $CO_2$-EOR modeling. This workflow was used for capacity prediction and parameter optimization in $CO_2$-EOR. The following conclusions were drawn:

(1) Compared to traditional simulation and prediction methods, machine learning approaches can effectively handle reservoir data and address non-linear problems. By incorporating multiple factors such as geology and operations, they significantly improve the efficiency and accuracy of the models.

(2) The investigation of the correlation between various factors and the cumulative oil production reveals that, from a geological perspective, there is a strong linear correlation between the porosity and permeability with the $CO_2$-EOR cumulative oil production. From an injection/production parameter perspective, there is a strong linear correlation between the $CO_2$-WAG gas injection rate and the $CO_2$-EOR cumulative oil production.

(3) Different machine learning models exhibited varying performance results in predicting production. By comparing eight different production prediction models, it can be concluded that the extreme gradient boosting (XGBoost) model outperforms other machine learning models in terms of predictive performance. The XGBoost model achieved an $R^2$ score of 0.99 on the training set and 0.98 on the testing set.

(4) The cumulative oil production, daily oil production, and recovery factor under the $CO_2$-WAG method were significantly higher than those under the WF method. This finding suggests that $CO_2$-EOR technology has great potential in improving the recovery factor of oil reservoirs.

(5) During the optimization of the $CO_2$-EOR parameters, PSO was coupled with the trained XGBoost model. PSO efficiently searches the parameter space to find the optimal $CO_2$-EOR parameters that maximize the cumulative oil production, thus saving computational costs. The optimized parameters resulted in a higher cumulative oil production and recovery factor when compared to previous results.

**Author Contributions:** Conceptualization, M.G. and Z.L.; methodology, M.G. and W.L. (Weirong Li); software, S.Q.; validation, Z.L., W.L. (Wanlu Liu) and H.Y.; formal analysis, W.L. (Wanlu Liu); investigation, J.C.; resources, H.Y.; data curation, J.C.; writing—original draft preparation, S.Q.; writing—review and editing, M.G. and W.L. (Weirong Li); visualization, J.C.; supervision, M.G.; project administration, M.G.; funding acquisition, M.G. All authors have read and agreed to the published version of the manuscript.

**Funding:** This research was funded by the Major Science and Technology project of the CNPC in China (grant No. 2021ZZ01-03 and No. 2021ZZ01-06).

**Data Availability Statement:** The raw/processed data required to reproduce these findings cannot be shared at this time, as the data also forms part of an ongoing study.

**Acknowledgments:** The authors are grateful for the financial support of the CNPC in China.

**Conflicts of Interest:** The authors declare that the publication of this paper has no conflict of interest.

# References

1. Kondori, J.; Miah, M.I.; Zendehboudi, S.; Khan, F.; Heagle, D. Hybrid Connectionist Models to Assess Recovery Performance of Low Salinity Water Injection. *J. Pet. Sci. Eng.* **2021**, *197*, 107833. [CrossRef]
2. Guo, J.-X.; Huang, C.; Wang, J.-L.; Meng, X.-Y. Integrated Operation for the Planning of $CO_2$ Capture Path in CCS–EOR Project. *J. Pet. Sci. Eng.* **2020**, *186*, 106720. [CrossRef]
3. Allahyarzadeh Bidgoli, A.; Hamidishad, N.; Yanagihara, J.I. The Impact of Carbon Capture Storage and Utilization on Energy Efficiency, Sustainability, and Production of an Offshore Platform: Thermodynamic and Sensitivity Analyses. *J. Energy Resour. Technol.* **2022**, *144*, 112102. [CrossRef]
4. Song, C. Global Challenges and Strategies for Control, Conversion and Utilization of $CO_2$ for Sustainable Development Involving Energy, Catalysis, Adsorption and Chemical Processing. *Catal. Today* **2006**, *115*, 2–32. [CrossRef]
5. Jiang, J.; Rui, Z.; Hazlett, R.; Lu, J. An Integrated Technical-Economic Model for Evaluating $CO_2$ Enhanced Oil Recovery Development. *Appl. Energy* **2019**, *247*, 190–211. [CrossRef]

6. Kong, S.; Huang, X.; Li, K.; Song, X. Adsorption/Desorption Isotherms of $CH_4$ and $C_2H_6$ on Typical Shale Samples. *Fuel* **2019**, *255*, 115632. [CrossRef]
7. Huang, X.; Gu, L.; Li, S.; Du, Y.; Liu, Y. Absolute Adsorption of Light Hydrocarbons on Organic-Rich Shale: An Efficient Determination Method. *Fuel* **2022**, *308*, 121998. [CrossRef]
8. Tang, Y.; Hou, C.; He, Y.; Wang, Y.; Chen, Y.; Rui, Z. Review on Pore Structure Characterization and Microscopic Flow Mechanism of $CO_2$ Flooding in Porous Media. *Energy Technol.* **2021**, *9*, 2000787. [CrossRef]
9. Kulkarni, M.M.; Rao, D.N. Experimental Investigation of Miscible and Immiscible Water-Alternating-Gas (WAG) Process Performance. *J. Pet. Sci. Eng.* **2005**, *48*, 1–20. [CrossRef]
10. Karimaie, H.; Nazarian, B.; Aurdal, T.; Nøkleby, P.H.; Hansen, O. Simulation Study of $CO_2$ EOR and Storage Potential in a North Sea Reservoir. *Energy Procedia* **2017**, *114*, 7018–7032. [CrossRef]
11. Tang, R.; Wang, H.; Yu, H.; Wang, W.; Chen, L. Effect of Water and Gas Alternate Injection on $CO_2$ Flooding. *Fault Block Oil Gas Field* **2016**, *23*, 358–362. [CrossRef]
12. Sanchez, N.L. *Management of Water Alternating Gas (WAG) Injection Projects*; OnePetro: Richardson, TX, USA, 1999.
13. Christensen, J.R.; Stenby, E.H.; Skauge, A. Review of WAG Field Experience. *SPE Reserv. Eval. Eng.* **2001**, *4*, 97–106. [CrossRef]
14. Al-Bayati, D.; Saeedi, A.; Myers, M.; White, C.; Xie, Q.; Clennell, B. Insight Investigation of Miscible $SCCO_2$ Water Alternating Gas (WAG) Injection Performance in Heterogeneous Sandstone Reservoirs. *J. $CO_2$ Util.* **2018**, *28*, 255–263. [CrossRef]
15. Sun, X.; Liu, J.; Dai, X.; Wang, X.; Yapanto, L.M.; Zekiy, A.O. On the Application of Surfactant and Water Alternating Gas (SAG/WAG) Injection to Improve Oil Recovery in Tight Reservoirs. *Energy Rep.* **2021**, *7*, 2452–2459. [CrossRef]
16. Pancholi, S.; Negi, G.S.; Agarwal, J.R.; Bera, A.; Shah, M. Experimental and Simulation Studies for Optimization of Water–Alternating-Gas ($CO_2$) Flooding for Enhanced Oil Recovery. *Pet. Res.* **2020**, *5*, 227–234. [CrossRef]
17. Ren, B.; Duncan, I.J. Maximizing Oil Production from Water Alternating Gas ($CO_2$) Injection into Residual Oil Zones: The Impact of Oil Saturation and Heterogeneity. *Energy* **2021**, *222*, 119915. [CrossRef]
18. Khather, M.; Yekeen, N.; Al-Yaseri, A.; Al-Mukainah, H.; Giwelli, A.; Saeedi, A. The Impact of Wormhole Generation in Carbonate Reservoirs on $CO_2$-WAG Oil Recovery. *J. Pet. Sci. Eng.* **2022**, *212*, 110354. [CrossRef]
19. Ren, D.; Wang, X.; Kou, Z.; Wang, S.; Wang, H.; Wang, X.; Tang, Y.; Jiao, Z.; Zhou, D.; Zhang, R. Feasibility Evaluation of $CO_2$ EOR and Storage in Tight Oil Reservoirs: A Demonstration Project in the Ordos Basin. *Fuel* **2023**, *331*, 125652. [CrossRef]
20. Rodrigues, H.; Mackay, E.; Arnold, D.; Silva, D. *Optimization of $CO_2$-WAG and Calcite Scale Management in Pre-Salt Carbonate Reservoirs*; OnePetro: Richardson, TX, USA, 2019.
21. Sen, D.; Chen, H.; Datta-Gupta, A. Inter-Well Connectivity Detection in $CO_2$ WAG Projects Using Statistical Recurrent Unit Models. *Fuel* **2022**, *311*, 122600. [CrossRef]
22. Imani, G. Machine Learning-Assisted Prediction of Oil Production and $CO_2$ Storage Effect in $CO_2$-Water-Alternating-Gas Injection ($CO_2$-WAG). *Appl. Sci.* **2022**, *12*, 958. [CrossRef]
23. He, R.; Ma, W.; Ma, X.; Liu, Y. Modeling and Optimizing for Operation of $CO_2$-EOR Project Based on Machine Learning Methods and Greedy Algorithm. *Energy Rep.* **2021**, *7*, 3664–3677. [CrossRef]
24. Mohagheghian, E.; James, L.A.; Haynes, R.D. Optimization of Hydrocarbon Water Alternating Gas in the Norne Field: Application of Evolutionary Algorithms. *Fuel* **2018**, *223*, 86–98. [CrossRef]
25. You, J.; Ampomah, W.; Tu, J.; Morgan, A.; Sun, Q.; Wei, B.; Wang, D. Optimization of Water-Alternating-$CO_2$ Injection Field Operations Using a Machine-Learning-Assisted Workflow. *SPE Reserv. Eval. Eng.* **2022**, *25*, 214–231. [CrossRef]
26. Jaber, A.K. Genetic Algorithm to Optimize Miscible Water Alternate $CO_2$ Flooding in Heterogeneous Clastic Reservoir. *Arab. J. Geosci.* **2022**, *15*, 714. [CrossRef]
27. Dehghan, M.H.; Hamidi, F.; Salajegheh, M. Study of Linear Regression Based on Least Squares and Fuzzy Least Absolutes Deviations and Its Application in Geography. In Proceedings of the 2015 4th Iranian Joint Congress on Fuzzy and Intelligent Systems (CFIS), Zahedan, Iran, 9–11 September 2015; IEEE: Piscataway, NJ, USA, 2015; pp. 1–6.
28. Kavitha, S.; Varuna, S.; Ramya, R. A Comparative Analysis on Linear Regression and Support Vector Regression. In Proceedings of the 2016 Online International Conference on Green Engineering and Technologies (IC-GET), Coimbatore, India, 19 November 2016; IEEE: Piscataway, NJ, USA, 2016; pp. 1–5.
29. Dorugade, A.V. New Ridge Parameters for Ridge Regression. *J. Assoc. Arab. Univ. Basic Appl. Sci.* **2014**, *15*, 94–99. [CrossRef]
30. Wang, Y.; Xia, S.-T. Unifying Attribute Splitting Criteria of Decision Trees by Tsallis Entropy. In Proceedings of the 2017 IEEE International Conference on Acoustics, Speech and Signal Processing (ICASSP), New Orleans, LA, USA, 5–9 March 2017; IEEE: Piscataway, NJ, USA, 2017; pp. 2507–2511.
31. Gamal, H.; Alsaihati, A.; Elkatatny, S.; Haidary, S.; Abdulraheem, A. Rock Strength Prediction in Real-Time While Drilling Employing Random Forest and Functional Network Techniques. *J. Energy Resour. Technol.* **2021**, *143*, 093004. [CrossRef]
32. Brieman, L. Random Forests. *Mach. Learn.* **2001**, *45*, 5–32. [CrossRef]
33. Jordan, M.I.; Mitchell, T.M. Machine Learning: Trends, Perspectives, and Prospects. *Science* **2015**, *349*, 255–260. [CrossRef]
34. Chen, T.; Guestrin, C. XGBoost: A Scalable Tree Boosting System. In Proceedings of the 22nd ACM SIGKDD International Conference on Knowledge Discovery and Data Mining, San Francisco, CA, USA, 13 August 2016; Association for Computing Machinery: New York, NY, USA, 2016; pp. 785–794.
35. Peterson, L. K-Nearest Neighbor. *Scholarpedia* **2009**, *4*, 1883. [CrossRef]

36. Abiodun, O.I.; Jantan, A.; Omolara, A.E.; Dada, K.V.; Mohamed, N.A.; Arshad, H. State-of-the-Art in Artificial Neural Network Applications: A Survey. *Heliyon* **2018**, *4*, e00938. [CrossRef]
37. Kennedy, J.; Eberhart, R. Particle Swarm Optimization. In Proceedings of the ICNN'95—International Conference on Neural Networks, Perth, WA, Australia, 27 November–1 December 1995; IEEE: Piscataway, NJ, USA, 1995; Volume 4, pp. 1942–1948.
38. Xu, Y.; Zhao, X.; Chen, Y.; Yang, Z. Research on a Mixed Gas Classification Algorithm Based on Extreme Random Tree. *Appl. Sci.* **2019**, *9*, 1728. [CrossRef]
39. Krizhevsky, A.; Sutskever, I.; Hinton, G.E. ImageNet Classification with Deep Convolutional Neural Networks. In Proceedings of the Advances in Neural Information Processing Systems, Stateline, NV, USA, 3–6 December 2012; Curran Associates, Inc.: Red Hook, NY, USA, 2012; Volume 25.
40. Song, K.; Yan, F.; Ding, T.; Gao, L.; Lu, S. A Steel Property Optimization Model Based on the XGBoost Algorithm and Improved PSO. *Comput. Mater. Sci.* **2020**, *174*, 109472. [CrossRef]
41. Eberhart; Shi, Y. Particle Swarm Optimization: Developments, Applications and Resources. In Proceedings of the 2001 Congress on Evolutionary Computation, Seoul, Republic of Korea, 27–30 May 2001; IEEE Cat. No.01TH8546; IEEE: Piscataway, NJ, USA, 2001; Volume 1, pp. 81–86.
42. Andalib Sahnehsaraei, M.; Mahmoodabadi, M.J.; Taherkhorsandi, M.; Castillo-Villar, K.K.; Mortazavi Yazdi, S.M. A Hybrid Global Optimization Algorithm: Particle Swarm Optimization in Association with a Genetic Algorithm. In *Complex System Modelling and Control through Intelligent Soft Computations*; Zhu, Q., Azar, A.T., Eds.; Studies in Fuzziness and Soft Computing; Springer International Publishing: Cham, Switzerland, 2015; Volume 319, pp. 45–86. ISBN 978-3-319-12882-5.
43. Ottah, D.G.; Ikiensikimama, S.S.; Matemilola, S.A. Aquifer Matching With Material Balance Using Particle Swarm Optimization Algorithm—PSO. In Proceedings of the SPE Nigeria Annual International Conference and Exhibition, Lagos, Nigeria, 4 August 2015; p. SPE-178319-MS.

**Disclaimer/Publisher's Note:** The statements, opinions and data contained in all publications are solely those of the individual author(s) and contributor(s) and not of MDPI and/or the editor(s). MDPI and/or the editor(s) disclaim responsibility for any injury to people or property resulting from any ideas, methods, instructions or products referred to in the content.

Article

# Front Movement and Sweeping Rules of $CO_2$ Flooding under Different Oil Displacement Patterns

Xiang Qi [1,2], Tiyao Zhou [1,2], Weifeng Lyu [1,2,*], Dongbo He [1,2,3,*], Yingying Sun [1,2], Meng Du [1,2], Mingyuan Wang [1,2] and Zheng Li [1,2]

1. PetroChina Research Institute of Petroleum Exploration & Development, Beijing 100083, China; qixiang@petrochina.com.cn (X.Q.); zhoutiyao@petrochina.com.cn (T.Z.); sunyingying@petrochina.com.cn (Y.S.); dumeng22@petrochina.com.cn (M.D.); wangmy21@petrochina.com.cn (M.W.); lizheng21@petrochina.com.cn (Z.L.)
2. State Key Laboratory of Enhanced Oil and Gas Recovery, Beijing 100083, China
3. JiDong Oilfield of PetroChina, Tangshan 063200, China
* Correspondence: lweifeng@petrochina.com.cn (W.L.); 311@petrochina.com.cn (D.H.)

**Citation:** Qi, X.; Zhou, T.; Lyu, W.; He, D.; Sun, Y.; Du, M.; Wang, M.; Li, Z. Front Movement and Sweeping Rules of $CO_2$ Flooding under Different Oil Displacement Patterns. *Energies* **2024**, *17*, 15. https://doi.org/10.3390/en17010015

Academic Editor: Riyaz Kharrat

Received: 25 October 2023
Revised: 5 December 2023
Accepted: 14 December 2023
Published: 19 December 2023

**Copyright:** © 2023 by the authors. Licensee MDPI, Basel, Switzerland. This article is an open access article distributed under the terms and conditions of the Creative Commons Attribution (CC BY) license (https://creativecommons.org/licenses/by/4.0/).

**Abstract:** $CO_2$ flooding is a pivotal technique for significantly enhancing oil recovery in low-permeability reservoirs. The movement and sweeping rules at the front of $CO_2$ flooding play a critical role in oil recovery; yet, a comprehensive quantitative analysis remains an area in need of refinement. In this study, we developed 1-D and 2-D numerical simulation models to explore the sweeping behavior of miscible, immiscible, and partly miscible $CO_2$ flooding patterns. The front position and movement rules of the three $CO_2$ flooding patterns were determined. A novel approach to the contour area calculation method was introduced to quantitatively characterize the sweep coefficients, and the sweeping rules are discussed regarding the geological parameters, oil viscosity, and injection–production parameters. Furthermore, the Random Forest (RF) algorithm was employed to identify the controlling factor of the sweep coefficient, as determined through the use of out-of-bag (OOB) data permutation analysis. The results showed that the miscible front was located at the point of maximum $CO_2$ content in the oil phase. The immiscible front occurred at the point of maximum interfacial tension near the production well. Remarkably, the immiscible front moved at a faster rate compared with the miscible front. Geological parameters, including porosity, permeability, and reservoir thickness, significantly impacted the gravity segregation effect, thereby influencing the $CO_2$ sweep coefficient. Immiscible flooding exhibited the highest degree of gravity segregation, with a maximum gravity segregation degree (GSD) reaching 78.1. The permeability ratio was a crucial factor, with a lower limit of approximately 5.0 for reservoirs suitable for $CO_2$ flooding. Injection–production parameters also played a pivotal role in terms of the sweep coefficient. Decreased well spacing and increased gas injection rates were found to enhance sweep coefficients by suppressing gravity segregation. Additionally, higher gas injection rates could improve the miscibility degree of partly miscible flooding from 0.69 to 1.0. Oil viscosity proved to be a significant factor influencing the sweep coefficients, with high seepage resistance due to increasing oil viscosity dominating the miscible and partly miscible flooding patterns. Conversely, gravity segregation primarily governed the sweep coefficient in immiscible flooding. In terms of controlling factors, the permeability ratio emerged as a paramount influence, with a factor importance value (FI) reaching 1.04. The findings of this study can help for a better understanding of sweeping rules of $CO_2$ flooding and providing valuable insights for optimizing oil recovery strategies in the field applications of $CO_2$ flooding.

**Keywords:** $CO_2$ front; sweep coefficient; Random Forest; main controlling factor

## 1. Introduction

The escalating emissions of greenhouse gases, predominantly carbon dioxide ($CO_2$), have led to a myriad of environmental challenges, including global climate change [1].

Statistics reveal that a staggering 64% of environmental pollution originates from the increasing concentration of $CO_2$ in the atmosphere [2]. In response to these concerns, China has aimed to peak carbon emissions by 2030 and achieve carbon neutrality by 2060 [3]. Carbon capture, utilization, and storage (CCUS) have emerged as pivotal strategies in addressing the global climate crisis. This technology involves the capture and separation of $CO_2$ from industrial emissions, followed by its beneficial utilization, ultimately contributing to a reduction in overall $CO_2$ emissions [4]. CCUS represents a win–win solution, fostering resource utilization and effective emission reduction. Particularly in the field of petroleum engineering, the application of $CO_2$ flooding holds promising prospects in terms of significantly enhancing oil recovery in low-permeability reservoirs [5].

As reservoir exploration continues, the proportion of conventional medium and high-permeability reservoir reserves is gradually decreasing, while the discovery of low-permeability reservoir reserves is steadily increasing [6–9]. In comparison with water flooding, carbon dioxide ($CO_2$) possesses lower density and viscosity, enabling it to penetrate even the tiny pores, improving the micro-sweep coefficient [10–12]. When the pressure exceeds the minimum miscible pressure (MMP), $CO_2$ can eliminate the interfacial tension between oil and gas through multi-contact miscibility, thus significantly enhancing the oil displacement efficiency [13–17]. Consequently, $CO_2$ flooding holds promising prospects for effectively enhancing oil recovery in low-permeability reservoirs. In addition, $CO_2$ can also enhance oil recovery by promoting crude oil expansion, reducing crude oil density and viscosity, changing reservoir wettability, and increasing reservoir permeability [18–21]. According to statistics, the application of CCUS technology in low-permeability oil fields in China's Daqing and Jilin regions can enhance oil recovery rates by 10–25% [22]. In the United States, the Kelly–Snyder oil field in the Permian Basin implemented a $CO_2$ miscible flooding project in the SACROC block in 2002, resulting in a cumulative increase of $2456 \times 10^4$ tons of crude oil and an estimated recovery rate improvement of over 26% [23].

The oil recovery within a reservoir is a product of both the sweep coefficient and the displacement efficiency [24]. In the case of miscible $CO_2$ flooding, the theoretical oil displacement efficiency can attain a remarkable 100%. Therefore, precise quantitative characterization of the $CO_2$ sweep coefficient holds immense importance in evaluating the enhanced oil recovery potential of $CO_2$ flooding. Furthermore, in low-permeability reservoirs, the challenges of severe heterogeneity, gravity segregation, and viscous fingering can lead to issues such as gas channeling and a low sweep coefficient [25,26]. To address these challenges effectively, conformance control technologies like WAG flooding and foam flooding become necessary [27–29]. A quantitative analysis of the sweep coefficient serves as the foundational basis for evaluating the efficacy of these conformance control technologies.

Characterizing the sweep coefficient involves both experimental and numerical simulation approaches [30–33]. Concerning experimental simulation, Bergit et al. [34] used high-resolution micro-positron emission tomography (µPET) to analyze the distribution of $CO_2$ in the core with the change of the PV number during $CO_2$ injection, and the sweeping effect was characterized qualitatively. Wang et al. [35] used a high-resolution nuclear magnetic resonance method and normalized MI value to distinguish the sweeping range of $CO_2$ in the core and studied the sweep characteristics of miscible and immiscible flooding. The results showed that the $CO_2$ front of immiscible flooding is more unstable, which affects the sweep coefficient of $CO_2$ flooding. Duraid et al. [36] used X-ray CT imaging technology to identify the distribution of oil and water in the heterogeneous core after $CO_2$ flooding. The results showed that the proportion of residual oil in the low-permeability part is significantly higher than that in the high-permeability part, which indirectly characterized the sweeping rules of $CO_2$ flooding. The visualization experiment can reflect the microscopic sweep characteristics of $CO_2$ flooding, and the core experiment can reflect the effect of $CO_2$ flooding from the macroscopy, but the numerical simulation method is helpful to analyze the sweep rules in detail. Lewis et al. [37] established a reservoir numerical model of a one-quarter five-point well pattern to study water, gas, and WAG flooding sweep coefficients. The sweep coefficients were calculated assuming that the oil displacement efficiency of gas flooding was 100%. Hao et al. [38] established a numerical simulation model of miscible flooding, delineating the contour line with oil and gas interfacial tension equal to 0.1 dyn/cm as the rear edge of the miscible zone and taking the $CO_2$ content in the oil phase equal to 20% as the front edge of the miscible zone, thus obtaining the sweep coefficient of $CO_2$ miscible flooding. Li et al. [39] established a mathematical model of miscible flooding considering $CO_2$ diffusion and adsorption and defined the location where a dimensionless $CO_2$ concentration of $C/C_0 = 0.5$ serves as the miscible front and the place where $C/C_0 = 0.95$ serves as the gas phase front to study the migration rule of the $CO_2$. C is the concentration of $CO_2$, and $C_0$ is the initial concentration of $CO_2$, which is $0.6 \text{ g}\cdot\text{cm}^{-3}$. When the adsorption is weak, increased porosity and initial injection rate accelerate the mass transfer and diffusion process. The stronger the diffusion effect, the larger the miscible region and the earlier the gas breakthrough. Enhanced adsorption reduces the sweeping range of the miscible zone. Li et al. [40] established a numerical simulation model of $CO_2$ flooding with a five-spot well pattern and an inverted nine-spot well pattern and proposed that the location where the viscosity of crude oil drops to a certain extent is defined as the miscible front, thus studying the $CO_2$ flooding sweeping rules of different well patterns. The results showed that the sweep coefficient of the inverted nine-spot well pattern was higher than that of the five-spot well pattern.

The factors affecting the $CO_2$ sweep coefficient include geological factors, fluid properties, the injection–production relationship, and so on [24,41,42]. Analyzing the main controlling factors affecting the $CO_2$ sweep coefficient is propitious when choosing suitable conformance control technology to improve the application effect of $CO_2$ flooding. Limited research has delved into identifying the main controlling factor in terms of the $CO_2$ sweep coefficient, with prevalent methodologies encompassing fuzzy analysis and orthogonal analysis [43,44]. Li et al. [45] evaluated the influence degrees of multiple factors on gas channeling through fuzzy analysis and concluded that fractures and heterogeneity are the most critical factors affecting gas channeling. Cui et al. [46] used orthogonal analysis to establish that the main controlling factor affecting the $CO_2$ sweep coefficient is well distance and established the calculation formula of the $CO_2$ sweep coefficient using multiple nonlinear regression methods. However, it is important to note that the selection of the weight matrix in fuzzy analysis is somewhat subjective, potentially impacting the precision of the main controlling factor analysis. Moreover, the limited data samples employed in orthogonal studies can lead to specific errors when analyzing the primary controlling factors. Consequently, there exists a need for the further exploration of methodologies for analyzing the main controlling factor influencing the $CO_2$ flooding sweep coefficient.

According to the relationship between reservoir pressure and minimum miscible pressure (MMP), $CO_2$ flooding is traditionally categorized into miscible flooding (MF) and immiscible flooding (IMF). However, throughout the reservoir development process, the continuous injection of $CO_2$ induces dynamic changes in the reservoir pressure distribution, leading to alterations in the miscible state of $CO_2$ and crude oil. Based on the pressure distribution between injection and production wells, we proposed a more nuanced classification of $CO_2$ flooding into three distinct displacement patterns: miscible flooding (MF), partly miscible flooding (IMF), and immiscible flooding (IMF). The position and migration law of the miscible front and the immiscible front under three displacement patterns were studied. The quantitative characterization of the sweep coefficients across these patterns was achieved through the application of the contour area method. Additionally, we elucidated the impact of geological factors, injection–production parameters, and crude oil viscosity on the sweep coefficients. Utilizing the Random Forest algorithm, we identified the main controlling factor influencing the sweep coefficient in $CO_2$ flooding. The findings of this study can help for providing valuable insights for optimizing oil recovery strategies in the field applications of $CO_2$ flooding. Sections 4.1 and 4.2 are dedicated to the fitting of PVT (pressure–volume–temperature) and slim tube experiment results carried out in the Jilin low-permeability oil field. Based on the fitting results, 1-D and 2-D numerical simulation models of $CO_2$ flooding were developed. Section 4.3 focuses on pinpointing the locations of both miscible and immiscible fronts through an analysis of the fluid property variations between wells within the 1-D model. In Section 4.4, based on the contour area method, the $CO_2$ sweep coefficients for different flooding patterns within the 2-D model are calculated using MATLAB software. Furthermore, we introduce parameters such as GSD (gravity segregation degree) and $D_m$ (miscible degree), investigating the impact of various factors on the sweep coefficients, gravity segregation degree, and miscible degree in the three $CO_2$ flooding patterns. These factors are discussed from three aspects: geology, oil viscosity, and injection–production parameters. Section 4.5 employs the out-of-bag permutation method within the Random Forest algorithm to identify the main controlling factors influencing the $CO_2$ sweep coefficients. Finally, our study concludes with a presentation of the findings in Section 5.

## 2. Materials and Methods

### 2.1. Materials

The crude oil and formation water used in the experiments were sampled from the Jilin oil field, and the crude oil fractions are shown in Table 1. $CO_2$ was purchased from the Beijing Analytical Instrument Factory, and the purity of the $CO_2$ was 99.95 mol%.

**Table 1.** Crude oil composition.

| Comp. | mol. (%) | Comp. | mol. (%) | Comp. | mol. (%) | Comp. | mol. (%) |
|---|---|---|---|---|---|---|---|
| $sCO_2$ | 0.153 | C7 | 3.835 | C17 | 2.249 | C27 | 1.062 |
| $N_2$ | 2.818 | C8 | 5.131 | C18 | 1.999 | C28 | 1.024 |
| C1 | 16.193 | C9 | 4.225 | C19 | 1.921 | C29 | 0.936 |
| C2 | 3.938 | C10 | 3.897 | C20 | 1.764 | C30 | 0.905 |
| C3 | 3.224 | C11 | 3.36 | C21 | 1.604 | C31 | 0.7 |
| IC4 | 1.675 | C12 | 3.256 | C22 | 1.554 | C32 | 0.716 |
| NC4 | 2.978 | C13 | 3.271 | C23 | 1.431 | C33 | 0.548 |
| IC5 | 0.904 | C14 | 2.697 | C24 | 1.385 | C34 | 0.532 |
| NC5 | 2.594 | C15 | 2.746 | C25 | 1.232 | C35 | 0.476 |
| C6 | 2.431 | C16 | 2.208 | C26 | 1.153 | C36+ | 5.275 |

### 2.2. Apparatus

The PVT analyzer (purchased from DBR, Edmonton, AB, Canada) had a maximum working pressure of 103 MPa and a full operating temperature of 180 °C. The flow chart of the PVT analyzer is shown in Figure 1. The slim tube experimental device was formed in the laboratory. The tube length was 16.0 m, the inner diameter was 6.35 mm, the outer diameter was 9.60 mm, the porosity was 33%, and the permeability of the gas measurement was 3.2 mD. The flow chart of the slim tube experiment is shown in Figure 2.

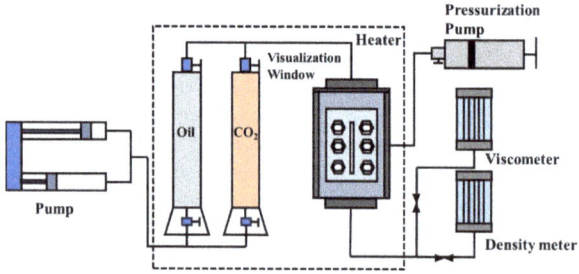

**Figure 1.** The diagram of the PVT analyzer.

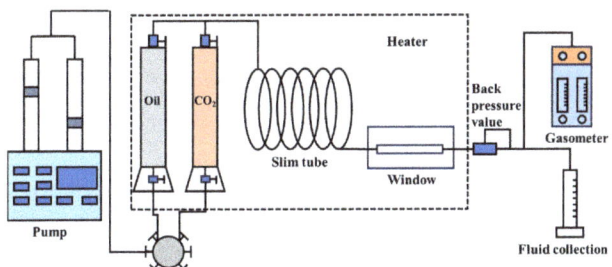

**Figure 2.** The diagram of the slim tube experiment.

*2.3. Methods*

2.3.1. PVT Experiments

1. Constant Composition Expansion experiments (CCE)

Constant composition expansion experiments were used to determine the relationship between the volume and pressure of crude oil with a constant mass at reservoir temperature to obtain parameters such as the saturation pressure, the compression coefficient, the relative volume, and the density of the formation fluids. The steps were as follows: the PVT analyzer was evacuated at a temperature of 97.3 °C, then a certain amount of crude oil was transferred to the PVT analyzer in a single phase at a constant temperature for more than 4 h, and the crude oil was pressurized to a pressure $P_1$, which was higher than the reservoir pressure. Above the saturation pressure, pressurization was carried out via stepwise depressurization at 1~2 MPa per step; below the saturation pressure, volume expansion was carried out at 1~20 cm$^3$ per step. After each depressurization step and expansion, the sample was stirred and stabilized thoroughly, and the pressure and volume were recorded. The experiment was stopped when the sample was expanded to twice the original sample.

2. Differential Liberation experiments (DL)

Differential liberation or multi-degassing experiments involved degassing crude oil at reservoir temperature in multi-graded pressure reductions to measure the relationship between oil and gas properties and composition with pressure. The crude oil was kept in a single phase, fully stirred to equilibrate, and then stabilized at saturation pressure. The sample was depressurized in three to five pressure steps, with each depressurization recording the sample volume and the volume of gas produced.

3. Swelling Tests (STs)

Swelling tests were used to measure the relationship between the density and viscosity of crude oil with the molar fraction of $CO_2$ by gradually increasing the molar fraction of the injected $CO_2$. First, the volume of the crude oil was tested at saturation pressure, and then a certain amount of $CO_2$ was injected into the crude oil at this pressure. The pressure increased until all the $CO_2$ was dissolved to test the saturation pressure, swelling factor, oil viscosity, and other parameters of the $CO_2$–crude oil system. The above experimental steps were repeated until the molar content of $CO_2$ in the system reached about 80%, which was when the experiment was stopped.

2.3.2. Slim Tube Experiments

The slim tube model was heated at the experimental temperature, and the slim tube was evacuated for more than 12 h. The pore volume was measured by injecting toluene into the tube model. The slim tube model was saturated with crude oil at the experimental pressure. $CO_2$ was injected at a constant rate of 15.00 cm$^3$/h to displace the crude oil. Whenever a certain amount of $CO_2$ was injected, the output oil and gas volumes, injection pressure, and back pressure were measured. Fluid phase and color changes were observed through a high-pressure observation window. The experiment was stopped after injecting 1.2 times the pore volume of $CO_2$. The above experiments were repeated, and the experimental pressure was changed to determine the minimum miscible pressure for $CO_2$ flooding.

2.3.3. Testing the Relative Permeability Curves Based on Unsteady-State Method

1.  Oil–water relative permeability curve

First, the absolute permeability ka of the core was measured. The core was then vacuumed to saturate the formation water, and the effective pore volume and porosity were measured. Under the back pressure of 24.5 MPa, the irreducible water saturation Swc was established by injecting formation oil to displace formation water at the injection rate of 0.2 mL/min, and the oil phase permeability under irreducible water saturation was measured. Keeping the same back pressure, the water flooding experiment was carried out at the injection rate of 0.2 mL/min. The data relating to the water breakthrough time, pressure difference, cumulative oil production, and cumulative liquid production were recorded. When the water cut at the core outlet reached 99.5%, the experiment was stopped. The relative permeabilities of the oil phase and water phase under different water saturations were measured, and the relative permeability curve was drawn.

2.  Gas–oil relative permeability curve

The core was saturated with formation water, and the water phase permeability was measured. The formation oil was injected at an injection rate of 0.2 mL/min to displace the formation water until the water saturation reached irreducible water saturation. The oil phase permeability under irreducible water saturation was measured. $CO_2$ was injected to displace oil at an injection rate of 0.2 mL/min, and the displacement pressure, oil production, and gas production at each time were recorded. When only gas was produced at the outlet of the core, the experiment was stopped, the relative permeabilities of the oil phase and gas phase under different gas saturations were measured, and the gas–oil relative permeability curve was finally drawn.

## 3. Numerical Simulation

*3.1. Simulation for Slim Tube Experiments and 1-D $CO_2$ Flooding*

3.1.1. Simulation for Slim Tube Experiments

The 1-D numerical simulation model was based on the parameters of the slim tube experiment. The cumulative $CO_2$ injection volume was 1.2 PV. The fluid parameters simulated for the slim tube experiments were obtained based on the fitting results of CCE, DL, and ST experiments with the WinProp module of the CMG software. The parameters of the thin tube model are shown in Table 2.

**Table 2.** Slim tube model parameters.

| Parameters | Values |
|---|---|
| Number of grids | $50 \times 1 \times 1$ |
| Grid size | 32 cm $\times$ 5.8 mm $\times$ 5.8 mm |
| Porosity/% | 0.33 |
| Permeability/mD | 3.2 |
| Injection pressure constraints/MPa | 35.0 |
| Gas injection rate/(PV/d) | 0.1 |

3.1.2. Simulation for 1-D $CO_2$ Flooding

The GEM module of CMG software was used to establish a 1-D numerical simulation model of $CO_2$ flooding, the EOS of which was fitted by using the WinProp module of CMG software. The values of the model parameters (porosity, permeability, saturation, and so on) were consistent with those in the Jilin oil field, and the processes of miscible flooding, immiscible flooding, and partly miscible flooding were simulated by setting different original reservoir pressures. To ensure the accuracy of the solution, the model was solved by using a fully implicit method. The relative permeability curves were obtained through the experiments, which are shown in Figure 3. The relative permeability of the oil, gas, and water phases was calculated using the Stone II method. The characterization of the miscible state was based on Coats' component model theory, which describes the miscible state through the relationship between the relative permeability and the interfacial tension of oil and gas. The reference interfacial tension $\sigma_{og}^{r}$ was set to 0.5 mN/m. When the interfacial tension $\sigma_{og}$ was higher than 0.5 mN/m, the interfacial tension did not affect the relative permeability curves. When the interfacial tension $\sigma_{og}$ was lower than 0.5 mN/m, oil started to be displaced in the form of miscible flooding. The relative permeability curve of miscible flooding was determined via interfacial tension interpolation. The interpolation equations are shown in Equations (1)–(3). The values of the 1-D model parameters are shown in Table 3.

$$k_{rom} = m \times k_{ro} + (1-m) \times \frac{k_{row}(S_w) + k_{rg}(S_g)}{2} \times \frac{S_o}{(1-S_w)} \quad (1)$$

$$k_{rgm} = m \times k_{rg} + (1-m) \times \frac{(k_{row}(S_W) + k_{rg}(S_g))}{2} \times \frac{S_g}{(1-S_w)} \quad (2)$$

$$m = \begin{cases} 1 & \sigma_{og} > \sigma_{og}^{r} \\ (\frac{\sigma_{og}}{\sigma_{og}^{r}})^n & \sigma_{og} \leq \sigma_{og}^{r} \end{cases} \quad (3)$$

where $k_{rom}$ and $k_{rgm}$ are the relative permeability of oil and gas after interpolation; $k_{ro}$ and $k_{rg}$ are the relative permeability of oil phase and gas before interpolation; and $S_w$, $S_g$, and $S_o$ are the saturation of water, gas, and oil, respectively. $k_{row}$ is the relative permeability of water in the oil–water relative permeability curve, which is a function of water saturation $S_w$; $k_{rg}$ is the relative permeability of gas in the oil–gas relative permeability curve, which is a function of gas saturation, $S_g$; $\sigma_{og}$ is the actual oil–gas interfacial tension in the model; $\sigma_{og}^{r}$ is the reference interfacial tension, which takes the value of 0.5 mN/m; and m is the interpolated value, the value of which depends on the relationship between $\sigma_{og}$ and $\sigma_{og}^{r}$. n is the index of the ratio of actual oil–gas interfacial tension $\sigma_{og}$ to the reference interfacial tension $\sigma_{og}^{r}$, which characterizes the transition speed of the relative permeability curve of immiscible flooding to miscible flooding.

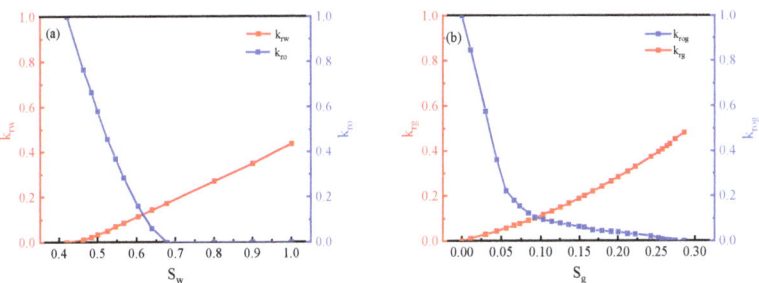

**Figure 3.** Relative permeability curve: (**a**) water/oil and (**b**) gas/oil.

*3.2. Simulation for 2-D CO$_2$ Flooding with Quarter Five-Spot Well Pattern*

A 2-D numerical simulation model of $CO_2$ flooding components was established using the GEM module of the CMG software, the EOS equations and relative permeability curves of which were consistent with the 1-D model. The model's porosity, permeability, and saturation were set with reference to the Jilin oil field. The miscible, immiscible, and partly miscible flooding processes were simulated by setting different original reservoir pressures. Continuous gas injection (CGI) was simulated with a constant gas injection period of 40 years. The model parameters are shown in Table 4.

Table 3. The 1-D numerical simulation parameters.

| Parameters | Values |
|---|---|
| Number of grids | 300 × 1 × 1 |
| Grid size | 1 m × 1 m × 1 m |
| Porosity/% | 0.13 |
| Temperature/°C | 97.3 |
| Permeability/mD | 4.5 |
| Initial formation pressure/MPa | 16.0 (immiscible flooding) 22.0 (partly miscible flooding) 26.0 (miscible flooding) |
| Gas injection rate/(m³/d) | 0.00195 (reservoir condition) |
| Well bottom pressure constraint/Mpa | 40.0 (upper limit) 10.0 (lower limit) |

Table 4. The 2-D numerical simulation parameters.

| Parameters | Values |
|---|---|
| Number of grids | 25 × 25 × 5 |
| Grid size | 8 m × 8 m × 0.6 m |
| Porosity/% | 0.13 |
| Temperature/°C | 97.3 |
| Permeability/mD | 4.5 |
| Initial formation pressure/Mpa | 16.0 (immiscible flooding) 22.0 (partly miscible flooding) 26.0 (miscible flooding) |
| Gas injection rate/(m³/d) | 0.77 (reservoir condition) |
| Well bottom pressure constraint/Mpa | 40.0 (upper limit) 10.0 (lower limit) |

Compared with the traditional numerical simulation model for $CO_2$ flooding, a series of improvements were made in this model: to enhance model convergence and solution accuracy, the model was solved using the fully implicit method instead of the traditional IMPSE method. In the iterative solution method, a nine-point difference format was used to suppress the grid orientation effect of $CO_2$ flooding instead of the conventional five-point difference format. For the general seepage equation $\Delta u = f$, the five-point differential and nine-point differential solution equations for the center point $u_{i,j}$ are shown in Equations (4) and (5), respectively. In Equations (4) and (5), f is a function of $(x_i, y_j)$, which represents the external driving force or source term at positions(i,j) in the differential equation.

$$\frac{1}{h^2}(u_{i-1,j} - 2u_{i,j} + u_{i+1,j}) + \frac{1}{h^2}(u_{i,j-1} - 2u_{i,j} + u_{i,j+1}) = f_{ij} \quad (4)$$

$$\frac{1}{h^2}(u_{i-1,j+1} + u_{i+1,j+1} + u_{i-1,j-1} + u_{i+1,j-1} + 4(u_{i,j+1} + u_{i,j-1} + u_{i-1,j} + u_{i+1,j}) - 20u_{i,j}) = f_{ij} \quad (5)$$

### 3.3. Solution of Sweep Coefficient for $CO_2$ Flooding

Parameters such as the $CO_2$ content in the oil and the interfacial tension of each grid obtained from the solution of the 2-D numerical model were imported into MATLAB R2022a by programming a data reading algorithm. The discrete grid data were expanded using linear or cubic interpolation methods. Contour maps of $CO_2$ content in oil and interfacial tension at different moments of $CO_2$ flooding were drawn. The contour lines corresponding to the miscible and immiscible front of $CO_2$ flooding were extracted. The polygon areas surrounded by the contour lines were solved, and the sweep coefficient at the corresponding time could be obtained compared with the total grid area. The mechanism is shown in Figure 4, and the sweep coefficient $E_{sweep}$ is calculated as shown in Equation (6).

$$E_{sweep} = \frac{S_{swept}}{S_{total}} \times 100\% \qquad (6)$$

where $S_{swept}$ and $S_{total}$ are the areas enclosed by the contour line of the $CO_2$ front and the coordinate axis and the total grid area, respectively.

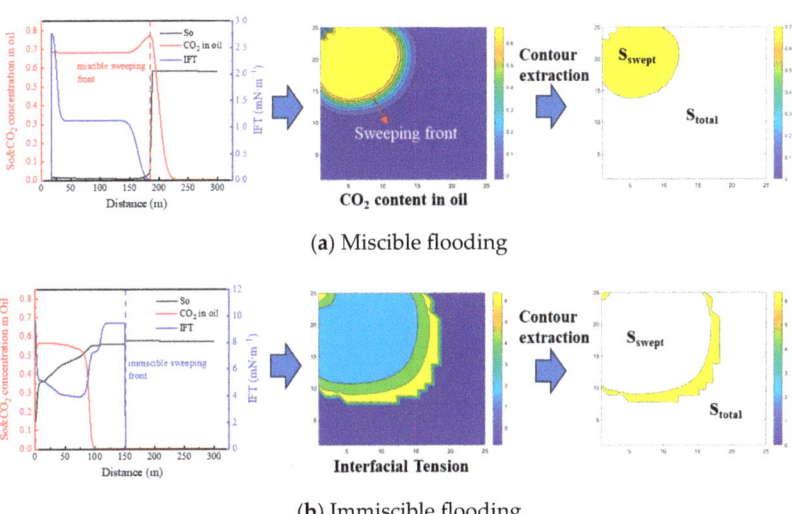

Figure 4. Schematic diagram of sweep coefficient calculation.

*3.4. Main Controlling Factor Assessment of $CO_2$ Flooding Based on Random Forest Algorithm*

Different ranges of values were assigned to each parameter affecting the sweep coefficients of $CO_2$ flooding, and 1000 sets of numerical simulation models with different combinations of parameters were formed using the Latin hypercube design method, which can make the values of each parameter satisfy the normal distribution. The range of values for each parameter is shown in Table 5. The sweep coefficients of $CO_2$ flooding for different parameter combinations were obtained by solving the 1000 sets of models. The Bootstrap sampling method was used to form 800 training sets and 200 prediction sets. The 800 training sets were used to construct multiple decision trees to develop a Random Forest model. Predictions were made on the prediction sets, and the ballot method was used to decide the regression or classification results. Different numbers of decision trees and leaves were set to form different Random Forest models and solve the prediction errors (MSE) of the different Random Forest models. The numbers of decision trees and leaves with minimum MSE were preferred as the parameters of the optimal Random Forest model. The principle of the Random Forest algorithm is shown in Figure 5. Based on the out-of-bag data permutation method (OOB), the optimized Random Forest model was used to evaluate the importance of each influencing factor (FI). The main steps were as follows: (1) For decision tree i, the prediction error rate $e_1$ of out-of-bag data was calculated. (2) After randomly replacing the observations of the influencing factor $X_j$, the decision tree was constructed again, and the prediction error rate $e_2$ of the out-of-bag data was calculated. (3) The differences between the two error rates were calculated and standardized to calculate the average $MDA_j$ in all decision trees, which was the factor importance value

($FI_j$). The larger the $FI_j$ value, the higher the importance of the parameter $X_j$. The calculation of $FI_j$ is shown in Equation (7) [47].

$$FI_j = MDA_j = \frac{1}{T}\sum_t^T \left[ \frac{1}{|D_t|}(\sum_{X_i^j \in D_t^j} \sum_k (R_k(X_i^j) - y_i^k)^2 - \sum_{X_i \in D_t} \sum_k (R_k(X_i) - y_i^k)^2 \right] \quad (7)$$

where T is the number of decision trees; $(X_i, y_i)$ is the training set; $X_i^j$ is the sample after a random exchange of the j-type influence parameter of $X_i$; $D_t$ is the out-of-bag sample set of the decision tree $_t$, and $D_t^j$ is the sample set formed after the j-th type of influence parameter exchange; and $R(X_i)$ is the predicted output of sample $X_i$. $y_i^k$ is the regression output of the k-type influence parameter under multi-objective regression.

Table 5. Range of values for each parameter.

| Parameters | Minimum Value | Maximum Value | Unit |
|---|---|---|---|
| Permeability | 0.5 | 100 | mD |
| Porosity | 0.05 | 0.2 | / |
| Reservoir thickness | 3.0 | 30.0 | m |
| Permeability ratio | 1.1 | 100 | / |
| Well spacing | 50 | 300 | M |
| Injection rate | 0.1 | 2.0 | m²/d (RC) |
| Oil viscosity | 0.74 | 33.52 | mPa·s |

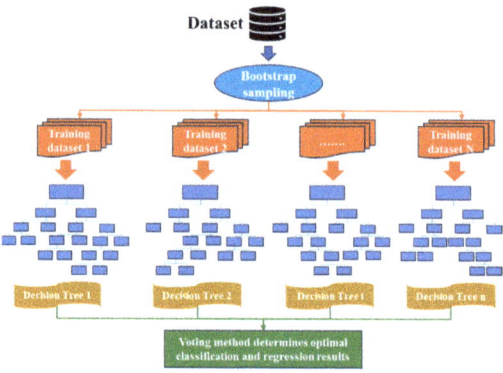

Figure 5. Schematic diagram of Random Forest algorithm.

## 4. Discussion

### 4.1. PVT Fitting and Establishment of EOS Equation for Reservoir Fluid

The saturation pressure $P_{sat}$, relative volume, oil phase density, viscosity, and swelling factor of reservoir fluid were measured by using CCE, DL, and ST experiments. The $C_{17+}$ component of the formation oil was divided into three components, $C_{17}$–$C_{26}$, $C_{27}$–$C_{39}$, and $C_{40+}$, and the $C_1$–$C_{16}$ components were merged into five components, $C_1$, $C_2$, $C_3$+$C_4$, $C_5$+$C_6$, and $C_7$–$C_{16}$. By adjusting the critical pressure $P_c$, critical temperature $T_c$, binary interaction coefficient, and other parameters of the heavy components, the fluid component model was obtained based on fitting the results of the PVT experiments. The results are shown in Figure 6. Some parameters of the EOS after PVT fitting are shown in Table 6. It can be seen from Figure 6 that the saturation pressure of the pure oil phase used in the experiment was 7.16 MPa, and the viscosity at saturation pressure was 1.74 mPa·s. When the mole fraction of injected $CO_2$ was 80%, the oil saturation pressure increased to 39.7 MPa, the viscosity decreased to 0.72 mPa·s, and the swelling factor increased to 1.8762. The swelling and viscosity reduction effects of $CO_2$ on crude oil were propitious to enhancing oil recovery. The experimental results were compared with the fitting results. It could be seen that the above component, the splitting and merging scheme for oil phase components, could accurately fit the results of PVT experiments, and the obtained EOS parameters could better simulate the high-temperature and high-pressure physical properties of crude oil in reservoirs.

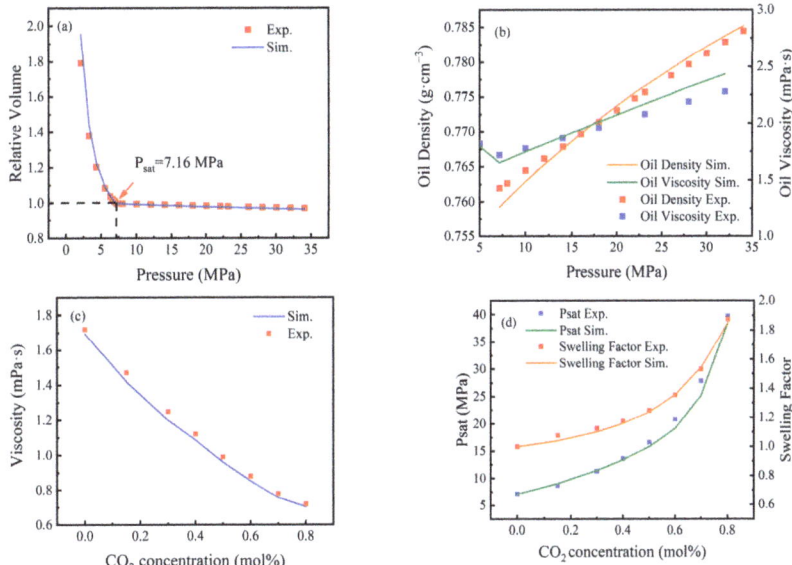

**Figure 6.** Results of PVT fitting: (**a**) relative volume fitting, (**b**) oil density and oil viscosity fitting, (**c**) swelling test fitting, and (**d**) saturation pressure and swelling factor fitting.

**Table 6.** Parameters of EOS after PVT fitting.

| Components | mol. (%) | $P_c$ (atm) | $T_c$ (K) | $M_w$ (g/mol) | $V_c$ (m$^3$/kmol) |
|---|---|---|---|---|---|
| $CO_2$ | 0.153 | 72.80 | 304.20 | 44.01 | 0.0940 |
| $N_2$ | 2.818 | 33.50 | 126.20 | 27.99 | 0.0898 |
| $C_1$ | 16.193 | 45.40 | 190.60 | 16.05 | 0.0990 |
| $C_2$ | 3.938 | 48.20 | 305.40 | 30.09 | 0.1480 |
| $C_3+C_4$ | 7.877 | 39.02 | 406.11 | 52.38 | 0.2371 |
| $C_5+C_6$ | 5.929 | 31.51 | 486.50 | 76.99 | 0.3335 |
| $C_7$-$C_{16}$ | 31.58 | 19.82 | 624.24 | 148.05 | 0.5701 |
| $C_{17}$-$C_{26}$ | 15.87 | 13.63 | 653.87 | 289.86 | 1.2768 |
| $C_{27}$-$C_{39}$ | 9.478 | 11.76 | 878.05 | 444.69 | 1.8926 |
| $C_{40+}$ | 6.164 | 10.60 | 989.31 | 716.21 | 3.2964 |

*4.2. Determination of Minimum Miscible Pressure Based on Slim Tube Experiment Simulation*

The $CO_2$ displacement experiment was carried out with a slim tube model at the formation temperature of 97.3 °C, and the displacement pressure ranged from 18 MPa to 30 MPa. The cumulative amount of injected $CO_2$ was 1.2 PV. The recovery factors of $CO_2$ flooding under different displacement pressures were calculated, and the results are shown in Figure 7a. When the displacement pressure was less than 22.10 MPa, the recovery factor of $CO_2$ flooding was low, which was an immiscible displacement process, and the recovery factor increased rapidly with the increase in displacement pressure. When the displacement pressure increased from 18 MPa to 23.11 MPa, the recovery degree rose from 76.89% to 94.53%. When the displacement pressure was higher than 22.10 MPa, the $CO_2$ flooding pattern turned to miscible flooding, and the increase in the recovery factor slowed down. When the displacement pressure increased from 23.11 MPa to 30.0 MPa, the recovery degree only increased from 94.53% to 96.08%. The sections of immiscible and miscible displacement were fitted, and the displacement pressure corresponding to the intersection of the fitting lines was 22.10 MPa. Based on the EOS equation formed in Section 4.1, a 1-D slim tube numerical model was established to simulate the results of the tube experiment. The results are shown in Figure 7b. The maximum fitting error of the recovery factor under immiscible flooding was about 7%, and the maximum fitting error of the recovery degree under miscible flooding was about 0.8%. The simulation results showed that the minimum miscible pressure was 23.30 MPa, and the error compared with the experimental results was 5.4%, which indicated that the EOS equation formed by PVT fitting could accurately reflect the characteristics of crude oil and could be used for subsequent numerical simulation work.

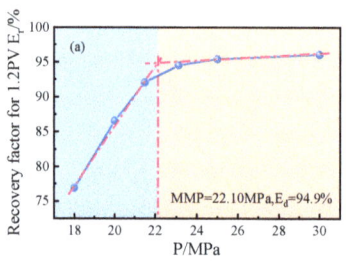 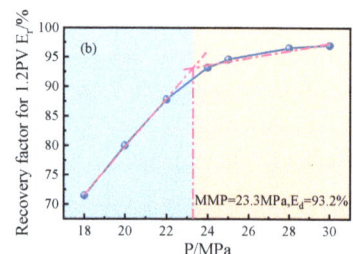

**Figure 7.** Results of slim tube experiment and slim tube experimental simulation: (**a**) slim tube experiment and (**b**) slim tube experimental simulation.

### 4.3. Front Movement Rules of $CO_2$ Flooding under Different Flooding Patterns

Currently, the classification of $CO_2$ flooding is mainly based on the relationship between original reservoir pressure and minimum miscible pressure, which is divided into miscible flooding, immiscible flooding, and near-miscible flooding. However, in the oil field development process, the reservoir pressure field undergoes dynamic changes due to the conduction between the injection and production wells, changing the miscible degree. Therefore, according to the miscible degree, $CO_2$ flooding was divided into three displacement patterns: (1) In miscible flooding (MF), the pressure between the injection and production wells was higher than the minimum miscible pressure, and the miscible degree was 1.0. (2) In partly miscible flooding (PMF), the original formation pressure of the reservoir was lower than the minimum miscible pressure, but the pressure near the injection well was higher than the minimum miscible pressure due to the reservoir energy enhancement of gas injection, thus forming a miscible flooding zone. The oil production of production wells decreased the reservoir pressure, and the immiscible flooding zone was formed within a certain range near the production well. The miscible degree of partly miscible flooding was between 0 and 1.0. (3) In immiscible flooding (IMF), the pressure between injection and production wells after gas injection was lower than the minimum miscible pressure, and the displacement process maintained immiscible flooding. The miscible degree of immiscible flooding was 0. By establishing a 1-D numerical simulation model, the distribution of $CO_2$ composition in the oil phase, oil saturation, and oil–gas interfacial tension between the injection and production wells under different displacement patterns of $CO_2$ flooding was studied, and the miscible/immiscible front position and front movement rules were determined. The results are shown in Figure 8. Figure 8a also shows the distribution of fluid properties between the injection and production wells in relation to miscible flooding. The oil saturation of miscible flooding maintained a meager value with increasing distance. Then, it increased rapidly to the initial oil saturation $S_{oi}$, showing the characteristics of a piston-like displacement. The interfacial tension increased rapidly with distance and then gradually decreased to the platform value. When the distance reached a certain value, it gradually decreased to 0. The $CO_2$ content in the oil phase increased rapidly and reached the platform value. The $CO_2$ content then increased gradually to the maximum value after increasing a certain distance and finally decreased to the original $CO_2$ content in the oil phase. According to the distribution of the above three parameters, the inter-well phase zone of miscible flooding was divided into a pure gas zone, a two-phase zone, a miscible zone, a diffusion zone, and an unswept zone. The gas zone was near the gas injection well area, where the oil saturation and interfacial tension were zero due to continuous $CO_2$ injection and miscibility. In the two-phase zone, the light components gradually evaporated into the gas phase, forming a rich gas that migrated to the production wells, significantly increasing the content of heavy components in crude oil at the miscible rare edge, thus forming an oil–gas two-phase zone. The crude oil in this zone was mainly residual oil with a high heavy component content. Due to the subsequent injection and dissolution of $CO_2$, the $CO_2$ content in this zone was significantly higher than the initial level. In the miscible zone, the rich gas in the miscible front contacted the crude oil through the condensate miscible effect, so the light component and $CO_2$ content in the oil phase increased significantly. The difference in the component content between oil and gas decreased, and the interfacial tension, therefore, gradually decreased and finally reached zero interfacial tension. The oil saturation suddenly changed to the initial oil saturation in this zone. In the diffusion zone, $CO_2$ entered the crude oil mainly through diffusion. As the distance increased, the diffusion effect gradually weakened, and the $CO_2$ content decreased to the initial level. The unswept zone only contained the pure oil phase. The interfacial tension was zero, the oil saturation was equal to the initial oil saturation, and the $CO_2$ content in the oil phase was equal to the initial value. The above analysis shows that the front of the miscible zone should be located at the maximum $CO_2$ content in the oil phase. Figure 8b shows the distribution of fluid properties between the injection and production wells in the case of immiscible flooding. With the increased distance, the oil saturation of immiscible flooding increased slowly to the initial oil saturation, forming a wide range of two-phase zones. The interfacial tension decreased rapidly with increasing distance, then increased sharply after a certain distance, and finally decreased to zero. The content of $CO_2$ in the oil phase decreased slightly at first and then decreased significantly to the initial level after a certain distance. Similarly, the injection–production inter-well phase zone of immiscible flooding was divided into two-phase, diffusion, and unswept zones. In the two-phase zone, due to the dissolution of $CO_2$ in the oil phase, the interfacial tension was reduced to 4 mN/m, and the $CO_2$ content in the oil phase was increased to about 0.55, which was lower than the $CO_2$ content in the case of miscible flooding. In the diffusion zone, the interfacial tension increased gradually due to the formation of a $CO_2$ concentration diffusion gradient. The interfacial tension reached the maximum at the junction of the diffusion zone and the unswept zone, defined as the immiscible front of immiscible flooding. Figure 8c shows the distribution of the fluid properties between injection and production wells in partly miscible flooding. The oil saturation of partly miscible flooding first maintained a meager value (residual oil saturation) with increasing distance. It then gradually increased to the initial oil saturation $S_{oi}$, but the increase amplitude was less than that of miscible flooding. The interfacial tension had two extreme values with the increase in distance. The $CO_2$ content in the oil phase gradually increased from the platform value of about 0.63 to the maximum value of 0.75. It then gradually decreased to the original $CO_2$ content in the oil phase. Unlike miscible and immiscible flooding, partly miscible flooding had both miscible and immiscible zones. In the miscible zone, due to condensate miscibility, the $CO_2$ content in the oil phase reached the maximum value, and the interfacial tension was reduced to zero, reaching the miscible state. In the immiscible zone, as the distance increased, the pressure between the injection and production wells decreased to less than the minimum miscible pressure, and the $CO_2$ in the oil phase re-evaporated into the gas phase, resulting in the gradual decrease in the $CO_2$ content in the oil phase and the rapid increase in interfacial tension. The displacement process changed from miscible flooding to immiscible flooding. The inter-well fluid zone divisions of the three displacement patterns are summarized in Figure 9. In addition, the gas injection volume was fixed at 0.3 HCPV, and the front movement positions of miscible flooding, partly miscible flooding, and immiscible flooding were compared, as shown in Figure 7d. The front movement of immiscible flooding was the fastest to reach the production well. The front of the immiscible zone in partly miscible flooding was located at 170 m, and the front of the miscible zone was located at 134 m, which indicated that the front movement velocity of the immiscible zone in partly miscible flooding was faster than that of the miscible zone. Therefore, immiscible flooding was more likely to cause gas channeling problems than miscible and partly miscible flooding.

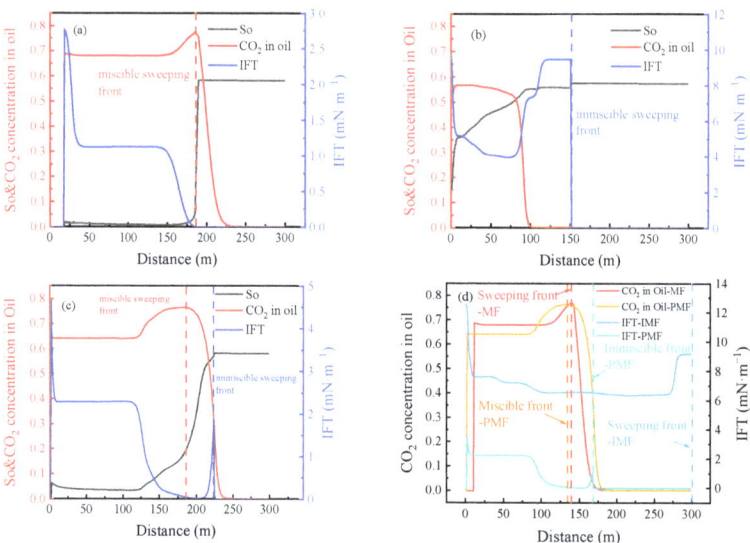

**Figure 8.** Distribution of fluid properties between wells: (**a**) miscible flooding (MF), (**b**) immiscible flooding (IMF), (**c**) partly miscible flooding (PMF), and (**d**) comparison chart (0.3 HCPV).

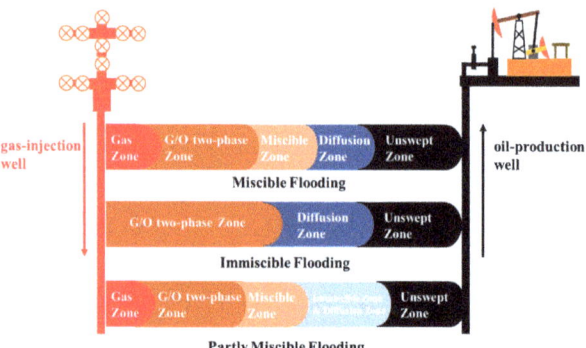

**Figure 9.** Division of inter-well fluid phase zones for three $CO_2$ displacement patterns.

*4.4. Front Movement Rules of $CO_2$ Flooding under Different Flooding Patterns*

According to the fluid phase zone division results of miscible, immiscible, and partly miscible flooding proposed in Section 4.3, the reservoir numerical simulation models of three displacement patterns were established based on the quarter five-point well pattern. The distribution of oil–gas interfacial tension and $CO_2$ content in the oil phase under different displacement patterns were output by MATLAB, and the contour map was formed. The effective sweep coefficient of $CO_2$ under different displacement patterns was solved by the program. In addition, the ratio of the maximum sweep coefficient $E_{smax}$ to the minimum sweep coefficient $E_{smin}$ between layers was defined as the gravity overlap degree (GSD), as shown in Equation (8), to characterize the strength of the gravity overlap during $CO_2$ flooding. The larger the GSD value was, the stronger the gravity overlap effect of the displacement process was. The ratio of the miscible zone swept volume $V_{s\text{-mis}}$ to the total swept volume $V_{s\text{-total}}$ of partly miscible flooding was defined as the miscible degree ($D_m$), as shown in Equation (9), to characterize the strength of the miscible degree in the process of partly miscible flooding. The larger the $D_m$, the stronger the influence of miscibility on the sweep coefficient in the partly miscible flooding, and the range of the miscible degree $D_m$ was 0~1.0. The effects of different factors on the sweep coefficient of $CO_2$ under three displacement patterns were studied from geological parameters, crude oil viscosity, and injection–production parameters.

$$GSD = \frac{E_{smax}}{E_{smin}} \tag{8}$$

$$D_m = \frac{V_{s\text{-mis}}}{V_{s\text{-total}}} \tag{9}$$

4.4.1. Geological Parameters

1. Permeability

The relationships between permeability and sweep coefficients of miscible flooding, partly miscible flooding, and immiscible flooding are compared in Figure 10a. The sweep coefficients of $CO_2$ under the three flooding patterns increased first and then decreased with increased permeability. However, the sweep coefficient of miscible flooding was higher than that of partly miscible and immiscible flooding. When the permeability was low, the sweep coefficients of the three displacement patterns were low due to the high seepage resistance in the reservoir, which was the main controlling factor affecting the sweep coefficient. When the permeability increased to 2 mD, the sweep coefficient gradually increased due to decreased seepage resistance. The sweep coefficient of miscible flooding reached 74.46%, while those of the partly miscible and immiscible flooding were 73.50% and 63.62%, respectively. When the permeability further increased, the sweep coefficients of the three displacement patterns gradually decreased. This was because the viscosity and density of $CO_2$ were lower than that of crude oil, and it was easy to produce gravity overlap in the reservoir with low seepage resistance, move to the top of the reservoir, and produce the crude oil in the top layer. The gravity segregation degrees (GSD values) of the three displacement patterns were analyzed, and the results are shown in Figure 10a. The GSD values of the three displacement patterns increased with increased permeability. When the permeability reached 100 mD, the GSD values of miscible, immiscible, and partly miscible flooding were 12.47, 69.44, and 51.75, respectively. The gravity segregation degree of miscible flooding was the lowest, followed by partly miscible flooding, and immiscible flooding was the highest. This was because the miscible effect continuously enriched $CO_2$ in the displacement process, and its density and viscosity were gradually similar to those of crude oil, so gravity segregation was significantly weakened. Figure 10b shows the variation in the sweep coefficient and miscible degree of partly miscible flooding with permeability. When the permeability was 0.5 mD, due to the large seepage resistance and high gas injection pressure, the overall pressure of the reservoir was increased so that $CO_2$ displaced crude oil in the form of miscible flooding, and the miscible degree ($D_m$ value) was 1.0. With the increase in permeability, the seepage resistance decreased, and $CO_2$ flooding began to show the characteristics of partly miscible flooding, where both miscible and immiscible zones could be observed. When the permeability increased to 20 mD, the $D_m$ value decreased to 0.52. This was mainly because the increase in permeability led to enhanced overlap, and $CO_2$ easily channeled along the top of the reservoir, resulting in a continuous decrease in the formation pressure, thus reducing the miscible degree. When the permeability increased to 100 mD, the miscibility rose to 0.70. The increased permeability reduced the seepage resistance, and the migration velocity of the immiscible front was accelerated. After gas channeling, the sweeping volume of the immiscible zone decreased due to the weak ability of the immiscible zone to sweep oil in both sides of the main streamline, while the miscible front continued to move and displace the oil on both sides, thus increasing the miscible degree.

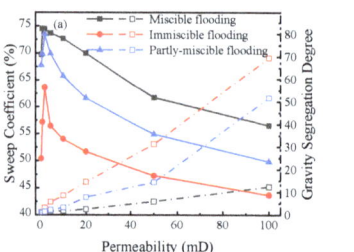
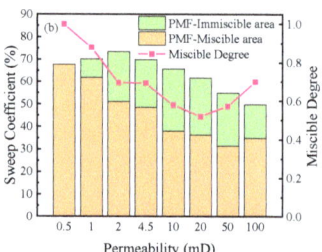

**Figure 10.** Relationship between permeability and sweep coefficient. (**a**) Comparison chart. (**b**) Sweep coefficient and miscible degree of partly miscible flooding (PMF).

2. Porosity

The relationships between the porosity and the sweep coefficient under three displacement patterns were compared, and the results are shown in Figure 11a. With the increase in porosity, the sweep coefficients of the three displacement patterns decreased gradually, and the sweep coefficients of miscible flooding and partly miscible flooding were higher than those of immiscible flooding. When the porosity was 0.03, the sweep coefficients of miscible, immiscible, and partly miscible flooding were 97.70%, 86.20%, and 92.13%, respectively. When the porosity increased to 0.3, the sweep coefficients of the three displacement patterns decreased to 50.46%, 42.98%, and 49.15%. The porosity affected the sweep coefficient, which was mainly attributed to the gravity segregation degree. It can be seen from Figure 10a that the GSD values of the three displacement patterns increased with the increase in porosity, and the GSD value of immiscible flooding was always greater than those of miscible flooding and partly miscible flooding. When the porosity was 0.3, the GSD value of immiscible flooding was 8.97, while the GSD values of miscible and partly miscible flooding were 1.93 and 13.57, respectively. The sweep coefficient and miscible degree of partly miscible flooding were analyzed, and the results are shown in Figure 11b. With the increase in porosity, the sweep coefficient of the immiscible zone of partly miscible flooding decreased significantly, but the sweep coefficient of the miscible zone increased slightly and then decreased slowly, and the sweep coefficient of the miscible zone decreased from 43.52% to 35.58%. The miscible degree rose from 0.47 to 0.72. On one hand, the increase in porosity accelerated the movement of the immiscible front and miscible front to production wells, but the moving velocity of the miscible front was slower. On the other hand, it was beneficial to the multi-contact miscibility in the movement of the miscible front, thus enhancing the miscible degree.

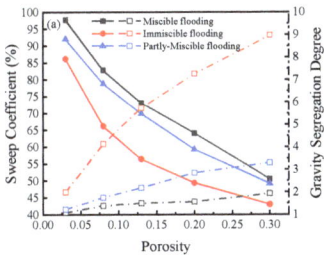
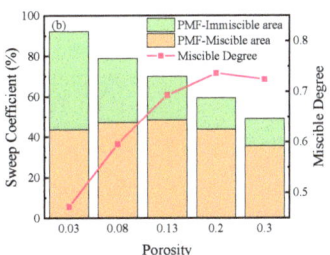

**Figure 11.** Relationship between porosity and sweep coefficient. (**a**) Comparison chart. (**b**) Sweep coefficient and miscible degree of partly miscible flooding (PMF).

3. Reservoir thickness

The influence of the reservoir thickness on $CO_2$ sweep coefficients of miscible flooding, partly miscible flooding, and immiscible flooding was studied by changing the reservoir thickness from 1.5 m to 10.0 m. The results are shown in Figure 12a. The sweep coefficients of $CO_2$ in the three displacement patterns decreased with the increase in reservoir thickness. When the reservoir thickness was 1.5 m, the sweep coefficients of miscible, partly miscible, and immiscible flooding were 86.85%, 82.75%, and 79.65%, respectively. When the reservoir thickness increased to 10.0 m, the sweep coefficients of the three displacement patterns decreased to 36.19%, 34.18%, and 33.01%, respectively. In addition, the GSD values of the three displacement patterns were 1.1~1.3, with a reservoir thickness of 1.5 m. When the reservoir thickness increased to 10.0 m, the GSD values of miscible flooding, partly miscible flooding, and immiscible flooding increased to 8.4, 16.4, and 78.1, respectively. The reservoir thickness had the most significant influence on immiscible flooding. Therefore, the variation in the gravity segregation degree caused by the change in the reservoir thickness was the critical factor affecting the sweep coefficient. The sweep coefficient and miscible degree of partly miscible flooding were analyzed, and the results are shown in Figure 12b. With the increase in the reservoir thickness, the sweep coefficients of miscible and immiscible zones of partly miscible flooding decreased gradually, but the miscible degree increased slightly from 0.62 to 0.74. This was mainly because the effect of the gravity segregation degree on the immiscible zone was stronger than that on the miscible zone. The increase in the reservoir thickness made the immiscible and miscible fronts move to the production well on the horizon and expand vertically simultaneously. The movement of the immiscible front was faster than that of the miscible front, so the immiscible zone quickly broke through to the production well in the high part of the reservoir. The sweeping of the miscible zone was more uniform, which ultimately improved the miscible degree.

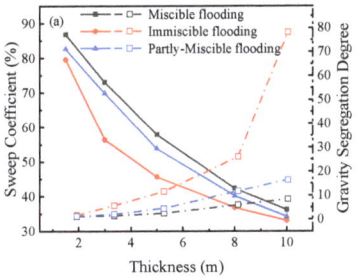
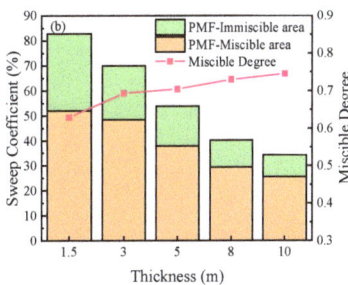

**Figure 12.** Relationship between thickness and sweep coefficient. (**a**) Comparison chart. (**b**) Sweep coefficient and miscible degree of partly miscible flooding (PMF).

4. Permeability Ratio

Two-layer heterogeneous models were established to study the influence of the permeability ratio on the sweep coefficients. The permeabilities of different layers changed in an anti-rhythmic pattern, the ratio of which ranged from 2 to 100. The results are shown in Figure 13a,b. The sweep coefficients of the whole layers and low-permeable zone decreased significantly under the three displacement patterns with the increased permeability ratio. When the permeability ratio was 10, the sweep coefficient of immiscible flooding was only 0.89%. In comparison, the sweep coefficients of miscible and partial miscible flooding were 4.8% and 2.8%, respectively, which were both lower than 5%. The result indicated the influence of heterogeneity on miscible flooding, and partly miscible flooding was weaker than immiscible flooding. Under different displacement patterns, the upper limit of the anti-rhythm reservoir permeability ratio suitable for applying $CO_2$ flooding was about 5.0.

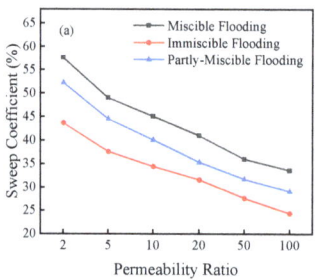
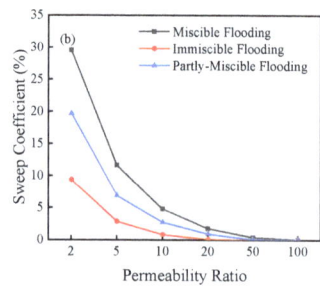

**Figure 13.** Relationship between permeability and sweep coefficient. (**a**) Total sweep coefficient and (**b**) sweep coefficient of low-permeable zone.

4.4.2. Injection–Production Parameters

1. Well spacing

The influence of injection–production well spacing on the $CO_2$ sweep coefficients of miscible flooding, partly miscible flooding, and immiscible flooding was studied in relation to a well spacing variation range of 100~300 m. The results are shown in Figure 14a. The sweep coefficients in the three displacement patterns decreased with the increase in well spacing. The increase in well spacing delayed the movement velocity of miscible and immiscible fronts to the production well in the horizontal direction. It significantly enhanced the gravity overlap of $CO_2$ in the vertical direction, so the sweep coefficients were reduced considerably. However, the reduced sweep coefficients of miscible and partly miscible flooding were mainly dominated by the weakening of the front movement velocity in the horizontal direction, and the gravity segregation degree did not change. When the well spacing increased to 300 m, the GSD values of miscible and partly miscible flooding increased from 1.0 and 1.2 to 2.1 and 3.6, respectively. The decrease in the sweep coefficient of immiscible flooding was mainly affected by gravity overlap, and the GSD value increased from 1.8 to 11.0. Figure 14b shows the influence of well spacing on the sweep coefficient and miscible degree of partly miscible flooding. Although the increase in well spacing reduced the total sweep coefficient of partly miscible flooding, the sweep coefficient of the miscible zone increased first and then decreased, and the miscible degree rose from 0.44 to 0.74. This was attributed to the delay in the miscible front moving through the production well. The effect of miscible flooding could be exploited adequately. When the well spacing was further increased, the gravity overlap of the miscible front was gradually enhanced, moving to the top of the reservoir and breaking through to the production well so that crude oil at the bottom of the reservoir could not be effectively displaced. When the well spacing was 200 m, it could not only ensure a higher sweep coefficient and miscible degree but also maximize the sweep coefficient of the miscible zone, giving full play to the role of miscible flooding.

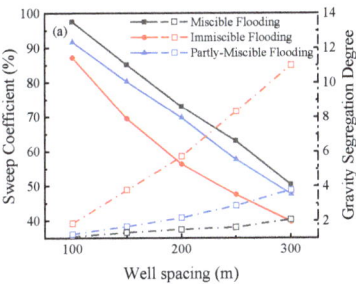
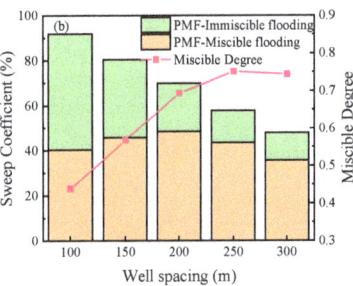

**Figure 14.** Relationship between well distance and sweep coefficient. (**a**) Comparison chart. (**b**) Sweep coefficient and miscible degree of partly miscible flooding (PMF).

2. Gas injection rate

The influence of the gas injection rate on the sweep coefficients of the three $CO_2$ flooding patterns was studied with the gas injection rate variation range of 85.98~517.5 m$^3$/d. The results are shown in Figure 15a. The sweep coefficients of $CO_2$ in the three $CO_2$ flooding patterns increased significantly first and then decreased slightly with an increasing gas injection rate. When the gas injection rate reached 517.5 m$^3$/d, the sweep coefficients of the three were about 70%. The GSD values of the three displacement patterns were significantly weakened with the increase in the gas injection rate. When the gas injection rate increased to 517.5 m$^3$/d, the GSD values of the three displacement patterns were about 1.01–1.07, which could achieve uniform displacement in each layer. The gas injection rate had two effects on the increasing $CO_2$ sweep coefficient. On one hand, the higher gas injection rate was beneficial to recovering reservoir energy, increasing the formation pressure to above the minimum miscible pressure, which made partly miscible flooding and immiscible flooding transition to miscible flooding. On the other hand, it promoted the seepage velocity of the miscible and immiscible fronts in the horizontal direction and inhibited vertical movement. However, the overhigh gas injection rate will accelerate gas channeling in the production wells. Hence, the sweep coefficients of the three displacement patterns decreased slightly with an increasing gas injection rate. Figure 15b shows the influence of the gas injection rate on the sweep coefficient and miscible degree of partly miscible flooding. When the gas injection rate was lower than 170.81 m$^3$/d, the sweep coefficient of the miscible zone and miscible degree were almost zero. When the gas injection rate exceeded 261.61 m$^3$/d, the sweep coefficient of the miscible zone reached 70%, and the miscible degree increased from 0.69 to 1.0. The miscible degree and sweep coefficient could be enhanced by appropriately increasing the gas injection rate for partly miscible and immiscible flooding.

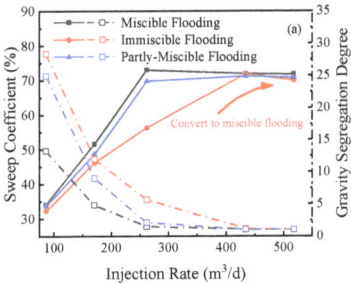
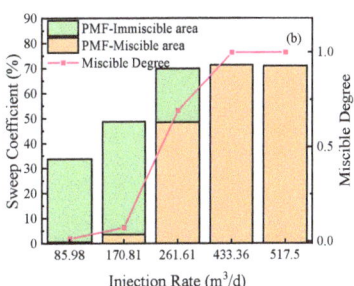

**Figure 15.** Relationship between injecting rate and sweep coefficients. (**a**) Comparison chart. (**b**) Sweep coefficient and miscible degree of partly miscible flooding (PMF).

### 4.4.3. Crude Oil Viscosity

The influence of the crude oil viscosity on the sweep coefficients of the three $CO_2$ flooding patterns was studied in relation to a gas injection rate variation range of 0.74~12.18 mPa·s. The results are shown in Figure 16a. The sweep coefficients of miscible and partly miscible flooding decreased with the increase in crude oil viscosity. When the viscosity of crude oil increased to 12.18 mPa·s, the sweep coefficients of miscible flooding and partly miscible flooding decreased from 80.42% and 78.98% to 64.42% and 49.92%, respectively. However, the gravity segregation degrees of miscible and partly miscible flooding remained almost unchanged with increasing crude oil viscosity, which were 1.40 and 1.60, respectively. Therefore, the decreases in the sweep coefficients of miscible flooding and partly miscible flooding were mainly due to the high seepage resistance caused by the increase in crude oil viscosity, and the influence of gravity overlap on the sweep coefficient was relatively small. The sweep coefficient of immiscible flooding decreased first and then increased with the increase in crude oil viscosity, and the GSD value rose first and then decreased. When the viscosity of crude oil was 4.98 mPa·s, the sweep coefficient of immiscible flooding reached the minimum value of 53%, and the GSD value increased to 5.27. When the viscosity of crude oil rose to 12.18 mPa·s, the sweep coefficient of immiscible flooding increased to 61.89%, and the GSD value decreased to 2.05. The change of the sweep coefficient of immiscible flooding had an excellent corresponding relationship with the variation in the gravity segregation degree. The influence of crude oil viscosity on the sweep coefficient and miscible degree of partly miscible flooding is shown in Figure 16b. The miscible degree of partly miscible flooding decreased first and then increased with the increase in crude oil viscosity and finally dropped from 0.78 to 0.71. When the viscosity of crude oil increased by 12.18 mPa·s, the sweep coefficient of the miscible zone decreased from 61.34% to 44.45%. In contrast, the sweep coefficient of the immiscible zone first increased to 22.59% and then dropped to 17.76%. The increase in crude oil viscosity made the multi-contact miscibility between $CO_2$ and crude oil more difficult, decreasing the sweep coefficient of the miscible zone. In addition, increased crude oil viscosity inhibited the horizontal movement of the immiscible zone and intensified the vertical overlap, thus decreasing the miscible degree. However, the overhigh viscosity of crude oil strengthened the horizontal viscous fingering, making the immiscible zone break through the production well more easily along the horizontal direction.

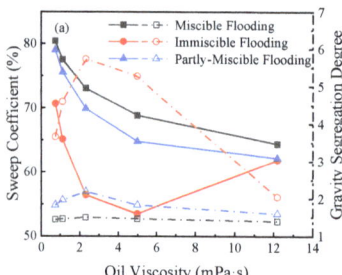 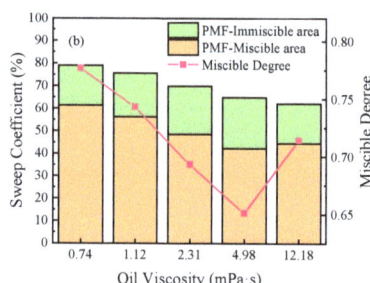

**Figure 16.** Relationship between oil viscosity and sweep coefficient. (**a**) Comparison chart. (**b**) Sweep coefficients and miscible degree of partly miscible flooding (PMF).

### 4.5. Analysis of the Main Controlling Factors of $CO_2$ Flooding Sweep Coefficient Based on the Random Forest Algorithm

Based on the single-factor analysis of the $CO_2$ flooding sweep coefficient in Section 4.4, the parameter variation ranges of different influencing factors were set, and 1000 sets of $CO_2$ flooding numerical simulation models were formed by using a Latin hypercube design. The Bootstrap sampling method was used to create 800 training sets and 200 testing sets. The Random Forest model was established by constructing multiple decision trees using the 800 training sets. The testing sets were predicted, and the voting method determined the prediction results. The number of decision trees and the number of leaves in the Random Forest model were optimized to obtain the optimal model parameters. The optimized Random Forest model was used to calculate the decrease rate MDA of the prediction accuracy before and after the permutation of the out-of-bag data of a certain influencing factor (such as porosity, permeability, etc.), which was used as a parameter to evaluate the importance degree (FI) of the influencing factor. The larger the FI value, the greater the influence of the factors on the sweep coefficient. The parameter optimization results of Random Forest are shown in Figure 17a. Under the condition of the same number of leaves, the mean square error MSE of the prediction results decreased rapidly with the increase in the number of decision trees. However, when the number of decision trees increased to 500, MSE almost did not change. Increasing the number of decision trees will lead to overfitting the model, so the optimal number of decision trees was 500. In addition, under the same number of decision trees, MSE increased with the increase in the number of leaves. Thereby, the optimal number of leaf nodes was five. The importance degree of the influencing factors of the $CO_2$ flooding sweep coefficient is shown in Figure 17b. The factors affecting the sweep coefficient of $CO_2$ flooding were ranked in order of importance degree (FI) as follows: permeability ratio, well spacing, reservoir thickness, gas injection rate, porosity, permeability, and crude oil viscosity. The main controlling factor affecting the sweep coefficients of $CO_2$ flooding was the permeability ratio, the importance degree (FI) of which reached 1.04. In the field application of $CO_2$ flooding, attention should be paid to the problem that the low-permeability layer caused by the high-permeability ratio cannot be effectively utilized and the gravity overlap problem of $CO_2$ flooding in thick reservoirs.

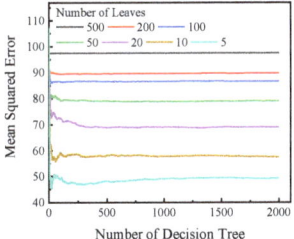
(**a**) Optimization of Random Forest parameters

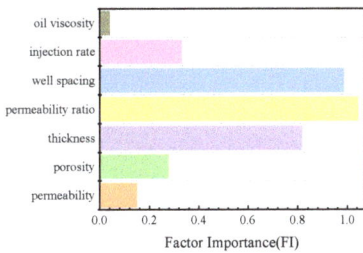
(**b**) Factor importance of $CO_2$ flooding (FI)

**Figure 17.** Analysis results of main controlling factors of $CO_2$ flooding sweep coefficient.

## 5. Conclusions

Based on the fitting results of the PVT and slim tube experiments in the Jilin oil field, the 1-D and 2-D reservoir numerical simulation models of $CO_2$ flooding were developed to analyze the movement rules relating to the $CO_2$ front in miscible, partly miscible, and immiscible flooding, along with the influence of different factors on the sweep coefficient. The main controlling factors affecting the sweep coefficient of $CO_2$ flooding were determined using the out-of-bag data permutation method based on the Random Forest algorithm.

(1) The miscible front was located at the point of maximum $CO_2$ content in the oil phase, and the immiscible front was located at the point of maximum interfacial tension near the production well. The movement speed of the immiscible front was faster than that of the miscible front, so gas channeling was more likely to occur in immiscible flooding. Comparatively, miscible and partly miscible flooding yielded higher sweep coefficients than immiscible flooding, underscoring the advantageous effect of miscibility in achieving uniform $CO_2$ sweeping within the reservoir.

(2) Regarding geological factors, the increases in porosity, permeability, and reservoir thickness intensified the gravity overlap effect and reduced the sweep coefficients of $CO_2$ under the three displacement patterns. The increase in the permeability ratio also notably reduced the sweep coefficient, albeit to a lesser extent, in miscible and partly miscible flooding when compared with immiscible flooding. The upper limit of the reservoir

permeability ratio suitable for applying $CO_2$ flooding under different displacement patterns was about 5. Concerning the injection–production parameters, the increase in well spacing amplified the seepage resistance in the horizontal direction, leading to reduced sweep coefficients in the cases of miscible and partly miscible flooding. In immiscible flooding, the reduction in the sweep coefficients was predominantly attributed to gravity overlap. Elevating the gas injection rate enhanced the miscibility degree in partly miscible flooding and immiscible flooding while inhabiting the gravity overlap, thus enhancing the sweep coefficient of $CO_2$ flooding. Crude oil viscosity primarily governed the sweep coefficients in miscible and partly miscible flooding by elevating the seepage resistance, whereas in immiscible flooding, gravity overlap took precedence in determining the sweep coefficient.

(3) The main controlling factor affecting the sweep coefficient of $CO_2$ flooding was the permeability difference, and the factor importance FI reached 1.04. In the field application of $CO_2$ flooding, attention should be paid to the problem of ineffective displacement in low-permeable zones in heterogeneous reservoirs.

**Author Contributions:** Conceptualization, X.Q. and T.Z.; methodology, X.Q.; software, X.Q.; validation, W.L., Y.S. and M.D.; formal analysis, W.L.; investigation, Z.L.; resources, X.Q.; data curation, X.Q. and M.W.; writing—original draft preparation, X.Q.; writing—review and editing, W.L.; visualization, D.H.; supervision, D.H.; project administration, D.H.; funding acquisition, W.L. All authors have read and agreed to the published version of the manuscript.

**Funding:** This research was funded by PetroChina's major scientific and technological project "Research and demonstration of key technologies for large-scale carbon dioxide capture, flooding and storage in the whole industrial chain", subject 03 "Study on CCUS geological fine description for oil displacement and key technologies in reservoir engineering" (2021ZZ01-03).

**Data Availability Statement:** The raw/processed data required to reproduce these findings cannot be shared at this time as the data also form part of an ongoing study.

**Acknowledgments:** The authors are grateful for the financial support of the CNPC in China.

**Conflicts of Interest:** Author Dongbo He was employed by the company JiDong Oilfield of PetroChina. The remaining authors declare that the research was conducted in the absence of any commercial or financial relationship that could be construed as a potential conflict of interest.

# References

1. McLaughlin, H.; Littlefield, A.A.; Menefee, M.; Kinzer, A.; Hull, T.; Sovacool, B.K.; Bazilian, M.D.; Kim, J.; Griffiths, S. Carbon capture utilization and storage in review: Sociotechnical implications for a carbon reliant world. *Renew. Sustain. Energy Rev.* **2023**, *177*, 113215. [CrossRef]
2. Kumar, N.; Augusto Sampaio, M.; Ojha, K.; Hoteit, H.; Mandal, A. Fundamental aspects, mechanisms and emerging possibilities of $CO_2$ miscible flooding in enhanced oil recovery: A review. *Fuel* **2022**, *330*, 125633. [CrossRef]
3. Zhao, C.; Ju, S.; Xue, Y.; Ren, T.; Ji, Y.; Chen, X. China's energy transitions for carbon neutrality: Challenges and opportunities. *Carbon Neutrality* **2022**, *1*, 7. [CrossRef]
4. Zhao, K.; Jia, C.; Li, Z.; Du, X.; Wang, Y.; Li, J.; Yao, Z.; Yao, J. Recent Advances and Future Perspectives in Carbon Capture, Transportation, Utilization, and Storage (CCTUS) Technologies: A Comprehensive Review. *Fuel* **2023**, *351*, 128913. [CrossRef]
5. Yuan, S.; Ma, D.; Li, J.; Zhou, T.; Ji, Z.; Han, H. Progress and prospects of carbon dioxide capture, EOR-utilization and storage industrialization. *Pet. Explor. Dev.* **2022**, *49*, 955–962. [CrossRef]
6. Yue, P.; Zhang, R.; Sheng, J.J.; Yu, G.; Liu, F. Study on the Influential Factors of $CO_2$ Storage in Low Permeability Reservoir. *Energies* **2022**, *15*, 344. [CrossRef]
7. Tsopela, A.; Bere, A.; Dutko, M.; Kato, J.; Niranjan, S.C.; Jennette, B.G.; Hsu, S.Y.; Dasari, G.R. $CO_2$ injection and storage in porous rocks: Coupled geomechanical yielding below failure threshold and permeability evolution. *Pet. Geosci.* **2022**, *28*, 1. [CrossRef]
8. Haeri, F.; Myshakin, E.M.; Sanguinito, S.; Moore, J.; Crandall, D.; Gorecki, C.D.; Goodman, A.L. Simulated $CO_2$ storage efficiency factors for saline formations of various lithologies and depositional environments using new experimental relative permeability data. *Int. J. Greenh. Gas Control* **2022**, *119*, 103720. [CrossRef]
9. Wei, J.; Zhou, X.; Zhou, J.; Li, J.; Wang, A. Experimental and simulation investigations of carbon storage associated with $CO_2$ EOR in low-permeability reservoir. *Int. J. Greenh. Gas Control* **2021**, *104*, 103203. [CrossRef]
10. Li, L.; Su, Y.; Sheng, J.J.; Hao, Y.; Wang, W.; Lv, Y.; Zhao, Q.; Wang, H. Experimental and Numerical Study on $CO_2$ Sweep Volume during $CO_2$ Huff-n-Puff EOR Process in Shale Oil Reservoirs. *Energy Fuels* **2019**, *33*, 4017–4032. [CrossRef]
11. Wang, X.; Gu, Y. Oil Recovery and Permeability Reduction of a Tight Sandstone Reservoir in Immiscible and Miscible $CO_2$ Flooding Processes. *Ind. Eng. Chem. Res.* **2011**, *50*, 2388–2399. [CrossRef]
12. Bikkina, P.; Wan, J.; Kim, Y.; Kneafsey, T.J.; Tokunaga, T.K. Influence of wettability and permeability heterogeneity on miscible $CO_2$ flooding efficiency. *Fuel* **2016**, *166*, 219–226. [CrossRef]
13. Zargar, G.; Bagheripour, P.; Asoodeh, M.; Gholami, A. Oil-$CO_2$ minimum miscible pressure (MMP) determination using a stimulated smart approach. *Can. J. Chem. Eng.* **2015**, *93*, 1730–1735. [CrossRef]
14. Han, L.; Gu, Y. Optimization of Miscible $CO_2$ Water-Alternating-Gas Injection in the Bakken Formation. *Energy Fuels* **2014**, *28*, 6811–6819. [CrossRef]
15. Li, L.; Zhou, X.; Su, Y.; Xiao, P.; Chen, Z.; Zheng, J. Influence of Heterogeneity and Fracture Conductivity on Supercritical $CO_2$ Miscible Flooding Enhancing Oil Recovery and Gas Channeling in Tight Oil Reservoirs. *Energy Fuels* **2022**, *104*, 103203. [CrossRef]
16. Guo, Y.; Liu, F.; Qiu, J.; Xu, Z.; Bao, B. Microscopic transport and phase behaviors of $CO_2$ injection in heterogeneous formations using microfluidics. *Energy* **2022**, *256*, 124524. [CrossRef]
17. Alhosani, A.; Lin, Q.; Scanziani, A.; Andrews, E.; Blunt, M.J. Pore-scale characterization of carbon dioxide storage at immiscible and near-miscible conditions in altered-wettability reservoir rocks. *Int. J. Greenh. Gas Control* **2021**, *105*, 103232. [CrossRef]
18. Seyyedi, M.; Sohrabi, M. Assessing the Feasibility of Improving the Performance of $CO_2$ and $CO_2$-WAG Injection Scenarios by CWI. *Ind. Eng. Chem. Res.* **2018**, *57*, 11617–11624. [CrossRef]

19. Liu, Y.; Rui, Z. A Storage-Driven $CO_2$ EOR for a Net-Zero Emission Target. *Engineering* **2022**, *18*, 79–87. [CrossRef]
20. Al-Bayati, D.; Saeedi, A.; Myers, M.; White, C.; Xie, Q.; Clennell, B. Insight investigation of miscible $SCCO_2$ Water Alternating Gas (WAG) injection performance in heterogeneous sandstone reservoirs. *J. CO2 Util.* **2018**, *28*, 255–263. [CrossRef]
21. Ren, B.; Littlefield, J.; Jia, C.; Duncan, I. Impact of Pressure-Dependent Interfacial Tension and Contact Angle on Capillary Trapping and Storage of $CO_2$ in Saline Aquifers. In Proceedings of the SPE Annual Technical Conference and Exhibition, San Antonio, TX, USA, 2–7 October 2023.
22. Song, X.; Wang, F.; Ma, D.; Gao, M.; Zhang, Y. Progress and prospect of carbon dioxide capture, utilization and storage in CNPC oilfields. *Pet. Explor. Dev.* **2023**, *50*, 229–244. [CrossRef]
23. Kalteyer, J. A Case Study of SACROC $CO_2$ Flooding in Marginal Pay Regions: Improving Asset Performance. In Proceedings of the SPE Improved Oil Recovery Conference, Virtual, 31 August–4 September 2020.
24. Zhao, X.; Liao, X. Evaluation Method of $CO_2$ Sequestration and Enhanced Oil Recovery in an Oil Reservoir, as Applied to the Changqing Oilfields, China. *Energy Fuels* **2012**, *26*, 5350–5354. [CrossRef]
25. Chen, X.; Li, Y.; Tang, X.; Qi, H.; Sun, X.; Luo, J. Effect of gravity segregation on $CO_2$ flooding under various pressure conditions: Application to $CO_2$ sequestration and oil production. *Energy* **2021**, *226*, 120294. [CrossRef]
26. Al Hinai, N.M.; Saeedi, A.; Wood, C.D.; Myers, M.; Valdez, R.; Sooud, A.K.; Sari, A. Experimental Evaluations of Polymeric Solubility and Thickeners for Supercritical $CO_2$ at High Temperatures for Enhanced Oil Recovery. *Energy Fuels* **2018**, *32*, 1600–1611. [CrossRef]
27. Sun, X.; Long, Y.; Bai, B.; Wei, M.; Suresh, S. Evaluation and Plugging Performance of Carbon Dioxide-Resistant Particle Gels for Conformance Control. *SPE J.* **2020**, *25*, 1745–1760. [CrossRef]
28. Li, Z.; Su, Y.; Li, L.; Hao, Y.; Wang, W.; Meng, Y.; Zhao, A. Evaluation of $CO_2$ storage of water alternating gas flooding using experimental and numerical simulation methods. *Fuel* **2022**, *311*, 122489. [CrossRef]
29. Alsumaiti, A.M.; Hashmet, M.R.; Alameri, W.S.; Antodarkwah, E. Laboratory Study of $CO_2$ Foam Flooding in High Temperature, High Salinity Carbonate Reservoirs Using Co-injection Technique. *Energy Fuels* **2018**, *32*, 1416–1422. [CrossRef]
30. Song, Y.; Yang, W.; Wang, D.; Yang, M.; Jiang, L.; Liu, Y.; Zhao, Y.; Dou, B.; Wang, Z. Magnetic resonance imaging analysis on the in-situ mixing zone of $CO_2$ miscible displacement flows in porous media. *J. Appl. Phys.* **2014**, *115*, 401–410. [CrossRef]
31. Wang, H.; Tian, L.; Chai, X.; Wang, J.; Zhang, K. Effect of pore structure on recovery of $CO_2$ miscible flooding efficiency in low permeability reservoirs. *J. Pet. Sci. Eng.* **2021**, *208*, 109305. [CrossRef]
32. Chen, M.; Cheng, L.; Cao, R.; Lyu, C.; Rao, X. Carbon dioxide transport in radial miscible flooding in consideration of rate-controlled adsorption. *Arab. J. Geosci.* **2020**, *13*, 1–11. [CrossRef]
33. Coats, K.H.; Smith, B.D. Dead-End Pore Volume and Dispersion in Porous Media. *Soc. Pet. Eng. J.* **1964**, *4*, 73–84. [CrossRef]
34. Brattekås, B.; Haugen, M. Explicit tracking of $CO_2$-flow at the core scale using micro-Positron Emission Tomography (μPET). *J. Nat. Gas Sci. Eng.* **2020**, *77*, 103268. [CrossRef]
35. Wang, S.; Jiang, L.; Cheng, Z.; Liu, Y.; Song, Y. Experimental study on the $CO_2$-decane displacement front behavior in high permeability sand evaluated by magnetic resonance imaging. *Energy* **2021**, *217*, 119433. [CrossRef]
36. Duraid, A.B.; Ali, S.; Quan, X.; Myers, M.B.; Cameron, W. Influence of Permeability Heterogeneity on Miscible $CO_2$ Flooding Efficiency in Sandstone Reservoirs: An Experimental Investigation. *Transp. Porous Media* **2018**, *125*, 341–356. [CrossRef]
37. Lewis, E.; Dao, E.; Mohanty, K.K. Sweep Efficiency of Miscible Floods in a High-Pressure Quarter Five-Spot Model. *SPE J.* **2006**, *13*, 432–439. [CrossRef]
38. Hao, Y.; Li, J.; Kong, C.; Guo, Y.; Lv, G.; Chen, Z.; Wei, X. Migration behavior of $CO_2$-crude oil miscible zone. *Pet. Sci. Technol.* **2021**, *39*, 959–971. [CrossRef]
39. Li, J.; Cui, C.; Wu, Z.; Wang, Z.; Wang, Z.; Yang, H. Study on the migration law of $CO_2$ miscible flooding front and the quantitative identification and characterization of gas channeling. *J. Pet. Sci. Eng.* **2022**, *218*, 110970. [CrossRef]
40. Li, N.; Tian, J.; Ren, Z. The research on spread rule of $CO_2$ miscible region in low permeability reservoir. *Well Test.* **2014**, *23*, 2023101670. [CrossRef]
41. Al-Abri, A.; Sidiq, H.; Amin, R. Mobility ratio, relative permeability and sweep efficiency of supercritical $CO_2$ and methane injection to enhance natural gas and condensate recovery: Coreflooding experimentation. *J. Nat. Gas Sci. Eng.* **2012**, *9*, 166–171. [CrossRef]
42. Perrin, J.C.; Benson, S. An Experimental Study on the Influence of Sub-Core Scale Heterogeneities on $CO_2$ Distribution in Reservoir Rocks. *Transp. Porous Media* **2010**, *82*, 93–109. [CrossRef]
43. Lu, Y.; Liu, R.; Wang, K.; Tang, Y.; Cao, Y. A study on the fuzzy evaluation system of carbon dioxide flooding technology. *Energy Sci. Eng.* **2021**, *9*, 239–255. [CrossRef]
44. Bai, S.; Song, K.P.; Yang, E.L. Optimization of water alternating gas injection parameters of $CO_2$ flooding based on orthogonal experimental design. *Spec. Oil Gas Reserv.* **2011**, *20*, 48–52. [CrossRef]
45. Chenglong, L. Gas Channeling Influencing Factors and Patterns of $CO_2$-flooding in Ultra-Low Permeability Oil Reservoir. *Spec. Oil Gas Reserv.* **2018**, *25*, 82–86. [CrossRef]

46. Cui, C.; Yan, D.; Yao, T.; Wang, J.; Zhang, C.; Wu, Z. Prediction method of migration law and gas channeling time of $CO_2$ flooding front: A case study of Gao 89-1 Block in Shengli Oilfield. *Reserv. Eval. Dev.* **2022**, *12*, 741–747+763. [CrossRef]
47. Sinha, U.; Dindoruk, B.; Soliman, M.Y. Prediction of $CO_2$ Minimum Miscibility Pressure MMP Using Machine Learning Techniques. In Proceedings of the SPE-Improved Oil Recovery Conference, Virtual, 31 August–4 September 2020.

**Disclaimer/Publisher's Note:** The statements, opinions and data contained in all publications are solely those of the individual author(s) and contributor(s) and not of MDPI and/or the editor(s). MDPI and/or the editor(s) disclaim responsibility for any injury to people or property resulting from any ideas, methods, instructions or products referred to in the content.

Article

# Experimental Study on Carbon Dioxide Flooding Technology in the Lunnan Oilfield, Tarim Basin

Zangyuan Wu [1,2], Qihong Feng [1,*], Yongliang Tang [2], Daiyu Zhou [2] and Liming Lian [3]

1 School of Petroleum Engineering, China University of Petroleum (East China), Qingdao 266580, China; wuzy-tlm@petrochina.com.cn
2 PetroChina Tarim Oilfield Company, Korla 841000, China; tangyl-tlm@petrochina.com.cn (Y.T.); zdy-tlm@petrochina.com.cn (D.Z.)
3 Research Institute of Petroleum Exploration and Development, Beijing 100083, China; lianliming@petrochina.com.cn
* Correspondence: fengqihong@upc.edu.cn

**Abstract:** The Lunnan Oilfield in the Tarim Basin is known for its abundant oil and gas resources. However, the marine clastic reservoir in this oilfield poses challenges due to its tightness and difficulty in development using conventional water drive methods. To improve the recovery rate, this study focuses on the application of carbon dioxide flooding after a water drive. Indoor experiments were conducted on the formation fluids of the Lunnan Oil Formation, specifically investigating gas injection expansion, thin tube, long core displacement, oil and gas phase permeability, and solubility. By injecting carbon dioxide under the current formation pressure, the study explores the impact of varying amounts of carbon dioxide on crude oil extraction capacity, high-pressure physical parameters of crude oil, and phase characteristics of formation fluids. Additionally, the maximum dissolution capacity of carbon dioxide in formation water is analyzed under different formation temperatures and pressures. The research findings indicate that the crude oil extracted from the Lunnan Oilfield exhibits specific characteristics such as low viscosity, low freezing point, low-medium sulfur content, high wax content, and medium colloid asphaltene. The measured density of carbon dioxide under the conditions of the oil group is 0.74 g/cm$^3$, which closely matches the density of crude oil. Additionally, the viscosity of carbon dioxide is 0.0681 mPa·s, making it well-suited for carbon dioxide flooding. With an increase in the amount of injected carbon dioxide, the saturation pressure and gas-oil ratio of the crude oil also increase. As the pressure rises, carbon dioxide dissolves rapidly into the crude oil, resulting in a gradual increase in the gas-oil ratio, expansion coefficient, and saturation pressure. As the displacement pressure decreases, the degree of carbon dioxide displacement initially decreases slowly, followed by a rapid decrease. Moreover, an increase in the injection rate of carbon dioxide pore volume leads to a rapid initial improvement in oil-displacement efficiency, followed by a slower increase. Simultaneously, the gas-oil ratio exhibits a slow increase initially, followed by a rapid rise. Furthermore, as the displacement pressure increases, the solubility of carbon dioxide in water demonstrates a linear increase. These research findings provide valuable theoretical data to support the use of carbon dioxide flooding techniques for enhancing oil recovery.

**Keywords:** Lunnan oilfield; carbon dioxide flooding; gas injection expansion; oil and gas phase seepage

Citation: Wu, Z.; Feng, Q.; Tang, Y.; Zhou, D.; Lian, L. Experimental Study on Carbon Dioxide Flooding Technology in the Lunnan Oilfield, Tarim Basin. *Energies* **2024**, *17*, 386. https://doi.org/10.3390/en17020386

Academic Editor: Mofazzal Hossain

Received: 31 October 2023
Revised: 6 December 2023
Accepted: 14 December 2023
Published: 12 January 2024

**Copyright:** © 2024 by the authors. Licensee MDPI, Basel, Switzerland. This article is an open access article distributed under the terms and conditions of the Creative Commons Attribution (CC BY) license (https://creativecommons.org/licenses/by/4.0/).

## 1. Introduction

With the rapid increase in the development of conventional oil reservoirs around the world, the reserves and annual production of conventional oil reservoirs and high-quality oil reservoirs are continuously decreasing [1]. As the reserves and production of tight oil and gas are increasing year by year, it has become the reservoir for global oil and natural gas production [2]. Tight reservoirs are characterized by poor physical properties, small pore throats, and threshold pressure gradients [3]. Through the high injection pressure, the water absorption capacity of the tight reservoir is poor, which leads to the slow development of water injection in oil wells [4]. The Lunnan Oilfield in the Tarim Basin is a low-porosity and low-permeability reservoir with a small pore throat radius, strong heterogeneity, large mud influence, and the development of micro-fractures, which makes overall development difficult [5]. By interacting carbon dioxide with tight oil, the oil displacement mechanism of viscosity reduction, expansion, and miscibility is achieved [6]. In order to improve reservoir recovery, the impact of carbon dioxide gas flooding on recovery has become a hot research topic [7].

At present, research on carbon dioxide flooding for enhanced oil recovery has yielded some results. Varfolomeev et al. studied the oil field production method of advanced water injection and direct injection. Taking the Weibei Oilfield as the research object, they carried out indoor experiments on ultra-low permeability reservoir cores and studied water flooding and carbon dioxide based on nuclear magnetic resonance data. The microscopic oil displacement mechanism of water flooding, conversion from water flooding to gas injection, etc., is concluded. The water flooding oil displacement efficiency in the Weibei ultra-low permeability reservoir is the lowest, only 36%, and the carbon dioxide oil flooding efficiency reaches 55%. The oil displacement efficiency of water flooding to gas flooding is the highest, reaching 60%, so the oil displacement efficiency of water flooding to carbon dioxide injection is the best [8]. Pal et al. studied the effect of carbon dioxide flooding, analyzed the displacement characteristics of carbon dioxide injection in tight oil reservoirs from a micro-scale, and concluded that carbon dioxide flooding effectively activates the crude oil stored in tight pores, and carbon dioxide breaks through quickly under a pressure of 10 MPa. The crude oil in the small pores did not move. Under the pressure of 24 MPa, the breakthrough of carbon dioxide gradually slowed down, the oil saturation decreased linearly after displacement, the oil in the pores was activated, and the recovery rate was high [9]. Ansari et al. studied the changes in the physical properties of crude oil and rocks through the effects of gas injection and expansion of crude oil and formation water dissolution of gas on the characteristics of the reservoir rocks. They concluded that after the injection of carbon dioxide, the viscosity and density of the reservoir crude oil gradually decreased. The volume of crude oil expanded to 45%, and the permeability of oil and gas was 6.5% higher than that of oil and water. Heavy components such as $C_{16}$-$C_{35}$ in crude oil gradually increased, and the asphaltene content reached 86%. After carbon dioxide was injected into the reservoir, it gradually dissolved in the formation water so that under the action of weakly acidic formation water, the permeability and porosity of the reservoir gradually increased, and the seepage capacity continued to become stronger [10]. Novak et al. studied the impact of carbon dioxide flooding on the recovery of fault block oil reservoirs. By developing a three-dimensional geological model of fault block oil reservoirs, they simulated the geomorphology of the oil reservoir and used phase analysis software to construct a fault block reservoir geomorphology. Based on the physical property model of the block oil reservoir and conducted carbon dioxide oil displacement experiments, it was concluded that the recovery effect at a well spacing of 250 m is relatively good. When the seepage pressure exceeds 30 MPa, the oil displacement efficiency is high, and the produced gas-oil ratio is the highest [11]. Yan et al. studied the impact of carbon dioxide flooding on the production of the ultra-low permeability Changqing Oilfield. Due to the low production efficiency of early water flooding, the development production was declining year by year. Through carbon dioxide flooding, the daily liquid production volume and daily oil production of the oil wells were increased. The amount has increased significantly, and the water content is gradually decreasing; compared with the water flooding injection pressure, the carbon dioxide injection pressure has not increased significantly and does not change with the injection amount, indicating that the reservoir has good carbon dioxide injection and good air-absorbing capabilities [12].

Judging from the recovery degree, the recovery degree of the homogeneous section is high, reaching 59.53%, but it is necessary to carry out tertiary oil recovery because of the thick reservoir and large remaining oil reserves in the homogeneous section [13].

After carbon dioxide displacement measures are implemented in reservoirs, there are four mechanisms for the change of rock wettability, namely, adsorption and flocculation of asphaltene, ion bridging, repulsion and attraction of net charges, and the influence of formation water.

The exploration of enhanced oil recovery technology by injecting carbon dioxide into a clastic reservoir in the Tarim Oilfield began in 2018. In 2021, in combination with the needs of carbon source conditions, miscible conditions, and the goal of "double carbon," the Tarim Oilfield selected the No.2 well area of the Lunnan Oilfield to carry out pilot tests of carbon dioxide flooding and storage. This practice has proven that carbon dioxide flooding technology can prolong the life of an oil field for more than 10 years, and the cooperative development of oil displacement, extraction, and carbon storage has special advantages. Well No.2 in the Lunnan Oilfield is a realistic potential area for CCUS-EOR development, which has the basis for integrated design of geological engineering.

The Lunnan Oilfield in the Tarim Basin is rich in oil and gas resources and is a tight oil reservoir. Conventional water flooding is difficult to develop, and the recovery rate can be effectively improved by injecting carbon dioxide after water flooding. This paper takes the Lunnan Oilfield in the Tarim Basin as the research object and carries out gas injection expansion, thin tube, long core displacement, oil and gas phase permeability, and solubility laboratory experiments. By injecting carbon dioxide into the formation, the impact of the injection amount of carbon dioxide on the extraction capacity of crude oil, the high-pressure physical property parameters of crude oil, and the phase characteristics of formation fluids were studied. The maximum solubility capacity of carbon dioxide in formation water under different formation temperatures and pressures was analyzed, which provides theoretical support for carbon dioxide flooding to improve oil recovery.

## 2. Research Methods

*2.1. Geological Characteristics*

The Tarim Basin is divided into eight first-order structural units, namely, the Kuqa Depression, Tabei Uplift, Northern Depression, Central Uplift, Southwest Depression, Tangguzibasi Depression, Tarnum Uplift, and Southeast Depression from north to south. The structural position of the study area belongs to the Akkule Uplift in the middle part of the Tabei Uplift, also known as the Lunnan Ancient Uplift and the Lunnan Low Uplift. It is bordered by the Luntai Fault in the north, adjacent to the Mangar Depression and the Shuntuoguole Uplift in the south, and the Caohu Depression and the Halahatang Depression in the east and west, respectively. The Lunnan Ancient Uplift covers an area of 4420 $km^2$ and consists of seven secondary structural units, namely, the northern slope zone, the Lunnan fault barrier zone, the central slope zone, the Santamu fault barrier zone, the southern slope zone, the western slope zone, and the eastern slope zone. Lunnan buried hill was formed by long-term evolution under the control of unified structural conditions of the Tabei Uplift and mainly experienced Gary.

There are four evolutionary stages: the formation period of the Dongbi Uplift, the formation period of the Hercynian anticline, the Indosinian-Yanshan fault activity period, and the Himalayan structure finalization period.

The surface of the Lunnan Oilfield is flat, with an average altitude of about 930 m. Except for the tamarisk and Achnatherum vegetation in a local area, no other plants have been developed [14].

The Lunnan Oilfield belongs to the Lunnan Low Bulge Lunnan Fault Zone tectonics of the Tarim Basin Tabei Uplift. It is an almost east-west trending long-axis anticline, which appears as an irregular skirt extending into the central depression, forming a set of fan delta-braided river delta-lacustrine facies sediments [15]. Well grain size analysis by borehole coring indicates that the lithology of the Lunnan Oil formation is predominantly a medium sandstone composed of quartz, feldspar, lithic debris, and miscellaneous matrix. The rock type is mainly feldspathic lithic sandstone, and the interstitial materials are mainly Kaolinite and mud. The content of potassium feldspar in the reservoir that easily reacts with carbon dioxide is between 9% and 25%. The reservoir sandstone has low compositional maturity and structural maturity [16].

Research has shown that the sand layer group studied in this article belongs to the subfacies of discernible delta plain, with the development of braided distributary channels and channel sand bar sedimentary microfacies. Sandstone is currently in the early stage of diagenesis B to the early stage of the late stage of diagenesis A, with the late stage of diagenesis A being the main stage. Residual intergranular pores and dissolution pores are the main spaces for oil and gas accumulation, with strong intra-layer and planar heterogeneity and weak interlayer heterogeneity. It is a typical medium porosity and medium permeability reservoir. The reservoir evaluation results indicate that the main reservoir types are Class II (good) reservoirs with medium to fine throats, followed by Class I (good) reservoirs with large to coarse throats and large to medium to fine throats, and Class III (medium) reservoirs with small to medium throats.

Injecting $CO_2$ into the formation to form weak acids can improve the permeability of the reservoir by interacting with rocks such as calcite. In sandstone reservoirs, $CO_2$ reacts to form precipitates such as kaolinite and clay minerals, which can also block pore throats and cause damage to the reservoir. When the temperature is high, the viscosity of crude oil will decrease, and the solubility of $CO_2$ in crude oil will also decrease, making it less prone to phase mixing. When the temperature is lower, the higher the viscosity of crude oil, the greater the seepage resistance, and the more severe the fingering phenomenon during the displacement process. The solubility of $CO_2$ will increase with the increase of reservoir pressure, and the lower the interfacial tension between oil and gas, the easier it is for mixing to occur. The oil displacement efficiency will also increase with the enhancement of $CO_2$ extraction capacity for light hydrocarbons.

The crude oil Lunnan Oilfield has good properties with low viscosity, low freezing point, low-medium sulfur content, high wax content, and medium colloid asphaltene; the methane content in natural gas is low, with a maximum of 88% and generally 60~84%; the salinity of formation water is $1.9 \times 10^5$ mg/L, the chloride ion content is $1.1 \times 10^5$ mg/L, and the formation water type is $CaCl_2$ type. Table 1 shows the fluid properties of the Lunnan Oilfield reservoir [16].

Table 1. Reservoir fluid properties of the Lunnan Oilfield.

| Crude Oil | | | | | | | Natural Gas | | Formation Water | |
|---|---|---|---|---|---|---|---|---|---|---|
| Ground Density (g/cm$^3$) | Initial Boiling Point (°C) | Viscosity 50 °C (mPa.s) | Wax Content (%) | Sulfur Content (%) | Freezing Point (°C) | Colloidal Asphaltene (%) | Proportion (g/cm$^3$) | Methane Content (%) | Total Mineralization ($10^5$ mg/L) | Chloride Ion Content ($10^5$ mg/L) |
| 0.8632 | 79.77 | 13.41 | 7.0 | 0.53 | −14.1 | 13.69 | 0.6919 | 78.23 | 1.9 | 1.1 |

The experimental cores in this paper were selected from the JIV1 sandstone reservoir core in the central depression of the Lunnan Oilfield in the Tarim Basin, and the depths of the cores ranged from 4554 m to 4583 m (See Table 2).

Table 2. Core parameter table of different injection pressure displacement experiments.

| Core Number | Length (cm) | Diameter (cm) | Porosity (%) |
|---|---|---|---|
| 1 | 5.06 | 2.48 | 19.38 |
| 2 | 5.73 | 2.51 | 14.51 |
| 3 | 5.05 | 2.52 | 16.65 |
| 4 | 4.83 | 2.47 | 17.80 |
| 5 | 5.29 | 2.53 | 15.60 |
| 6 | 7.08 | 2.51 | 16.52 |
| 7 | 7.25 | 2.49 | 17.23 |

## 2.2. Experimental Methods

The experiment adopts the national standard "Oil and Gas Reservoir Fluid Physical Property Analysis Method" [17]; the experimental instrument adopts the oil and gas reservoir fluid analysis instrument, the maximum pressure can reach 150 MPa, and the maximum temperature is 200 °C (See Figure 1).

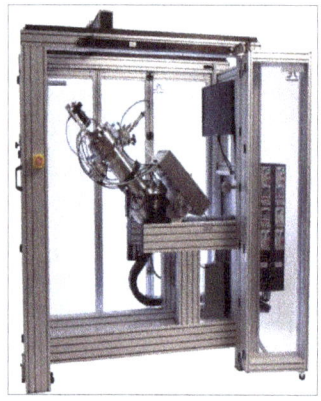

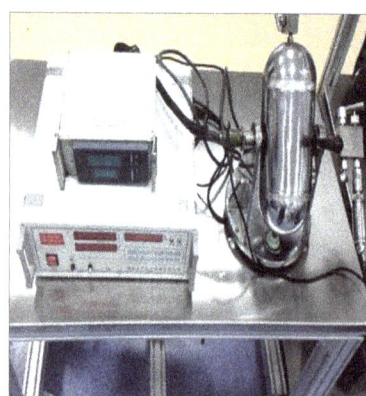

**Figure 1.** Oil and gas reservoir fluid analysis instrument.

Experiment procedure:

(1) In the experiment, the temperature was set to 128 °C, the pressure was set to 45.4 MPa, pure carbon dioxide was used as the injection gas, and on-site formation water was used as the water sample. By transferring an appropriate amount of mixed oil samples into a high-temperature and high-pressure sampler and adding excess carbon dioxide, the temperature is kept constant at the formation temperature, the pressure is kept at the bubble point pressure, and the oil is stabilized after sufficient stirring [18]. The excess carbon dioxide from the upper part of the sampler is discharged under constant pressure. The fluid in the sampler is the formation of a crude oil sample, which is used to carry out experiments.

(2) During the experiment, the pressure and temperature data in the reaction kettle were continuously collected and recorded through a data collector. At the same time, the flow rate of carbon dioxide gas and the time of chemical reaction were recorded. Based on the data collected from the experiment, physical parameters such as density and viscosity of carbon dioxide gas are calculated.

(3) After extracting and drying the core, measure the gas permeability of the core, sort them in order according to the permeability, connect the core model, and measure the porosity of the saturated formation water in the core model. Oil is injected into the water-saturated core to drive water with oil until there is no water flow at the outlet of the core, and the irreducible water saturation of the core model is calculated.

(4) Connect the cylinder to the container to ensure that the gas enters the container smoothly. Set a pressure sensor and a temperature sensor on the cylinder, connect it to the back pressure valve at the outlet end of the core, and add a predetermined pressure; adjust a certain displacement pressure, and carbon dioxide gas is injected into the core to conduct carbon dioxide gas flooding tight oil experiments until no more crude oil is produced at the outlet end, thereby measuring the cumulative volume of displaced oil and gas and the cumulative volume of injected gas.

(5) Pump the formation oil into the thin tube core until no water comes out from the outlet end, and calculate the irreducible water saturation and original oil saturation in the thin tube model; maintain the back pressure during the experiment, determine

the displacement pressure, conduct the displacement experiment under a certain displacement pressure, and record the pump volume and oil production parameters.

(6) By changing the displacement pressure and repeating the experiment, the gas drive recovery rate under different displacement pressures is measured. The minimum pressure corresponding to the gas drive recovery rate of 90–95% is recorded as the minimum miscible pressure [19].

(7) Place the core sample in the permeability measuring device to ensure that the sample can seep smoothly. A pressure sensor and a temperature sensor are provided on the permeability measurement device to monitor pressure and temperature changes during the seepage process.

(8) In the experiment, the permeability was set to 0.1 mD~1000 mD, the porosity was 5~30%, and the pressure gradient was set to 0.1 MPa/m~10 MPa/m. Oil and gas were injected from the oil and gas source into the core sample, and the pressure sensors and temperature sensors monitor the pressure and temperature of oil and gas, and the flow meter monitors the seepage rate of oil and gas. At the same time, data such as the permeability, pressure, and temperature of oil and gas in core samples are continuously collected, and the injection time and amount of oil and gas are recorded through a data collector.

(9) Put the prepared solution into a constant temperature bath, heat it to the set temperature, and stir the solution with a magnetic stirrer to ensure uniformity of the solution; during the dissolution process, monitor the solution in real-time with a pH meter and conductivity meter and monitor the concentration changes of dissolved substances in the solution through spectrophotometer and other instruments, and continuously collect and record the concentration, pH value, conductivity, dissolution time and amount of dissolved substances in the solution (See Figure 2).

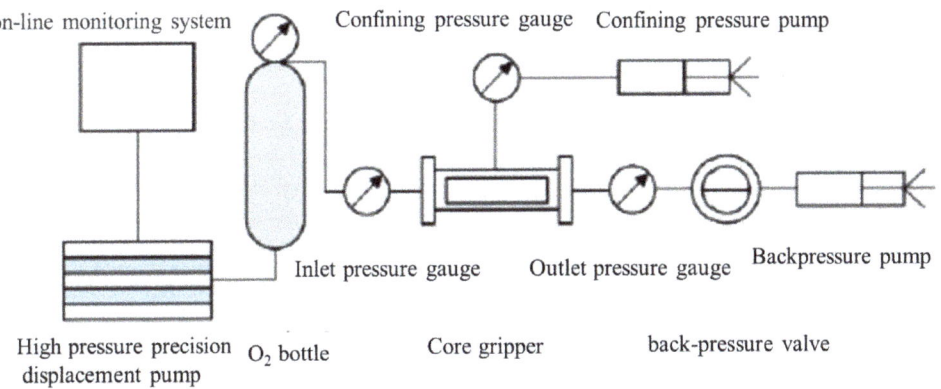

**Figure 2.** Flow chart of saturated crude oil core displacement experiment.

## 3. Results

*3.1. High-Temperature and High-Pressure Physical Properties of Carbon Dioxide*

By recording the physical property changes of carbon dioxide under different temperatures and pressures, the high-temperature and high-pressure physical properties of carbon dioxide under reservoir conditions are determined. Table 3 shows the crude oil sampling results of the Lunnan oil group. It can be seen from Table 3 that under oil group conditions, the density of carbon dioxide is 0.74 g/cm$^3$, which is basically close to the density of crude oil, but its viscosity is 0.0681 mPa·s, which improves the mobility ratio of crude oil and has good adaptability to carbon dioxide flooding [20]. Through a chromatographic analysis of the oil and gas sample composition and well flow composition, the well flow composition of the formation fluid and the single-degassing experimental data were obtained (Table 4).

Table 3. Lunnan oil group crude oil sampling results.

| Gas/Oil Ratio ($m^3/m^3$) | Coefficient of Expansion | Volume Coefficient | Bubble Point Pressure (MPa) | Crude Oil Density ($g/cm^3$) | Crude Oil Viscosity under Formation Pressure (mPa·s) |
|---|---|---|---|---|---|
| 42.52 | 1 | 1.14 | 12.06 | 0.79 | 3.6 |

Table 4. Formation fluid well flow components composition.

| Component Name | Well Flow Molar Composition (%) | Component Name | Well Flow Molar Composition (%) |
|---|---|---|---|
| $N_2$ | 1.15 | $nC_4$ | 1.73 |
| carbon dioxide | 0.88 | $iC_5$ | 1.34 |
| $C_1$ | 29.89 | $nC_5$ | 1.36 |
| $C_2$ | 2.15 | $C_6$ | 4.40 |
| $C_3$ | 1.33 | $C_{7+}$ | 55.11 |
| $iC_4$ | 0.64 | $C_{7+}$ molecular weight | 362.20 |

*3.2. Gas Injection Expansion Characteristics*

By transferring the prepared formation fluid sample into the PVT instrument, after the formation temperature stabilizes for 2 h, add an appropriate amount of pressurized formation crude oil sample and stir it thoroughly for 2 h to make the sample into a homogeneous single-phase state. Then, slowly reduce the pressure, measure its bubble point, and perform a single removal test to test the amount of dissolved gas in crude oil and the density and viscosity of the fluid [21]. After the test is completed, continue to inject carbon dioxide into the oil sample according to the above experimental method, increase stirring to make the sample into a homogeneous single-phase state, then reduce the pressure to measure its new bubble point and conduct a single removal test. Through multiple consecutive additions of injected gas, the gas injection expansion results of the Lunnan Oilfield formation were obtained (See Table 5).

Table 5. Gas injection expansion results of formations in the Lunnan Oilfield.

| Injected Gas Mole Fraction (Decimal) | Gas/Oil Ratio ($m^3/m^3$) | Coefficient of Expansion | Volume Coefficient | Saturation Pressure (MPa) | Surface Degassed Crude Oil Density ($g/cm^3$) | Viscosity under Formation Pressure (mPa·s) |
|---|---|---|---|---|---|---|
| 0 | 42.52 | 1 | 1.14 | 12.06 | 0.79 | 3.62 |
| 0.39 | 126.63 | 1.1 | 1.34 | 19.72 | 0.81 | 1.98 |
| 0.56 | 206.56 | 1.2 | 1.52 | 25.50 | 0.82 | 1.48 |
| 0.67 | 316.56 | 1.3 | 1.76 | 30.65 | 0.82 | 1.27 |
| 0.82 | 594.68 | 1.4 | 2.37 | 39.68 | 0.82 | 1.05 |

As the amount of injected carbon dioxide increases, the saturation pressure of crude oil increases from 12.06 MPa to 39.68 MPa, and the gas-oil ratio increases from 42.52 $m^3/m^3$ to 594.68 $m^3/m^3$, indicating that the sample has a strong impact on carbon dioxide. The dissolving ability of crude oil is strong. As the pressure increases, carbon dioxide will quickly dissolve into crude oil, and its gas-oil ratio, expansion coefficient, and saturation pressure will gradually increase. Due to the dissolved carbon dioxide in crude oil and the extraction of crude oil by carbon dioxide, the viscosity gradually decreased from 3.62 mPa·s to 1.05 mPa·s, and the expansion coefficient increased from 1 to 1.4. Therefore, after carbon dioxide is injected into the oil reservoir, it achieves good expansion and viscosity reduction effects.

## 3.3. Minimum Miscibility Pressure

This article uses configured formation fluid samples to test the minimum miscible pressure of carbon dioxide and formation crude oil to determine whether the reservoir has achieved miscible carbon dioxide flooding. The degree of recovery is achieved by injecting gas under different experimental pressures [22]. Figure 3 shows the degree of $CO_2$ flooding recovery under different pressures. As the displacement pressure continues to decrease, the degree of carbon dioxide displacement decreases slowly at first and then decreases rapidly. When the displacement pressure exceeds 27.5 MPa, the degree of carbon dioxide displacement exceeds 90%. Therefore, the minimum miscible pressure for carbon dioxide injection is 27.35 MPa. However, the current formation pressure is 49.88 MPa, which can achieve the miscibility of carbon dioxide and the formation of crude oil.

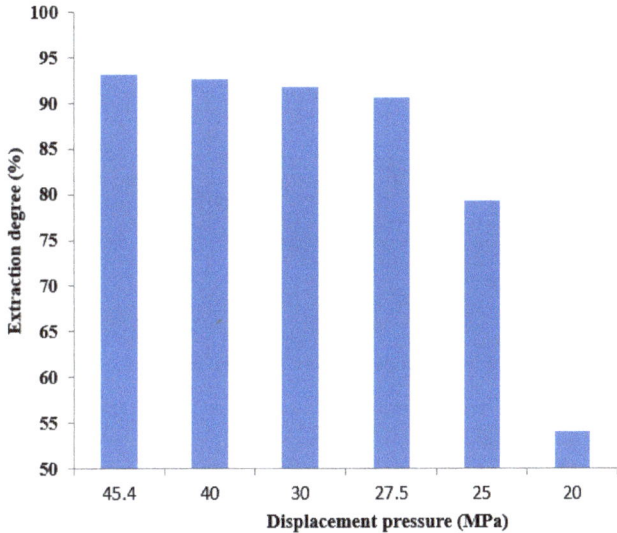

**Figure 3.** Carbon dioxide injection and recovery degree under different pressures.

## 3.4. Long Core Displacement Characteristics

In this paper, core samples with a length of 6 cm were spliced to prepare cores of the required length and filter paper was placed between two adjacent cores to effectively reduce the end effect [23]. In order to characterize the physical properties and water content of the Lunnan oil reservoir, a long core displacement experiment was conducted by testing the porosity and permeability data of the core. The displacement experiments of continuous gas flooding after water flooding, water and gas alternation, periodic gas injection, and carbon dioxide + hydrocarbon gas slug were completed. The core length of the Lunnan Oil Formation water flooding followed by different displacement methods of carbon dioxide injection is 100 cm, the diameter is 3.8 cm, the permeability is 45 mD, and the porosity is 17%. Taken from wild outcrops, they are prepared by bonding, pressing, and wire cutting. Figure 4 shows the long core displacement experimental core of carbon dioxide injected after water flooding in the Lunnan Oil Formation.

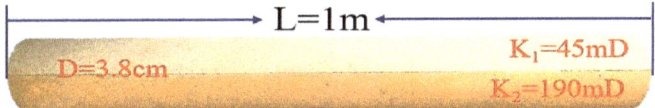

**Figure 4.** Carbon dioxide injection long core displacement experimental core after water flooding in the Lunnan Oil Formation.

The long core displacement experimental results were obtained through continuous carbon dioxide gas injection (See Figure 5). After water flooding in the Lunnan oil formation, as the injection rate of carbon dioxide pore volume increases, the displacement efficiency increases first rapidly and then slowly, with a turning point of 0.8 HCPV. The gas-oil ratio increased slowly at first and then rapidly, with the turning point being 0.6 HCPV.

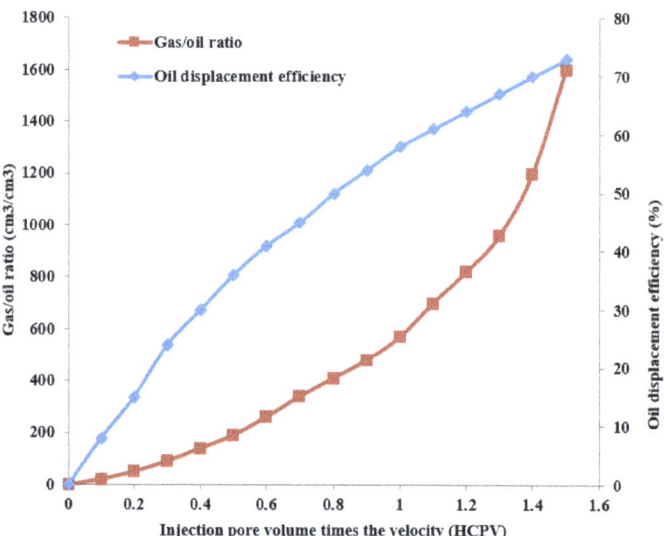

**Figure 5.** Experimental results of continuous carbon dioxide gas flooding after water flooding in the Lunnan Oil Formation.

A new and accurate calculation method has been proposed to predict the vapor-liquid equilibrium of $CO_2$ binary mixtures, which is not entirely dependent on experimental data [24]. Carbon capture and storage (CCS) has become a promising way to solve this challenge. It is estimated that in heavy industry, CCS can reduce emissions by 25–67% [25]. The saltwater layer has a $CO_2$ injection well and 200 GRs, which have different uncertain petrophysical characteristics. UML framework can be used to select RGRs and capture the whole uncertain domain. The calculation cost related to scheme testing, decision-making, and development planning of $CO_2$ storage sites under geological uncertainty is significantly reduced [26]. The carbon dioxide flooding in this paper can also achieve the purpose of carbon sequestration.

### 3.5. Oil and Gas Phase Permeability Characteristics

Based on the expansion experiment, thin tube experiment, and long core displacement experiment, the phase permeability characteristics of carbon dioxide injection gas flooding in the reservoir were obtained. Table 6 shows the characteristics of the reservoir fluid components. During the carbon dioxide flooding process, miscibility, mass transfer, extraction, expansion, and other functions greatly increase the fluidity of crude oil. The water flooding efficiency of the Lunnan oil formation is close to the limit. After conversion to carbon dioxide flooding development, on the one hand, the co-permeability interval is increased, and the fluidity of crude oil is increased; on the other hand, the residual oil saturation and irreducible water saturation are further reduced, releasing the movable space for oil, gas, and water, and the residual oil saturation is reduced from 30% to 12%.

Table 6. Reservoir fluid component characteristics.

| Components Name | Molecular Weight (g/mol) | Equation Coefficient | Equation Coefficient | Critical Temperature (K) | Critical Pressure (b) | Critical Volume | Critical Z Factor | Volume Offset Coefficient | Eccentric Factor |
|---|---|---|---|---|---|---|---|---|---|
| $N_2$ | 28.01 | 0.45 | 0.078 | 126.20 | 33.50 | 90.00 | 0.29 | −0.13 | 0.04 |
| $CO_2$ | 44.01 | 0.45 | 0.078 | 320.08 | 102.96 | 94.00 | 0.36 | 0 | 0.34 |
| $C_1$ | 16.04 | 0.45 | 0.078 | 277.00 | 212.70 | 98.00 | 0.91 | −0.022 | 0.0055 |
| $C_2$~$C_5$ | 53.34 | 0.45 | 0.078 | 397.86 | 39.43 | 237.45 | 0.28 | −0.065 | 0.18 |
| $C_6$~$C_{10}$ | 110.76 | 0.45 | 0.078 | 578.48 | 35.15 | 447.39 | 0.33 | 0.0 | 0.32 |
| $C_{11+}$ | 180.03 | 0.45 | 0.078 | 666.99 | 23.52 | 728.18 | 0.31 | 0.0001 | 0.58 |
| $C_{17+}$ | 365.77 | 0.45 | 0.078 | 859.91 | 12.43 | 1458.12 | 0.25 | 0.0002 | 1.09 |

Completely miscible $CO_2$ displacement can well displace various components of crude oil, including heavy components, and the difference of oil sample family components in different displacement stages is very small, and the miscible effect is excellent. After $CO_2$ flooding with large injection, the heavy components in the ultra-low permeability tight reservoir are well recovered, and the final recovery degree of heavy components is similar to that in the medium and low permeability reservoir, which shows that $CO_2$ flooding with large injection is a practical and effective displacement scheme to improve the recovery degree of ultra-low permeability tight reservoir. This makes $CO_2$ flooding very suitable for oil displacement and exploitation in the Lunnan Oilfield.

### 3.6. Water Solubility Characteristics

By obtaining aqueous solutions containing supersaturated carbon dioxide under different pressures and carrying out a single degassing experiment on saturated carbon dioxide water samples after the pressure stabilizes, the ability of water samples to dissolve carbon dioxide under different pressure conditions is measured. Figure 6 shows the variation pattern of sample solubility with pressure. The solubility of carbon dioxide in water increases with the increase of pressure, basically showing a linear relationship. Under the current formation pressure conditions, the solubility of carbon dioxide in formation water is 24 $m^3/m^3$.

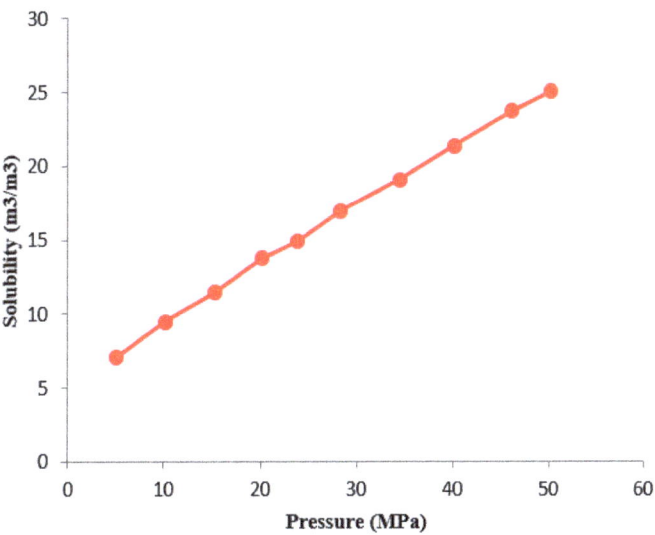

**Figure 6.** Solubility changes with pressure.

After carbon dioxide displacement, the wettability of reservoir rocks turns to be lipophilic. Most of the actual reservoirs are strongly hydrophilic, but after carbon dioxide displacement, the reservoirs can become weakly hydrophilic. After the injection of dissolved $CO_2$, the reservoir pressure and the reservoir pressure impact range continuously increase. After the injection is stopped, the reservoir pressure quickly recovers to its original value, ensuring the feasibility and safety of dissolved $CO_2$ storage. The pH value is mainly affected by pressure, water–rock reactions, and diffusion. Injecting dissolved $CO_2$ can preserve the reservoir for a long time, ensuring safety. Illite minerals undergo dissolution, while kaolinite minerals undergo precipitation. The total dissolution of minerals in the reservoir near the injection well is greater than the total precipitation, which increases the porosity of the reservoir. In the later stage of injection, the porosity of various positions in the reservoir no longer changes. The increase in reservoir porosity increases the permeability of the reservoir, alleviates the increase in reservoir pressure, and ensures the feasibility and safety of dissolved $CO_2$ storage.

## 4. Discussion

The heterogeneity of continental reservoirs in China is strong, and the fractures in low permeability reservoirs are developed. At the same time, the types of reservoirs in China are complex, and they are difficult to develop. The research of $CO_2$ oil recovery technology in China started late, and it was not until the end of the 1950s that China began to study the topic of $CO_2$ flooding. Then, in 1963, China first conducted research on enhancing oil recovery by $CO_2$ flooding in the Daqing Oilfield and then conducted the pilot tests of $CO_2$ injection in 1966 and 1969, respectively. In 1980, the pilot test of $CO_2$ miscible flooding was carried out in the Shaxia Reservoir of Pucheng Oilfield, and the purpose of dewatering flooding was achieved. $CO_2$ flooding is developing rapidly. In 2014, $CO_2$ flooding was carried out in the Jingbian Oilfield, a typical low-permeability reservoir. The final results show that $CO_2$ water-gas alternation can make the oil displacement efficiency reach 77.3%, and a stable production increase can be achieved by using $CO_2$ flooding technology.

In this paper, the study of $CO_2$ flooding, the determination of minimum miscible pressure in the Lunnan sandstone reservoir in the Tarim Basin, the study of reservoir fluid phase state after $CO_2$ injection, and the feasibility of $CO_2$-EOR technology is possible because injected carbon dioxide can reduce the viscosity of crude oil, increase the fluidity of crude oil, reduce the interfacial tension between oil and water, and expand the volume of crude oil, thus improving the oil recovery. The conclusion of this experiment is that with the increase in pressure, the injection capacity of $CO_2$ increases. The injection capacity of $CO_2$ is much greater than that of water injection. Compared with water injection, pure $CO_2$ flooding can improve oil recovery.

## 5. Conclusions

Injecting carbon dioxide under current formation pressure conditions, the impact of different amounts of carbon dioxide injection on crude oil extraction capacity, high-pressure physical properties of crude oil, and formation fluid phase characteristics varies. The following conclusions are ultimately drawn:

(1) The crude oil from the Lunnan Oilfield has low viscosity, low solidification point, low medium sulfur content, high wax content, and medium colloidal asphaltene. The carbon dioxide density measured under oil group conditions is 0.74 $g/cm^3$, which is similar to the density of crude oil. But its viscosity is 0.0681 mPa·s, which has good applicability for carbon dioxide flooding.

(2) As the amount of carbon dioxide injected increases, the saturation pressure of crude oil increases from 12.06 MPa to 39.68 MPa. The gas–oil ratio has increased from 42.52 $m^3/m^3$ to 594.68 $m^3/m^3$. The gas–oil ratio, expansion coefficient, and saturation pressure gradually increase. As the displacement pressure continues to decrease, the carbon dioxide displacement efficiency begins to slowly decrease and then rapidly decline. As the injection rate of carbon dioxide pore volume increases, the growth rate first increases and then slows down, reaching a turning point at 0.8 HCPV., The oil–gas ratio also exhibits a turning point occurring at 0.6 HCPV. The solubility of

carbon dioxide in the formation of water is determined to be 24 m$^3$/m$^3$. Beneficial for subsequent oil recovery and carbon storage.

**Author Contributions:** Conceptualization, Z.W.; Methodology, Q.F., Y.T., D.Z. and L.L.; Software, Z.W., Q.F. and Y.T.; Validation, Z.W., Y.T., D.Z. and L.L.; Formal analysis, Z.W., Q.F. and Y.T.; Investigation, Z.W., Y.T. and D.Z.; Resources, Z.W., Q.F. and D.Z.; Data curation, Z.W., Q.F. and L.L.; Writing – original draft, Z.W., Q.F., Y.T., D.Z. and L.L.; Writing – review & editing, Z.W., Q.F., Y.T., D.Z. and L.L.; Project administration, Q.F. All authors have read and agreed to the published version of the manuscript.

**Funding:** This work was supported by Major Science and Technology Project of CNPC (Grant No. ZD2019-183-007).

**Data Availability Statement:** The figures and tables used to support the findings of this study are included in the article.

**Acknowledgments:** The authors would like to show sincere thanks to those techniques who have contributed to this research.

**Conflicts of Interest:** Author Zangyuan Wu, Yongliang Tang and Daiyu Zhou were employed by the company PetroChina Tarim Oilfield Company. The remaining authors declare that the research was conducted in the absence of any commercial or financial relationships that could be construed as a potential conflict of interest. The authors declare that they have no known competing financial interests or personal relationships that could have appeared to influence the work reported in this paper. Author Zangyuan Wu, Yongliang Tang and Daiyu Zhou are employee of CNPC, who provided funding and teachnical support for the work. The funder had no role in the design of the study; in the collection, analysis, or interpretation of data, in the writing of the manuscript, or in the decision to publish the results.

# References

1. Al-Obaidi, D.A.; Al-Mudhafar, W.J.; Al-Jawad, M.S. Experimental evaluation of Carbon Dioxide-Assisted Gravity Drainage process (CO$_2$-AGD) to improve oil recovery in reservoirs with strong water drive. *Fuel* **2022**, *324*, 124409. [CrossRef]
2. Guerra, A.; McElligott, A.; Du, C.Y.; Marić, M.; Rey, A.D.; Servio, P. Dynamic viscosity of methane and carbon dioxide hydrate systems from pure water at high-pressure driving forces. *Chem. Eng. Sci.* **2022**, *252*, 117282. [CrossRef]
3. Liu, M.; Yang, X.; Wen, J.; Wang, H.; Feng, Y.; Lu, J.; Wang, J. Drivers of China's carbon dioxide emissions: Based on the combination model of structural decomposition analysis and input-output subsystem method. *Environ. Impact Assess. Rev.* **2023**, *100*, 107043. [CrossRef]
4. Kalam, S.; Olayiwola, T.; Al-Rubaii, M.M.; Amaechi, B.I.; Jamal, M.S.; Awotunde, A.A. Carbon dioxide sequestration in underground formations: Review of experimental, modeling, and field studies. *J. Pet. Explor. Prod.* **2021**, *11*, 303–325. [CrossRef]
5. Lv, Z.; Qiao, K.; Chu, F.; Yang, L.; Du, X. Experimental study of divalent metal ion effects on ammonia escape and carbon dioxide desorption in regeneration process of ammonia decarbonization. *Chem. Eng. J.* **2022**, *435*, 134841. [CrossRef]
6. Sun, X.; Cai, J.; Li, X.; Zheng, W.; Wang, T.; Zhang, Y. Experimental investigation of a novel method for heavy oil recovery using supercritical multithermal fluid flooding. *Appl. Therm. Eng.* **2021**, *185*, 116330. [CrossRef]
7. Chuah, L.F.; Bokhari, A.; Asif, S.; Klemeš, J.J.; Dailin, D.J.; El Enshasy, H.; Yusof, A.H.M. A review of performance and emission characteristic of engine diesel fuelled by biodiesel. *Chem. Eng. Trans.* **2022**, *94*, 1099–1104.
8. Varfolomeev, M.A.; Yuan, C.; Bolotov, A.V.; Minkhanov, I.F.; Mehrabi-Kalajahi, S.; Saifullin, E.R.; Shaihutdinov, D.K. Effect of copper stearate as catalysts on the performance of in-situ combustion process for heavy oil recovery and upgrading. *J. Pet. Sci. Eng.* **2021**, *207*, 109125. [CrossRef]
9. Pal, N.; Zhang, X.; Ali, M.; Mandal, A.; Hoteit, H. Carbon dioxide thickening: A review of technological aspects, advances and challenges for oilfield application. *Fuel* **2022**, *315*, 122947. [CrossRef]
10. Ansari, K.B.; Gaikar, V.G.; Trinh, Q.T.; Khan, M.S.; Banerjee, A.; Kanchan, D.R.; Danish, M. Carbon dioxide capture over amine functionalized styrene divinylbenzene copolymer: An experimental batch and continuous studies. *J. Environ. Chem. Eng.* **2022**, *10*, 106910. [CrossRef]
11. Novak Mavar, K.; Gaurina-Međimurec, N.; Hrnčević, L. Significance of enhanced oil recovery in carbon dioxide emission reduction. *Sustainability* **2021**, *13*, 1800. [CrossRef]
12. Yan, H.; Zhang, J.; Li, B.; Zhu, C. Crack propagation patterns and factors controlling complex crack network formation in coal bodies during tri-axial supercritical carbon dioxide fracturing. *Fuel* **2021**, *286*, 119381. [CrossRef]
13. Zhou, M.; Feng, J.; Jiang, T.; Liu, J. Preliminary study on tertiary oil recovery of high temperature and high salinity reservoirs in Tarim Oilfield. *Xinjiang Pet. Geol.* **2010**, *2*, 59–62.

14. Wen, B.; Shi, Z.; Jessen, K.; Hesse, M.A.; Tsotsis, T.T. Convective carbon dioxide dissolution in a closed porous medium at high-pressure real-gas conditions. *Adv. Water Resour.* **2021**, *154*, 103950. [CrossRef]
15. Bagheri, H.; Hashemipour, H.; Rahimpour, E.; Rahimpour, M.R. Particle size design of acetaminophen using supercritical carbon dioxide to improve drug delivery: Experimental and modeling. *J. Environ. Chem. Eng.* **2021**, *9*, 106384. [CrossRef]
16. Wang, Y.; Dong, Y.; Zhang, L.; Chu, G.; Zou, H.; Sun, B.; Zeng, X. Carbon dioxide capture by non-aqueous blend in rotating packed bed reactor: Absorption and desorption investigation. *Sep. Purif. Technol.* **2021**, *269*, 118714. [CrossRef]
17. Ringrose, P.S.; Furre, A.K.; Gilfillan, S.M.; Krevor, S.; Landrø, M.; Leslie, R.; Zahid, A. Storage of carbon dioxide in saline aquifers: Physicochemical processes, key constraints, and scale-up potential. *Annu. Rev. Chem. Biomol. Eng.* **2021**, *12*, 471–494. [CrossRef]
18. Yang, L.; Wang, Y.; Lian, Y.; Dong, X.; Liu, J.; Liu, Y.; Wu, Z. Rational planning strategies of urban structure, metro, and car use for reducing transport carbon dioxide emissions in developing cities. *Environ. Dev. Sustain.* **2023**, *25*, 6987–7010. [CrossRef]
19. Mao, F.; Li, Z.; Zhang, K. A comparison of carbon dioxide emissions between battery electric buses and conventional diesel buses. *Sustainability* **2021**, *13*, 5170. [CrossRef]
20. Vaz, S., Jr.; de Souza, A.P.R.; Baeta, B.E.L. Technologies for carbon dioxide capture: A review applied to energy sectors. *Clean. Eng. Technol.* **2022**, *8*, 100456. [CrossRef]
21. Lyu, Q.; Tan, J.; Li, L.; Ju, Y.; Busch, A.; Wood, D.A.; Hu, R. The role of supercritical carbon dioxide for recovery of shale gas and sequestration in gas shale reservoirs. *Energy Environ. Sci.* **2021**, *14*, 4203–4227. [CrossRef]
22. Jia, W.; Jia, X.; Wu, L.; Guo, Y.; Yang, T.; Wang, E.; Xiao, P. Research on regional differences of the impact of clean energy development on carbon dioxide emission and economic growth. *Humanit. Soc. Sci. Commun.* **2022**, *9*, 25. [CrossRef]
23. Voumik, L.C.; Ridwan, M.; Rahman, M.H.; Raihan, A. An investigation into the primary causes of carbon dioxide releases in Kenya: Does renewable energy matter to reduce carbon emission? *Renew. Energy Focus* **2023**, *47*, 100491. [CrossRef]
24. Motie, M.; Bemani, A.; Soltanmohammadi, R. On The Estimation of Phase Behavior Of $CO_2$-Based Binary Systems Using ANFIS Optimized By GA Algorithm. In *Fifth $CO_2$ Geological Storage Workshop*; European Association of Geoscientists & Engineers: Utrecht, The Netherlands, 2018; Volume 2018, pp. 1–5.
25. Mahjour, S.K.; Faroughi, S.A. Risks and uncertainties in carbon capture, transport, and storage projects: A comprehensive review. *Gas Sci. Eng.* **2023**, *119*, 205117. [CrossRef]
26. Mahjour, S.K.; Faroughi, S.A. Selecting representative geological realizations to model subsurface $CO_2$ storage under uncertainty. *Int. J. Greenh. Gas Control.* **2023**, *127*, 103920. [CrossRef]

**Disclaimer/Publisher's Note:** The statements, opinions and data contained in all publications are solely those of the individual author(s) and contributor(s) and not of MDPI and/or the editor(s). MDPI and/or the editor(s) disclaim responsibility for any injury to people or property resulting from any ideas, methods, instructions or products referred to in the content.

*Article*

# Simulations of $CO_2$ Dissolution in Porous Media Using the Volume-of-Fluid Method

Mohammad Hossein Golestan * and Carl Fredrik Berg

PoreLab, Department of Geoscience and Petroleum, Norwegian University of Science and Technology, NTNU, 7031 Trondheim, Norway; carl.f.berg@ntnu.no
* Correspondence: mohammad.h.golestan@ntnu.no

**Abstract:** Traditional investigations of fluid flow in porous media often rely on a continuum approach, but this method has limitations as it does not account for microscale details. However, recent progress in imaging technology allows us to visualize structures within the porous medium directly. This capability provides a means to confirm and validate continuum relationships. In this study, we present a detailed analysis of the dissolution trapping dynamics that take place when supercritical $CO_2$ ($scCO_2$) is injected into a heterogeneous porous medium saturated with brine. We present simulations based on the volume-of-fluid (VOF) method to model the combined behavior of two-phase fluid flow and mass transfer at the pore scale. These simulations are designed to capture the dynamic dissolution of $scCO_2$ in a brine solution. Based on our simulation results, we have revised the Sherwood correlations: We expanded the correlation between Sherwood and Peclet numbers, revealing how the mobility ratio affects the equation. The expanded correlation gave improved correlations built on the underlying displacement patterns at different mobility ratios. Further, we analyzed the relationship between the Sherwood number, which is based on the Reynolds number, and the Schmidt number. Our regression on free parameters yielded constants similar to those previously reported. Our mass transfer model was compared to experimental models in the literature, showing good agreement for interfacial mass transfer of $CO_2$ into water. The results of this study provide new perspectives on the application of non-dimensional numbers in large-scale (field-scale) applications, with implications for continuum scale modeling, e.g., in the field of geological storage of $CO_2$ in saline aquifers.

**Keywords:** porous media; dissolution; $CO_2$ geological storage; pore-scale simulations; dissolution trapping; Sherwood correlation

**Citation:** Golestan, M.H.; Berg, C.F. Simulations of $CO_2$ Dissolution in Porous Media Using the Volume-of-Fluid Method. *Energies* **2024**, *17*, 629. https://doi.org/10.3390/en17030629

Academic Editors: Yongbin Wu, Weifeng Lv, Tiyao Zhou and Xiaoqing Lu

Received: 12 December 2023
Revised: 15 January 2024
Accepted: 24 January 2024
Published: 28 January 2024

**Copyright:** © 2024 by the authors. Licensee MDPI, Basel, Switzerland. This article is an open access article distributed under the terms and conditions of the Creative Commons Attribution (CC BY) license (https://creativecommons.org/licenses/by/4.0/).

## 1. Introduction

Single and multiphase flow in porous media is the underlying foundation for numerous industrial and natural processes. Researchers have addressed porous media flow at different spatial scales, including the pore scale and the continuum scale. Traditionally, the continuum scale was the dominant experimental approach due to limitations in the measurements inside individual pores. However, recent advancements in imaging technology have enabled the scientific community to capture underlying processes within the void space of the porous medium. Additionally, information from pore scale imaging can be used for the validation of numerical models [1,2]. At the continuum scale, we only retain effective properties of the pore scale details [3]. By construction, effective properties do not capture the full pore scale physics. Therefore, they might lose out on details that are significant for continuum scale effects. One approach for retaining more pore-scale physics is multi-scale modeling, which incorporates pore-scale results into the continuum-scale simulations [4].

Both single and multiphase flow might occur coincidentally with the transfer of species. In a single-phase flow situation, the transport of a dissolved species depends

One of the challenges in research concerning porous media is the limited ability to directly observe the processes occurring within it. Previous investigations primarily relied on scaled-up mass transfer methods [43,44].

The rate of interphase mass transfer is commonly expressed in terms of the Sherwood number ($Sh$), which is a function of fluid transport and porous media properties. The interfacial mass transfer coefficient, $k_f$, can be expressed as a function of the non-dimensional Sherwood ($Sh$) number defined as [33]:

$$Sh = \frac{k_f d_m}{D_m} \qquad (2)$$

where $D_m$ is the molecular diffusion coefficient [m$^2$/s], while $d_m$ is the mean grain diameter of the porous medium [m]. The Sherwood number is derived from continuum scale experiments while assuming a uniform distribution of fluids in the porous medium. However, pore-scale structure properties such as mean grain diameter, pore size distribution, and wettability have indisputable effects on fluid displacement. Therefore, pore-scale investigations can provide an in-depth perspective on the application of Sherwood number in large-scale (field-scale) applications.

Note that conventionally, due to the challenges in experimental detection of the fluid interface, the mass transfer coefficient and interfacial area are considered as lumped parameters within the analysis. Consequently, a combined mass transfer coefficient ($ka$), derived from the product of the mass transfer coefficient ($k$) and the specific interfacial area ($a$), was used. The significant limitation of relying on a combined mass transfer coefficient is the reliance on porous media characteristics, particularly particle size, as the specific interfacial area cannot be excluded. Consequently, in prior research, the mass transfer model was formulated using a modified Sherwood number instead of the conventional Sherwood number:

$$Sh' = \frac{ka d_m^2}{D_m} \qquad (3)$$

where, as before, $d_m$ is an effective particle size and $D_m$ is the diffusion coefficient of the trapped phase to the flowing phase. In this article, we will differentiate these terms individually by capturing the interfacial area, and thereby provide a more accurate description of the process. The capability of pore-scale investigations in incorporating pore-scale properties can refine the calculated Sherwood number for further use in continuum-based simulations.

## 3. Mathematical Model

In this research, we tried to clarify the correlation between the Sherwood number and interfacial mass transfer from a bottom-top perspective. We applied direct pore-scale modeling on three different porous media under different flow conditions. The porous media were generated by Particula [45], which generates three-dimensional packings of particles with predetermined size distributions, simulating both spherical and non-spherical particles with regular and irregular shapes. Then, the surface mesh of each packing (stl files) was converted to a binary Cartesian grid. These binary grids were imported into the voxelToFoam module of the poreFoam solver [46] to generate the volume mesh required by the computational fluid dynamics simulator.

*3.1. Two-Phase Flow Modeling*

The handling of two-phase, multi-component mass transfer at the pore scale was carried out through two sets of equations—the volume-of-fluid (VOF) formulation for the flow of the multi-phase flow, and the concentration equation with a mass transfer at the fluid–fluid interface (and its extended formulation at the solid boundary). The Navier–Stokes equations were solved on a regular grid using the second-order projection method to simulate multi-phase flows. The pressure was found from the divergence of the temporary solution, and the velocity was corrected by adding the gradient of the pressure. The marker function that identifies different fluids was updated differently in various multi-phase simulation methods.

The VOF method uses unified equations for momentum and continuity to represent the movement of fluid phases across the computational domain, linked with the volume fraction $\alpha_i$ of phase $i$. The unified equations are given as:

$$\frac{\nabla(\rho u)}{\partial t} + \nabla \cdot (\rho u \times u) = \mu \nabla^2 u - \nabla p + \rho g + F_{st} \quad (4)$$

$$\nabla \cdot u = 0 \quad (5)$$

where u represents the velocity field, $p$ signifies the pressure field, $\rho$ stands for the fluid density, $\mu$ represents fluid viscosity, and $F_{st}$ denotes the surface tension force, which can be expressed as:

$$F_{st} = \sigma k n \delta \quad (6)$$

Here, $\sigma$ represents interfacial tension, k denotes interface curvature (computed as k = $-\nabla \cdot$n), where n is the interface normal determined by n = $\frac{\nabla \alpha}{|\nabla \alpha|}$, and $\delta$ is a Dirac delta function positioned on the interface. To incorporate the surface tension effect, it must be converted into a volumetric force. This transformed force can then be applied as a term similar to a body force within the momentum equation. This conversion process is accomplished through the continuum-surface-force (CSF) method developed by Brackbill et al. [47]:

$$F_{st,CSF} = -\sigma \nabla \cdot \left( \frac{\nabla \alpha}{\|\nabla \alpha\|} \right) \nabla \alpha \quad (7)$$

The CSF model can cause inaccuracies when calculating surface tension forces at low capillary numbers (Ca < 0.01) due to unwanted artificial currents. [47–49]. Raeni et al. [46] developed a model called the sharp surface force (SSF) to reduce the impact of artificial currents. The SSF model uses smoothed and sharpened indicator functions to calculate the curvature and the Dirac delta function ($\delta$).

The evolution of the indicator function $\alpha_i$ for each phase, where $i$ = 1 or 2, is governed by an advection equation formulated as

$$\frac{\partial \alpha_i}{\partial t} + \nabla \cdot (\alpha_i u) - \nabla \cdot (\alpha_i (1 - \alpha_i) u_r) = 0 \quad (8)$$

where $u_r$ represents the relative velocity existing between the two phases [50]. The third component in this equation is an artificial compression factor utilized to enhance the sharpness of the interface, thereby improving the precision of the interface representation. Importantly, this extra term exclusively operates within the interface area and does not impact the solution beyond this zone. The indicator function ($\alpha$) is determined at cell centers and extrapolated linearly to solid boundaries. To calculate normal vectors at the interface, the indicator function is smoothed by interpolating values between cell centers ($c$) and faces ($f$):

$$\alpha_s = C \left\langle \langle \alpha \rangle_{c \to f} \right\rangle_{c \to f} + (1 - C) \alpha \quad (9)$$

The coefficient $C$ is set to a value of 0.5, as specified by [46]. This modified indicator function, denoted as $\alpha_s$, is employed to determine the normal direction at the centers of the grid cells. When dealing with solid boundaries, adjustments are made to the direction of the normal vector at the interface ($n_{sb}$) to accommodate solid adhesion. This correction is carried out as follows:

$$n_{sb} = n_w \cos \theta + s_w \sin \theta \quad (10)$$

Here, $\theta$ represents the contact angle of the fluid-fluid interface with the solid boundary, while $n_w$ and $s_w$ denote unit vectors normal and tangential to the solid boundary, respectively.

Ultimately, the smoothed surface tension at the central points of the faces is computed as:

$$F_{st,SSF} = \sigma \nabla \cdot \left( \frac{\nabla \alpha_s}{\|\nabla \alpha_s\|} \right) \nabla \alpha_{sh} \tag{11}$$

Here, $\alpha_{sh}$ represents a refined indicator function as defined by Raeini et al. [46] as:

$$\alpha_{sh} = \frac{1}{1 - C_{sh}} \left[ \min\left( \max\left( \alpha, \frac{C_{sh}}{2} \right), 1 - \frac{C_{sh}}{2} \right) - \frac{C_{sh}}{2} \right] \tag{12}$$

In our simulations, we set $C_{sh}$ to a value of 0.5, which serves as a sharpening coefficient. When $C_{sh}$ is equal to 0, it returns $\alpha$ (as in the CSF model), and when it is set to 1, it results in an extremely sharp representation of the interface, which is unstable.

### 3.2. Mass Transfer Modeling

In this section, we outline how chemical mass transfer is incorporated into the VOF framework. We describe the conservation equation for the local concentration $C_{i,k}$ of the chemical species $k$ in each phase $i$:

$$\frac{\partial C_{i,k}}{\partial t} + \nabla \cdot (u_i C_{i,k}) = \nabla \cdot (J_{i,k}) + R_{i,k} \tag{13}$$

Here, $u_i$ represents the velocity of phase $i$ locally. $J_{i,k}$ is the diffusion mass flux, determined by Fick's law and represented as $J_{i,k} = -D_{i,k} \nabla C_{i,k}$, where $D_{i,k}$ is the molecular diffusion coefficient of component $k$ in phase i. In this study, we disregard the reaction term, $R_{i,k}$. The continuity of fluxes and chemical potentials at the fluid–fluid interface is expressed as adherence to the jump condition, as:

$$\begin{aligned} (C_{i,k}(u_i - u_I) - D_{i,k} \nabla C_{i,k}) \cdot \mathbf{n} = 0 \\ C_{2,k} = H_k C_{1,k} \end{aligned} \tag{14}$$

Here, $u_I$ is the interface velocity and $H_k$ is Henry's constant. [51]. When Henry's law conditions are not met, mass transfer occurs between fluid phases to achieve equilibrium. To simplify, we introduce a global variable for the concentration of species $k$ in a two-phase flow as:

$$C_k = \alpha C_{1,k} + (1 - \alpha C_{2,k}) \tag{15}$$

In line with the continuous species transfer (CST) model as introduced by Haroun et al. [38] and Deising et al. [52], the concentration equation within the VOF framework can be expressed as a single-field equation as follows:

$$\begin{aligned} \frac{\partial C_k}{\partial t} + \nabla \cdot (u C_k) = \nabla \cdot (D_h \nabla C_k + \Phi_k) + R_k \\ \Phi_k = -D_h \frac{C_k(1 - H_k)}{\alpha + H_k(1 - \alpha)} \nabla \alpha \\ R_k = \alpha R_{1,k} + (1 - \alpha) R_{2,k} \end{aligned} \tag{16}$$

where $D_h$ is the harmonic mean diffusion coefficient as

$$D_h = \frac{D_1 D_{2,k}}{\alpha D_{2,k} + (1 - \alpha) D_{1,k}} \tag{17}$$

Deising et al. [52] reported that the harmonic mean diffusivity is a more suitable option than the arithmetic mean. In Equation (15), the CST term $\Phi_k$ arises from concentration variations that cause an extra flux at the fluid–fluid interface. This term characterizes the thermodynamic equilibrium at the fluid–fluid interface. When Henry's coefficient equals one ($C_{1,k} = C_{2,k}$), the CST term disappears, resulting in Equation (15) transforming into the standard advection-diffusion equation without the impact of solubility. Maes and Soulaine [26] introduced an enhanced continuous species transfer (C-CST) model by adjusting the advection term within the CST framework. This modification is aimed at improving the accuracy of predictions in situations characterized by high Péclet numbers, where advective mass transfer prevails over diffusion. The C-CST representation of the concentration equation is as follows:

$$\frac{\partial C_k}{\partial t} + \nabla \cdot (uC_k) + \nabla \cdot \left( \frac{C_j - H_j C_j}{H_j + \alpha(1 - H_j)} \alpha(1-\alpha) u_r \right) = \nabla \cdot (D_h \nabla C_k + \Phi_k) + R_k \quad (18)$$

Given the demonstrated superior robustness of C-CST compared to the CST framework, we have incorporated it into the VOF framework for simulating species mass transport.

## 4. Simulations

We utilized the governing equations within the OpenFOAM toolbox to simulate coupled two-phase flow and mass transport [53]. We used interFoam solver in OpenFOAM, equipped with the SSF interfacial tension model, to model two-phase fluid flow. Equation (15) for chemical species conservation was integrated into the GeoChemFoam module by incorporating the hydrodynamics solver with the multicomponent mass transfer model [26].

In our simulations, we generated three distinct sets of densely random-packed grains. These packs consisted of polydisperse grains with no friction. One of these grain packs was composed of non-spherical grains, aiming to represent a real rock sample, while the other two packs were made up of spherical grains. Each pack contained 1000 grains and was enclosed within a cubic container. The grains had a density of 2.65 g/cm$^3$, which corresponds to quartz minerals, and they exhibited no friction between each other. Figure 1 illustrates the structure of these generated rock samples, while Table 1 provides the relevant physical properties of each sample.

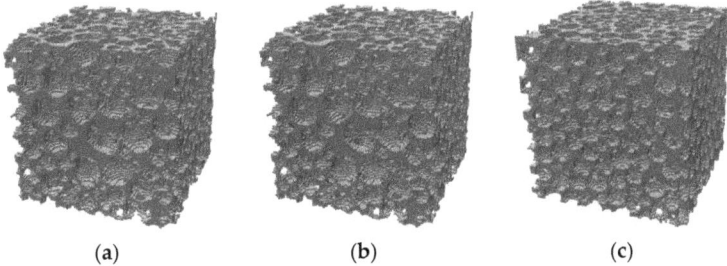

**Figure 1.** Pore geometry of the rock samples used in this study: (**a**) Realistic rock sample; (**b**) monodisperse sphere pack; (**c**) polydisperse sphere pack.

**Table 1.** Physical properties of the three rocks.

| Rock Sample | Porosity | Permeability (m$^2$) | Voxel Size (μm) | Side Length (m) | Tortuosity | Grain Surface Area (m$^2$) |
|---|---|---|---|---|---|---|
| Realistic rock sample | 0.29 | $2.82 \times 10^{-12}$ | 3.0 | 900 | 1.22 | $3.6474 \times 10^{-5}$ |
| Monodisperse sphere pack | 0.34 | $5.44 \times 10^{-12}$ | 3.0 | 900 | 1.19 | $3.5774 \times 10^{-5}$ |
| Polydisperse sphere pack | 0.34 | $4.11 \times 10^{-12}$ | 3.0 | 900 | 1.20 | $3.9047 \times 10^{-5}$ |

In this study, we conducted a sensitivity analysis to examine how fluid properties and flow conditions affect the interfacial mass transfer coefficient. In all our simulations, initially the pore geometries were saturated with water and subsequently injected supercritical carbon dioxide (scCO$_2$) into the domain to displace the water. We assumed that the pore geometries exhibited a water-wet characteristic with a contact angle of $\theta = 45$ degrees. To achieve our target capillary numbers defined as:

$$Ca = \frac{U\mu_{co2}}{\sigma} \tag{19}$$

we adjusted the inlet velocity of the non-wetting scCO$_2$ (displacing) phase defined as $U$ in Equation (19) accordingly.

Additionally, we varied the diffusivity coefficients of scCO$_2$ in water to attain different Peclet numbers, defined as the ratio of advective transport rate to diffusive transport rate:

$$Pe = \frac{Ud_m}{D_m} \tag{20}$$

where, as before, $d_m$ is an effective grain diameter of the porous medium [m] and $D_m$ is the molecular diffusion coefficient [m$^2$/s]. The Henry coefficient was held constant equal to 0.5 for all cases. We continued the simulations for several pore volumes beyond the breakthrough of the displacing phase. Figures 2 and 3 illustrate the evolution of saturation and concentration of scCO$_2$ for each of the rock samples.

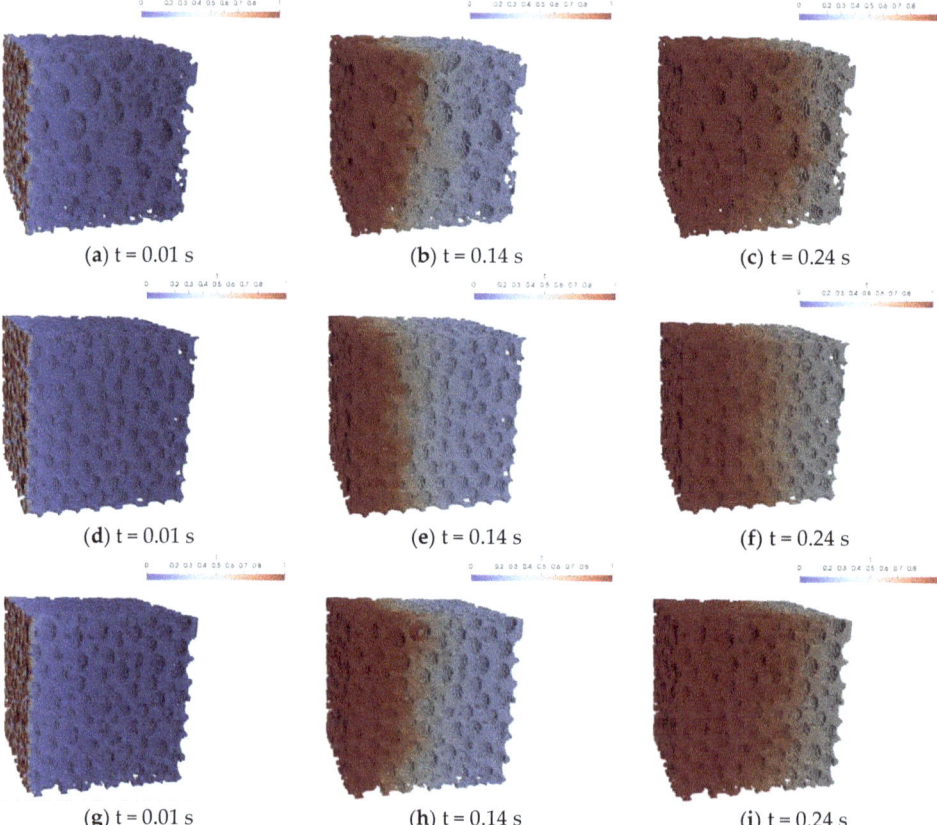

**Figure 2.** Changes in the concentration of dissolved CO$_2$ (red) in the brine phase (blue) over time. The porous medium is assumed to be water-wet with a contact angle of $\theta = 45°$. (**a**–**c**) Realistic rock sample; (**d**–**f**) monodisperse sphere pack; (**g**–**i**) polydisperse sphere pack.

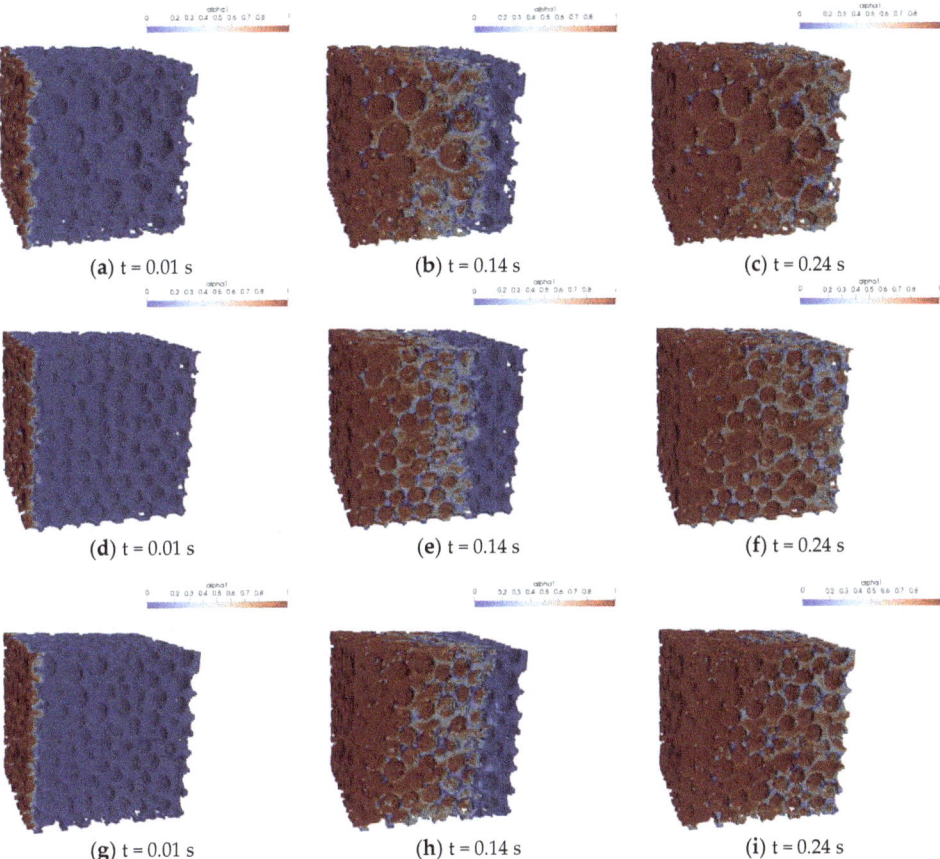

**Figure 3.** Changes in the distribution of the scCO$_2$ phase (red) and the brine phase (blue) within the pore space over time. (**a**–**c**) Realistic rock sample; (**d**–**f**) monodisperse sphere pack; (**g**–**i**) polydisperse sphere pack.

To construct the computational grid, we initially formed a background hexahedral mesh covering the entire domain. Then, we eliminated all grid cells located within the solid portion and produced a boundary-fitted mesh by iterative refining and aligning the grid cells with the solid region. The total number of computational grid cells for the realistic rock sample, monodisperse sphere pack, and polydisperse sphere pack after meshing were 305,462, 362,100, and 339,869, respectively. The governing equations for fluid flow and mass transport were discretized using a finite-volume method. These equations were then solved within the pore space of a porous medium to investigate the dissolution of scCO$_2$ into brine over time.

The pore space was initially saturated with brine and then scCO$_2$ was injected at a constant flow rate from the left boundary (Figure 3). We maintained a constant pressure condition at the outlet boundary. Additionally, we enforced a no-slip boundary condition at all other boundaries, including the interfaces between the fluid and solid phases. To accommodate the water-wet nature, we specified a receding contact angle $\theta$ of 45 degrees. We set the interfacial tension ($\sigma$) at a value of 34 mN/m. Initially, we assumed that the brine phase did not contain any dissolved CO$_2$ (dsCO$_2$). As we injected scCO$_2$ from the left boundary, it had the opportunity to dissolve into the brine.

In the current study, we did not consider mineral dissolution or precipitation as a factor. This process typically occurs at a very gradual pace in sandstone reservoirs and becomes significant only over geological timescales [54]. Nevertheless, it is worth noting that certain prior studies have explored the concept of mineral dissolution and its effect on $CO_2$ trapping, particularly during the injection phase. These findings are more relevant to carbonate reservoirs and have been discussed in studies such as those by Seyyedi et al. [55].

In this study, the influence of the composition of the brine in the aquifer on the dissolution of $CO_2$ was not considered. The existence of ions in the brine influences the dissolution rate of the $CO_2$ in the brine as studied by Lin et al. [40].

## 5. Results and Discussion

We conducted simulations focusing on $CO_2$ drainage and dissolution to investigate the underlying mechanisms of $CO_2$ dissolution and mass transfer within a porous medium saturated with brine. The procedure involved injecting supercritical $CO_2$ into the medium from the left side, with the porous medium initially saturated with brine that was free of dissolved $CO_2$ (referred to as fresh brine or $dsCO_2$-free brine, where $dsCO_2$ indicates the dissolved $CO_2$ in the brine phase). Through these simulations, we could observe and track the dissolution of $CO_2$ into the water phase when injecting $scCO_2$ into the porous medium.

Mobility is a factor that characterizes the flow of a particular phase within a multiphase system. The phase with the greatest mobility will be more mobile and dominate the flow. Mobility is defined as the ratio between the relative permeability of a phase and its viscosity. The simulations were conducted in two different mobility ratios, $M$, defined as the mobility of the displacing fluid ($scCO_2$) behind the front, $\lambda_{scCo2}$, divided by the mobility of the displaced fluid (water) ahead of the front, $\lambda_w$:

$$M = \frac{\lambda_{scCo2}}{\lambda_w} \qquad (21)$$

The mobility ratios $M = 1$ and $M = 0.1$ were selected to investigate the effect of fluid dynamics in our study.

### 5.1. Analysis of the Simulation Resutlts

The changes in the interfacial area for all our simulations are shown in Figure 4. As we see, the interfacial area for the simulations conducted in $M = 1$ was higher than the ones in $M = 0.1$. The higher mobility ratio led to more instability of the fluid interfaces, which again led to more viscous fingering and thereby more fluid–fluid interfacial area in the porous medium. Figure 5 displays the mass flux throughout the simulation for each rock sample. The mobility ratio had an impact on the mass flux for all the rock samples, with higher mobility ratios leading to increased mass flux. The mentioned viscous fingering provided the $scCO_2$ front with a larger fluid–fluid interfacial area, and thereby a greater potential for mass transfer. To account for the impact of the interfacial area on our mass flux results, we depicted the mass flux per interfacial area in Figure 6. We note that the high mobility ratio cases not only had more interfacial area, but also higher mass flux per interfacial area. As mentioned, at high mobility ratio $scCO_2$ will channel into the medium. We expect the higher mass flux per interfacial area was due to less dissolved $CO_2$ around the fingers, leading to a higher mass flux.

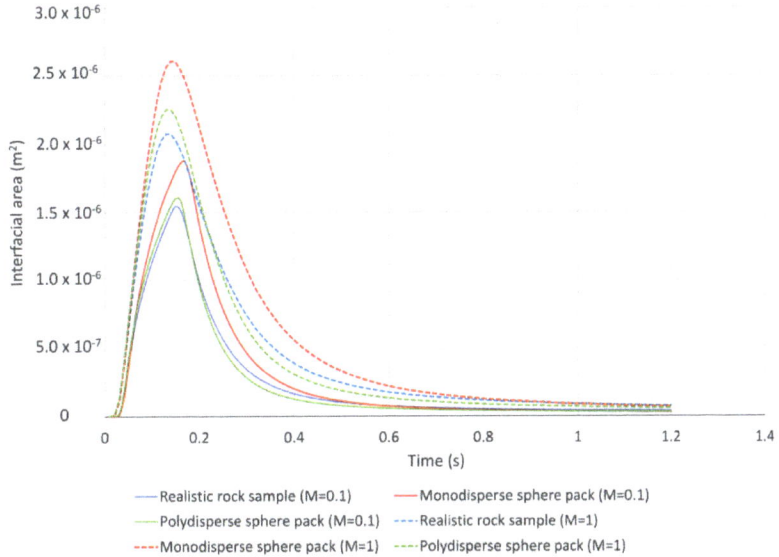

**Figure 4.** Changes in interfacial area between the scCO$_2$ and water, during the water drainage by scCO$_2$ in different rock samples.

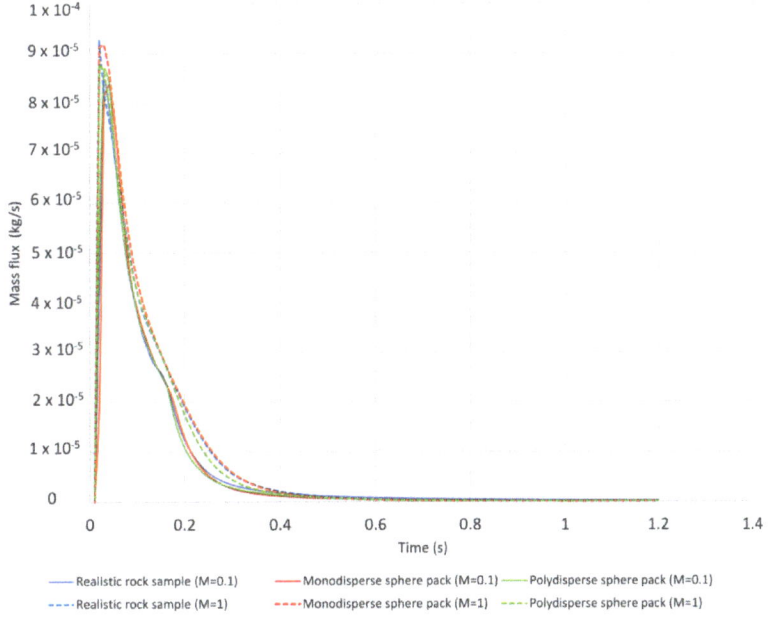

**Figure 5.** Evolution of CO$_2$ mass flux from the scCO$_2$ into the water during the water drainage by scCO$_2$ in different rock samples.

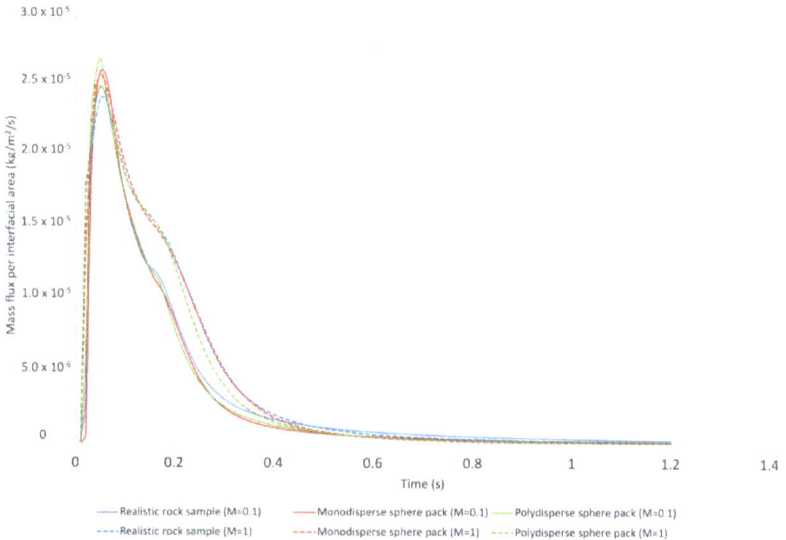

**Figure 6.** Evolution of CO$_2$ mass flux from the scCO$_2$ into the water (as dsCO$_2$) per interfacial area, during the water drainage by scCO$_2$ in different rock samples.

## 5.2. Estimation of the Mass Transfer Coefficien

In all our simulation cases, there was a strong correlation between mass flux per interfacial area and concentration differences for all rock samples, as depicted in Figure 7. Figure 8 shows how the total mass flux per interfacial area changed as the concentration difference ($HC_{co2} - C_w$) varied in all our simulations. For simplification, we investigated the simulations for the realistic rock sample only, which could be generalized to the other simulations. Figure 9 shows the changes in the total mass flux per interfacial area as a function of concentration difference ($HC_{co2} - C_w$), during the water drainage by scCO$_2$ in the realistic rock sample. We notice that, with an equivalent concentration difference, the total mass flux per interfacial area was greater when $M = 1$ in comparison to $M = 0.1$. In non-equilibrium upscaling models, the interfacial mass transfer coefficient is obtained as the slope of total mass flux per interfacial area versus the concentration difference $HC_{co2} - C_w$ [56]:

$$\Phi^T = k(HC_{co_2} - C_w) \tag{22}$$

where $k$ is the mass exchange coefficient (in m/s) and $\Phi^T$ is the interfacial mass flux.

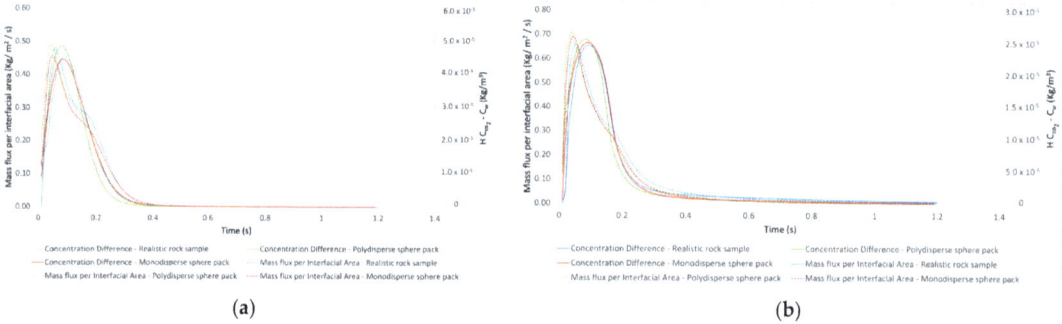

**Figure 7.** Total mass flux per interfacial area and concentration difference (HC$_{co2}$-C$_w$) during the displacement mechanism in mobility ratio = 1 (**a**) and mobility ratio = 0.1 (**b**).

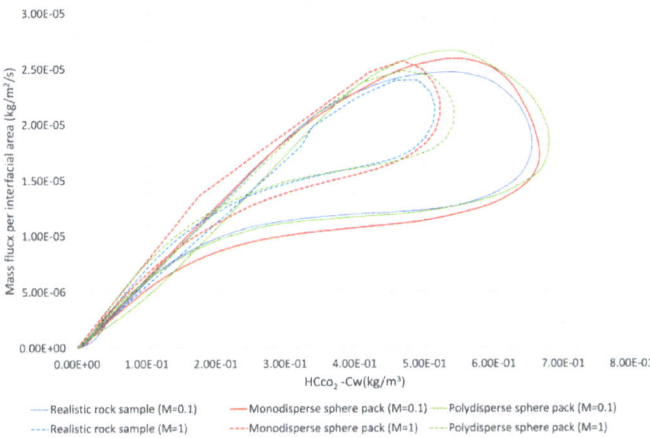

**Figure 8.** Changes in the total mass flux per interfacial area, concerning the concentration difference ($HC_{CO_2}$-$C_w$), during the water drainage by $scCO_2$ in different rock samples.

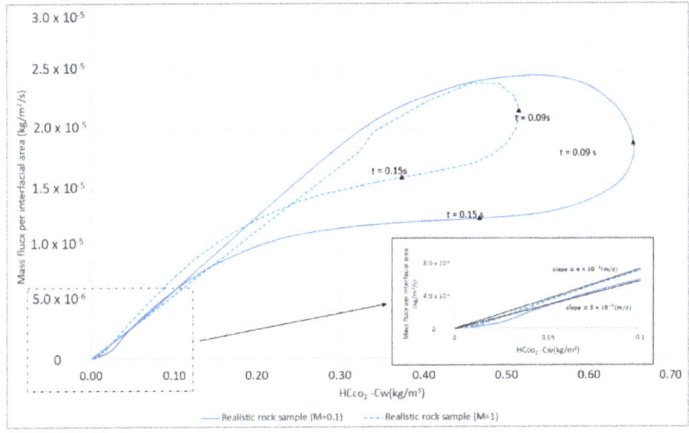

**Figure 9.** Changes in the total mass flux per interfacial area, concerning the concentration difference ($HC_{CO_2}$-$C_w$), during the water drainage by $scCO_2$ in the realistic rock sample. The triangles on the graphs specify three time steps in the simulations. The inset shows how the interfacial mass transfer was calculated for each of the simulations.

Following the breakthrough of the $scCO_2$, it is evident in Figure 8 that the flux can be approximated as linear relative to the concentration difference. The interfacial mass transfer coefficient ($k$) can subsequently be determined by evaluating the gradient of this function. In Figure 9, considering two different mobility ratios, $M = 1$ and $M = 0.1$, we derived values of $k = 4 \times 10^{-5}$ m/s and $k = 3 \times 10^{-5}$ m/s, respectively, for the realistic rock sample. The calculation of the interfacial mass transfer coefficient ($k$) for the remaining rock samples followed the same procedure.

### 5.3. Development of Sherwood Correlation with Peclet

To explore the relationship between mass exchange coefficients, mobility ratios, and Peclet numbers, the procedure was repeated across a range of diffusion coefficients spanning from $10^{-6}$ to $10^{-10}$ m²/s, for both mobility ratios $M = 1$ and $M = 0.1$. Subsequently, the mass exchange coefficients were plotted against the Peclet number on a logarithmic scale in Figure 10. The equations extracted from the trend line in Figure 10 are reported in Table 2. Based on the extracted equations, we see a clear effect of the mobility ratio: in the log–log plot there was a shift in intercept value, while the slope was fairly constant. This implies that the mobility should affect the multiplier in a function of interfacial mass transfer as a function of Peclet number. Therefore, we propose a generalized equation that describes interfacial mass transfer ($k$) as a function of Peclet number (Pe) including the mobility ratio ($M$) as:

$$k = \psi_1 \times M^{\psi_2} \times Pe^{\psi_3} \tag{23}$$

where $\psi_1$, $\psi_2$, and $\psi_3$ are constants and $M$ is the mobility ratio. We used a simple regression to estimate the free variables in Equation (23) and obtained the values $\psi_1 = 6.29 \times 10^{-5}$, $\psi_2 = -0.402$, and $\psi_3 = 0.128$. Thus, for our simulated data we have:

$$k = 6.29 \times 10^{-5} \times M^{0.128} \times Pe^{-0.402} \tag{24}$$

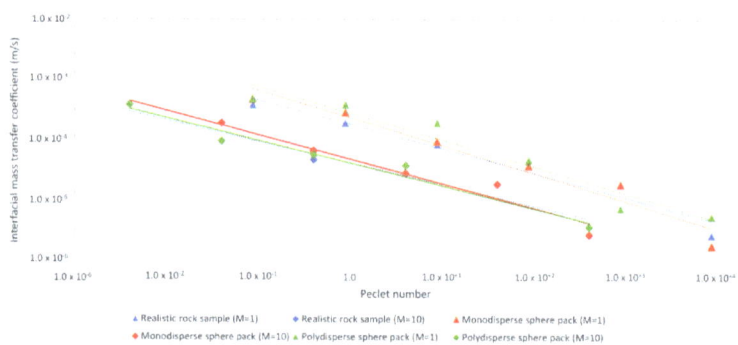

**Figure 10.** Interfacial mass transfer coefficient (k) measured after the breakthrough of scCO$_2$ versus Peclet number on a log–log scale for all the rock samples.

**Table 2.** The equations and corresponding R-squared values extracted from each of the trendlines in Figure 10.

| Rock Sample | M = 1 | M = 0.1 |
|---|---|---|
| Realistic rock sample | $k = 2 \times 10^{-4} Pe^{-0.3999}$<br>$R^2 = 0.989$ | $k = 4 \times 10^{-5} Pe^{-0.352}$<br>$R^2 = 0.9572$ |
| Monodisperse sphere pack | $k = 2 \times 10^{-4} Pe^{-0.455}$<br>$R^2 = 0.9431$ | $k = 5 \times 10^{-5} Pe^{-0.4}$<br>$R^2 = 0.9899$ |
| Polydisperse sphere pack | $k = 3 \times 10^{-4} Pe^{-0.437}$<br>$R^2 = 0.8156$ | $k = 4 \times 10^{-5} Pe^{-0.372}$<br>$R^2 = 0.9614$ |

In Figure 11 we have plotted the correlation between the extracted mass transfer coefficients from the simulation results versus the values given by Equation (24). The correlation is good, indicating the predictivity of our proposed equation.

Mass transfer is inherently a phenomenon that depends on the characteristics of the specific system in question. A practical approach for making experimental and small-scale numerical data more universally applicable is to represent them in a dimensionless style and establish relationships between the ratios of forces that impact the system. This allows us to predict that a system operating at a different scale but with equivalent force ratios is likely to yield comparable outcomes. Utilizing our correlation between the interfacial mass transfer coefficient and the Peclet number, we can replace the interfacial mass transfer coefficient from Equation (2) with the one in Equation (23), resulting in an equation for the Sherwood number as a function of mobility ratio ($M$), the Peclet number ($Pe$), the mean grain diameter of the porous medium ($d_m$), and the molecular diffusion coefficient ($D_m$).

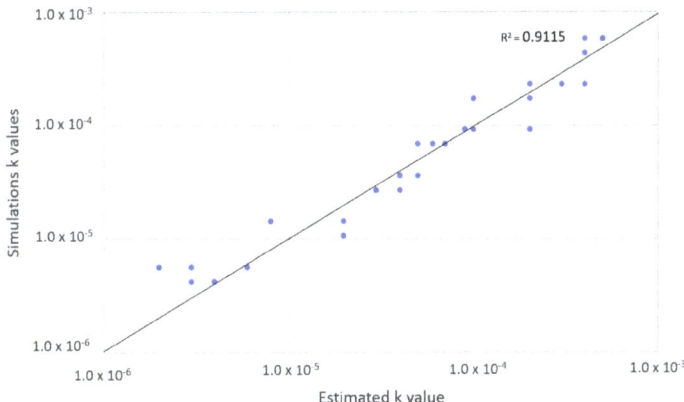

**Figure 11.** The correlation between the estimated and the calculated mass transfer coefficients.

$$Sh = \psi_1 \times M^{\psi_2} \times Pe^{\psi_3} \frac{d_m}{D_m} \quad (25)$$

By replacing the $d_m/D_m$ in Equation (25) with the one in the Peclet number definition (Equation (20)), we obtain the below equation:

$$Sh = \frac{\psi_1 \times M^{\psi_2} \times Pe^{(1+\psi_3)}}{U} \quad (26)$$

We carried out our simulations using a velocity of $U = 0.004$ (m/s)s. Considering the orders of magnitude of the constant values, mobility ratio, and velocity in Equation (26), we have a constant value approximately in the order of $1 \times 10^{-2}$.

For comparison, we tried to identify distinctive correlations between the Sherwood and Peclet numbers for each of the rock samples for our simulation results. Hence, we plotted the Sherwood number as a function of the Peclet number in Figure 12. The equations extracted for each of the rocks are shown in Table 3. As observed, the constants in the equations remain moderately consistent across various rock samples with diverse mean grain diameters. Therefore, we also considered general correlation derived from the trendline encompassing all the data in Figure 12, yielding the following equation with an R-squared value equal to 0.8341.

$$Sh = 0.0833 Pe^{0.5984} \quad (27)$$

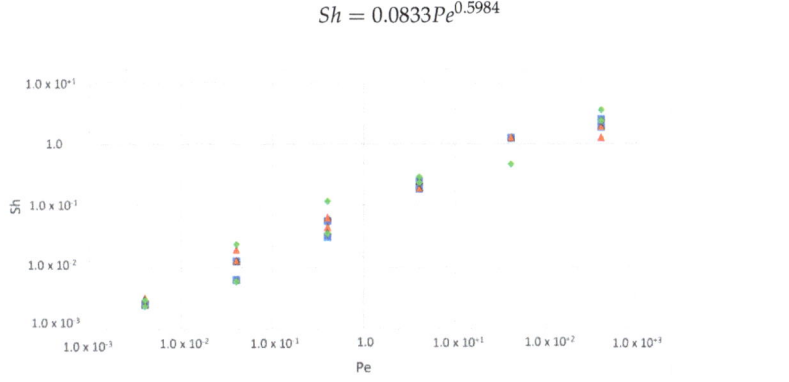

**Figure 12.** The correlation between Sherwood and Peclet numbers for each of the rock samples.

Table 3. Sherwood as a function of Peclet for each of the rock samples.

| Rock Sample | Realistic Rock Sample | Monodisperse Sphere Pack | Polydisperse Sphere Pack |
|---|---|---|---|
| $d_m(m)$ | $6.28 \times 10^{-5}$ | $6.40 \times 10^{-5}$ | $5.87 \times 10^{-5}$ |
| Sherwood and Peclet equation | $Sh = 0.0779 Pe^{0.6243}$ $R^2 = 0.8755$ | $Sh = 0.0879 Pe^{0.5727}$ $R^2 = 0.7462$ | $Sh = 0.0848 Pe^{0.5984}$ $R^2 = 0.9434$ |

The constant multiplier value in Equation (27) is similar to ($\psi_1 \times M^{\psi_2}/U$), derived in Equation (26). Furthermore, the exponent constant in Equation (27) is also similar to the $(1 + \psi_1)$ value we obtained in Equation (26).

Researchers have attempted to experimentally determine a relationship between the Sherwood number and the Peclet number [17,33,42,44]. The multiplier and exponent constants reported in the literature [17,33,44] are in good agreement with our findings. However, they were reported for NAPL interfacial mass transfer.

*5.4. Development of Sherwood Correlation with Reynold and Schmidt*

The Gilliland–Sherwood correlation is a common method for estimating mass transfer coefficient, considering both the stagnant film model and flow conditions:

$$Sh = \varphi_1 Re^{\varphi_2} Sc^{\varphi_3} \tag{28}$$

where $\varphi_1$, $\varphi_2$, and $\varphi_3$ are considered constants. Within this context, $\varphi_2$ reflects the relationship between the mass transfer system, fluid flow, and the particle size of the porous media, while $\varphi_3$ describes the connection between the mass transfer system and the characteristics of both the trapped phase and the flowing phase. Additionally, $\varphi_1$ accounts for other influencing factors, like the effects of dissolution fingering and two-stage dissolution [57–59]. The variable $Re$ is the Reynolds number:

$$Re = \frac{U d_m}{\nu_{scCO2}} \tag{29}$$

where U is the mean fluid velocity [m/s], $d_m$ is the geometrical mean diameter of the soil grains [m], $\nu$ is the kinematic viscosity of the displacing phase (scCO$_2$) [m$^2$/s]. The Reynolds number is a ratio that represents the relationship between the inertia and viscous forces of a moving fluid. The variable $Sc$ is the Schmidt number:

$$Sc = \frac{\mu_w}{\rho_w D_m} \tag{30}$$

The Schmidt number is a dimensionless quantity that compares the momentum diffusivity to the mass diffusivity. By using Equation (31), the mass transfer model for our simulations can be expressed as follows:

$$Sh = 0.004 Re^{0.67} Sc^{0.577} \tag{31}$$

The precision of the model is depicted in Figure 13 through a comparison of the Sherwood number obtained from the simulation results on the $x$-axis with the predicted Sherwood number on the $y$-axis. As indicated by the black line, this model can also reasonably predict the Sherwood number.

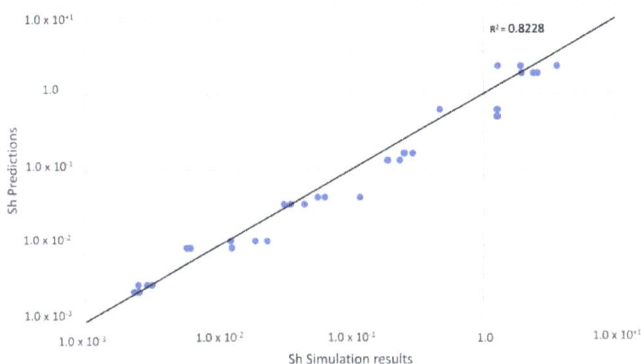

**Figure 13.** The Sherwood number obtained from the simulation versus the predicted Sherwood number, with the 1-1 line represented by the black line.

Figure 14 compares our mass transfer model with those previously reported in the literature. Equation (31) was developed for Reynolds numbers between 0.1 and 4 and Schmidt numbers ranging from 0.089 to 8900. In Figure 14, we have presented our model for Schmidt number 2000 as a solid black line. In earlier studies, the interfacial area was difficult to measure experimentally, leading researchers to report the modified Sherwood number instead. Table 4 provides a list of previous models that reported the Sherwood number models. The study by Donaldson et al. [60] explored the mass transfer of $O_2$ in porous media. Under the assumption of uniform sphere distribution equal to particle size, they approximated the interfacial area, and hence, the interfacial area was underestimated. Therefore, their model overestimates the Sherwood number compared to our work. According to our findings, the model created by Powers et al. [42] overestimates the Sherwood number in comparison to our model. Powers model was investigated for a solid–liquid system using naphthalene as the solid phase. This is because there is lower mass transfer in a solid–liquid system, while our model is designed for a $scCO_2$-water (liquid–liquid) system. The research paper by Patmonoaji et al. [61] proposes a model for a system with gas ($N_2$) and liquid (water). However, this model overestimates the Sherwood number values. The reason for this could be the difference in the nature of the gas used in their experiment ($N_2$) compared to ours ($scCO_2$). The closest model to our proposed model in the literature is given by Patmonoaji and Suekane [57]. Their study investigated mass transfer in a $CO_2$–water system, and as shown in Figure 14, it is consistent with our model and is widely accepted.

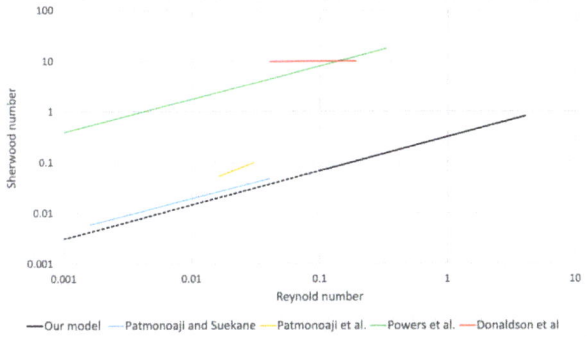

**Figure 14.** Comparison of the mass transfer model developed in this study with other works [42,57,60,61].

Table 4. Mass transfer models (Sherwood as a function of Reynolds and Schmidt) reported in the literature.

| Source | Model | Detail |
|---|---|---|
| Patmonoaji and Suekane [57] | $Sh = 0.386 Re^{0.645}$ | Schmidt number of trapped $CO_2$ gas at 532 and Reynolds numbers between 0.0016 and 0.04. |
| Patmonoaji et al. [61] | $Sh = 0.337 Re$ | Schmidt number of trapped $N_2$ gas at 534 and Reynolds numbers between 0.016 and 0.03. |
| Powers et al. [42] | $Sh = 36.8 Re^{0.654}$ | Schmidt number of trapped solid Naphthalene at 1250 and Reynolds numbers between 0.001 and 0.33. |
| Donaldson et al. [60] | $Sh = 2 + 0.6 Re^{1/2} + Sc^{1/3}$ | Schmidt number of trapped solid $N_2$ at 478 and Reynolds numbers between 0.04 and 0.19. |

## 6. Conclusions

We explored the temporary dissolution of $scCO_2$ as it was injected into porous media saturated with brine, simulating conditions akin to $CO_2$ geological storage in saline aquifers. Through detailed pore-scale simulations of multiphase and multicomponent flow, we accurately characterized the dynamic distribution of $scCO_2$ and brine phases, as well as the dissolution process of $scCO_2$ during the injection phase.

The coefficient governing mass transfer at the interface, necessary for modeling the dissolution rate, is typically described using Sherwood correlations linked to the system's properties. The primary contribution of this research is an updated Sherwood number, dependent on the Peclet number: we extended the relationship between the Sherwood and Peclet numbers, discovering how the mobility ratio influences the equation. Therefore, we argue that the impact of the mobility ratio should be included in this equation. We have also adjusted the relationship between the Sherwood number, which is based on the Reynolds number and Schmidt number.

Interfacial mass transfer in porous media models is limited due to interfacial area measurement challenges during the dissolution process. For our data, the inclusion of the mobility ratio significantly improved the predictability of mass transfer modeling that omits direct measurements of interfacial area. Our numerically derived mass transfer models based on Sherwood, Reynolds, and Schmidt numbers for the dissolution mass transfer in porous media could improve continuum scale modeling, e.g., modeling the dissolution of $scCO_2$ in water. We compared our mass transfer model with experimental models reported in the literature and found good agreement for interfacial mass transfer of $CO_2$ into water.

**Author Contributions:** Conceptualization, M.H.G. and C.F.B.; methodology, M.H.G.; software, M.H.G.; validation, M.H.G.; formal analysis, M.H.G. and C.F.B.; investigation, M.H.G.; resources, M.H.G.; data curation, M.H.G.; writing—original draft preparation, M.H.G.; writing—review and editing, C.F.B.; visualization, M.H.G.; supervision, C.F.B.; project administration, C.F.B.; funding acquisition, C.F.B. All authors have read and agreed to the published version of the manuscript.

**Funding:** This work was supported by the Research Council of Norway through its Center of Excellence funding scheme, project number 262644 (Porelab).

**Data Availability Statement:** Data are contained within the article.

**Conflicts of Interest:** The author declares no conflict of interest.

## References

1. Bultreys, T.; Boever, W.; De Cnudde, V. Imaging and image-based fluid transport modeling at the pore scale in geological materials: A practical introduction to the current state-of-the-art. *Earth-Sci. Rev.* **2016**, *155*, 93–128. [CrossRef]
2. Xiong, T.; Chen, M.; Jin, Y.; Zhang, W.; Shao, H.; Wang, G.; Long, E.; Long, W. A New Multi-Scale Method to Evaluate the Porosity and MICP Curve for Digital Rock of Complex Reservoir. *Energies* **2023**, *16*, 7613. [CrossRef]
3. Mobile, M.; Widdowson, M.; Stewart, L.; Nyman, J.; Deeb, R.; Kavanaugh, M.; Mercer, J.; Gallagher, D. In-situ determination of field-scale NAPL mass transfer coefficients: Performance, simulation and analysis. *J. Contam. Hydrol.* **2016**, *187*, 31–46. [CrossRef]

4. Carrillo, F.J.; Bourg, I.C.; Soulaine, C. Multiphase Flow Modelling in Multiscale Porous Media: An Open-Sourced Micro-Continuum Approach. *J. Comput. Phys. X* **2020**, *8*, 100073. [CrossRef]
5. Icardi, M.; Boccardo, G.; Marchisio, D.L.; Tosco, T.; Sethi, R. Pore-scale simulation of fluid flow and solute dispersion in three-dimensional porous media. *Phys. Rev. E* **2014**, *90*, 013032. [CrossRef] [PubMed]
6. Chen, L.; Kang, Q.; Robinson, B.A.; He, Y.-L.; Tao, W.-Q. Pore-scale modeling of multiphase reactive transport with phase transitions and dissolution-precipitation processes in closed systems. *Phys. Rev. E* **2013**, *87*, 043306. [CrossRef]
7. Agaoglu, B.; Scheytt, T.; Copty, N.K. Impact of NAPL architecture on interphase mass transfer: A pore network study. *Adv. Water Resour.* **2016**, *95*, 138–151. [CrossRef]
8. Niessner, J.; Hassanizadeh, S.M. Modeling Kinetic Interphase Mass Transfer for Two-Phase Flow in Porous Media Including Fluid–Fluid Interfacial Area. *Transp. Porous Media* **2009**, *80*, 329. [CrossRef]
9. Illangasekare, T.H.; Smits, K.M.; Fučík, R.; Davarzani, H. From Pore to the Field: Upscaling Challenges and Opportunities in Hydrogeological and Land–Atmospheric Systems. In *Pore Scale Phenomena*; World Scientific Series in Nanoscience and Nanotechnology; World Scientific, 2014; Volume 10, pp. 163–202. [CrossRef]
10. Chen, L.; Kang, Q.; Tang, Q.; Robinson, B.A.; He, Y.-L.; Tao, W.-Q. Pore-scale simulation of multicomponent multiphase reactive transport with dissolution and precipitation. *Int. J. Heat Mass Transf.* **2015**, *85*, 935–949. [CrossRef]
11. Ehlers, W.; Häberle, K. Interfacial Mass Transfer During Gas-Liquid Phase Change in Deformable Porous Media with Heat Transfer. *Transp. Porous Media* **2016**, *114*, 525–556. [CrossRef]
12. Chen, L.; Wang, M.; Kang, Q.; Tao, W. Pore scale study of multiphase multicomponent reactive transport during $CO_2$ dissolution trapping. *Adv. Water Resour.* **2018**, *116*, 208–218. [CrossRef]
13. Mwenketishi, G.T.; Benkreira, H.; Rahmanian, N. A Comprehensive Review on Carbon Dioxide Sequestration Methods. *Energies* **2023**, *16*, 7971. [CrossRef]
14. Pan, X.; Sun, L.; Huo, X.; Feng, C.; Zhang, Z. Research Progress on $CO_2$ Capture, Utilization, and Storage (CCUS) Based on Micro-Nano Fluidics Technology. *Energies* **2023**, *16*, 7846. [CrossRef]
15. Kokkinaki, A.; O'Carroll, D.M.; Werth, C.J.; Sleep, B.E. An evaluation of Sherwood–Gilland models for NAPL dissolution and their relationship to soil properties. *J. Contam. Hydrol.* **2013**, *155*, 87–98. [CrossRef] [PubMed]
16. Miller, C.T.; Christakos, G.; Imhoff, P.T.; McBride, J.F.; Pedit, J.A.; Trangenstein, J.A. Multiphase flow and transport modeling in heterogeneous porous media: Challenges and approaches. *Adv. Water Resour.* **1998**, *21*, 77–120. [CrossRef]
17. Agaoglu, B.; Copty, N.K.; Scheytt, T.; Hinkelmann, R. Interphase mass transfer between fluids in subsurface formations: A review. *Adv. Water Resour.* **2015**, *79*, 162–194. [CrossRef]
18. Held, R.J.; Celia, M.A. Pore-scale modeling and upscaling of nonaqueous phase liquid mass transfer. *Water Resour. Res.* **2001**, *37*, 539–549. [CrossRef]
19. Joekar-Niasar, V.; Hassanizadeh, S.M. Analysis of Fundamentals of Two-Phase Flow in Porous Media Using Dynamic Pore-Network Models: A Review. *Crit. Rev. Environ. Sci. Technol.* **2012**, *42*, 1895–1976. [CrossRef]
20. Parker, J.C.; Park, E. Modeling field-scale dense nonaqueous phase liquid dissolution kinetics in heterogeneous aquifers. *Water Resour. Res.* **2004**, *40*. [CrossRef]
21. Sainz-Garcia, A.A. Dynamics of Underground Gas Storage. Insights from Numerical Models for Carbon Dioxide and Hydrogen. Ph.D. Thesis, Université Toulouse 3 Paul Sabatier (UT3 Paul Sabatier), Toulouse, France, 2017.
22. Brusseau, M.L.; DiFilippo, E.L.; Marble, J.C.; Oostrom, M. Mass-removal and mass-flux-reduction behavior for idealized source zones with hydraulically poorly-accessible immiscible liquid. *Chemosphere* **2008**, *71*, 1511–1521. [CrossRef]
23. Schnaar, G.; Brusseau, M.L. Pore-Scale Characterization of Organic Immiscible-Liquid Morphology in Natural Porous Media Using Synchrotron X-ray Microtomography. *Environ. Sci. Technol.* **2005**, *39*, 8403–8410. [CrossRef] [PubMed]
24. Seyedabbasi, M.A.; Farthing, M.W.; Imhoff, P.T.; Miller, C.T. The influence of wettability on NAPL dissolution fingering. *Adv. Water Resour.* **2008**, *31*, 1687–1696. [CrossRef]
25. Dillard, L.A.; Blunt, M.J. Development of a pore network simulation model to study nonaqueous phase liquid dissolution. *Water Resour. Res.* **2000**, *36*, 439–454. [CrossRef]
26. Maes, J.; Soulaine, C. A new compressive scheme to simulate species transfer across fluid interfaces using the Volume-Of-Fluid method. *Chem. Eng. Sci.* **2018**, *190*, 405–418. [CrossRef]
27. Bear, J. *Dynamics of Fluids in Porous Media*; Dover publications: New York, NY, USA, 1988; ISBN 978-0-486-65675-5.
28. Costanza-Robinson, M.S.; Harrold, K.H.; Lieb-Lappen, R.M. X-ray Microtomography Determination of Air–Water Interfacial Area–Water Saturation Relationships in Sandy Porous Media. *Environ. Sci. Technol.* **2008**, *42*, 2949–2956. [CrossRef] [PubMed]
29. Kim, H.; Rao, P.S.C.; Annable, M.D. Gaseous Tracer Technique for Estimating Air–Water Interfacial Areas and Interface Mobility. *Soil Sci. Soc. Am. J.* **1999**, *63*, 1554–1560. [CrossRef]
30. Segura, L.A.; Toledo, P.G. Pore-Level Modeling of Isothermal Drying of Pore Networks Accounting for Evaporation, Viscous Flow, and Shrinking. *Dry. Technol.* **2005**, *23*, 2007–2019. [CrossRef]
31. Wang, Q.; Jia, Z.; Cheng, L.; Li, B.; Jia, P.; Lan, Y.; Dong, D.; Qu, F. Characterization of Flow Parameters in Shale Nano-Porous Media Using Pore Network Model: A Field Example from Shale Oil Reservoir in Songliao Basin, China. *Energies* **2023**, *16*, 5424. [CrossRef]
32. Luo, C.; Wan, H.; Chen, J.; Huang, X.; Cui, S.; Qin, J.; Yan, Z.; Qiao, D.; Shi, Z. Estimation of 3D Permeability from Pore Network Models Constructed Using 2D Thin-Section Images in Sandstone Reservoirs. *Energies* **2023**, *16*, 6976. [CrossRef]

33. Sarikurt, D.A.; Gokdemir, C.; Copty, N.K. Sherwood correlation for dissolution of pooled NAPL in porous media. *J. Contam. Hydrol.* **2017**, *206*, 67–74. [CrossRef]
34. van Genuchten, M.T.; Alves, W.J. *Analytical Solutions of the One-Dimensional Convective-Dispersive Solute Transport Equation*; United States Department of Agriculture, Economic Research Service: Washington, DC, USA, 1982.
35. Hunt, J.R.; Sitar, N.; Udell, K.S. Nonaqueous phase liquid transport and cleanup: 1. Analysis of mechanisms. *Water Resour. Res.* **1988**, *24*, 1247–1258. [CrossRef]
36. Al-Futaisi, A.; Patzek, T.W. Impact of wettability alteration on two-phase flow characteristics of sandstones: A quasi-static description. *Water Resour. Res.* **2003**, *39*. [CrossRef]
37. Latt, J.; Malaspinas, O.; Kontaxakis, D.; Parmigiani, A.; Lagrava, D.; Brogi, F.; Belgacem, M.B.; Thorimbert, Y.; Leclaire, S.; Li, S.; et al. Palabos: Parallel Lattice Boltzmann Solver. *Comput. Math. Appl.* **2021**, *81*, 334–350. [CrossRef]
38. Haroun, Y.; Legendre, D.; Raynal, L. Volume of fluid method for interfacial reactive mass transfer: Application to stable liquid film. *Chem. Eng. Sci.* **2010**, *65*, 2896–2909. [CrossRef]
39. Benson, S.M.; Bennaceur, K.; Cook, P.; Davison, J.; De Coninck, H.; Farhat, K.; Ramirez, A.; Simbeck, D.; Surles, T.; Verma, P.; et al. Carbon Capture and Storage. In *Global Energy Assessment (GEA)*; Johansson, T.B., Nakicenovic, N., Patwardhan, A., Gomez-Echeverri, L., Eds.; Cambridge University Press: Cambridge, UK, 2012; pp. 993–1068. ISBN 978-0-511-79367-7.
40. Liu, B.; Mahmood, B.S.; Mohammadian, E.; Khaksar Manshad, A.; Rosli, N.R.; Ostadhassan, M. Measurement of Solubility of $CO_2$ in NaCl, $CaCl_2$, $MgCl_2$ and $MgCl_2$ + $CaCl_2$ Brines at Temperatures from 298 to 373 K and Pressures up to 20 MPa Using the Potentiometric Titration Method. *Energies* **2021**, *14*, 7222. [CrossRef]
41. Wang, H.; Kou, Z.; Ji, Z.; Wang, S.; Li, Y.; Jiao, Z.; Johnson, M.; McLaughlin, J.F. Investigation of enhanced $CO_2$ storage in deep saline aquifers by WAG and brine extraction in the Minnelusa sandstone, Wyoming. *Energy* **2023**, *265*, 126379. [CrossRef]
42. Powers, S.E.; Abriola, L.M.; Dunkin, J.S.; Weber, W.J. Phenomenological models for transient NAPL-water mass-transfer processes. *J. Contam. Hydrol.* **1994**, *16*, 1–33. [CrossRef]
43. Miller, C.T.; Poirier-McNeil, M.M.; Mayer, A.S. Dissolution of Trapped Nonaqueous Phase Liquids: Mass Transfer Characteristics. *Water Resour. Res.* **1990**, *26*, 2783–2796. [CrossRef]
44. Powers, S.E.; Abriola, L.M.; Weber, W.J., Jr. An experimental investigation of nonaqueous phase liquid dissolution in saturated subsurface systems: Steady state mass transfer rates. *Water Resour. Res.* **1992**, *28*, 2691–2705. [CrossRef]
45. Ibrahim, M.A.A.; Kerimov, A.; Mukerji, T.; Mavko, G. Particula: A simulator tool for computational rock physics of granular media. *Geophysics* **2019**, *84*, F85–F95. [CrossRef]
46. Raeini, A.Q.; Blunt, M.J.; Bijeljic, B. Modelling two-phase flow in porous media at the pore scale using the volume-of-fluid method. *J. Comput. Phys.* **2012**, *231*, 5653–5668. [CrossRef]
47. Brackbill, J.U.; Kothe, D.B.; Zemach, C. A continuum method for modeling surface tension. *J. Comput. Phys.* **1992**, *100*, 335–354. [CrossRef]
48. Ubbink, O.; Issa, R.I. A Method for Capturing Sharp Fluid Interfaces on Arbitrary Meshes. *J. Comput. Phys.* **1999**, *153*, 26–50. [CrossRef]
49. Popinet, S.; Zaleski, S. A front-tracking algorithm for accurate representation of surface tension. *Int. J. Numer. Methods Fluids* **1999**, *30*, 775–793. [CrossRef]
50. Graveleau, M.; Soulaine, C.; Tchelepi, H.A. Pore-Scale Simulation of Interphase Multicomponent Mass Transfer for Subsurface Flow. *Transp. Porous Media* **2017**, *120*, 287–308. [CrossRef]
51. Henry, W.; Banks, J. III. Experiments on the quantity of gases absorbed by water, at different temperatures, and under different pressures. *Philos. Trans. R. Soc. Lond.* **1997**, *93*, 29–274. [CrossRef]
52. Deising, D.; Marschall, H.; Bothe, D. A unified single-field model framework for Volume-Of-Fluid simulations of interfacial species transfer applied to bubbly flows. *Chem. Eng. Sci.* **2016**, *139*, 173–195. [CrossRef]
53. Jasak, H.; Jemcov, A.; Tuković, Ž. OpenFOAM: A C++ Library for Complex Physics Simulations. 2007. Available online: www.openfoam.org/ (accessed on 23 January 2024).
54. Zhang, D.; Song, J. Mechanisms for Geological Carbon Sequestration. *Procedia IUTAM* **2014**, *10*, 319–327. [CrossRef]
55. Seyyedi, M.; Mahmud, H.K.B.; Verrall, M.; Giwelli, A.; Esteban, L.; Ghasemiziarani, M.; Clennell, B. Pore Structure Changes Occur During CO2 Injection into Carbonate Reservoirs. *Sci. Rep.* **2020**, *10*, 3624. [CrossRef]
56. Soulaine, C.; Debenest, G.; Quintard, M. Upscaling multi-component two-phase flow in porous media with partitioning coefficient. *Chem. Eng. Sci.* **2011**, *66*, 6180–6192. [CrossRef]
57. Patmonoaji, A.; Suekane, T. Investigation of CO2 dissolution via mass transfer inside a porous medium. *Adv. Water Resour.* **2017**, *110*, 97–106. [CrossRef]
58. Patmonoaji, A.; Hu, Y.; Zhang, C.; Suekane, T.; Patmonoaji, A.; Hu, Y.; Zhang, C.; Suekane, T. Dissolution Mass Transfer of Trapped Phase in Porous Media. In *Porous Fluids—Advances in Fluid Flow and Transport Phenomena in Porous Media*; IntechOpen: London, UK, 2021; ISBN 978-1-83962-712-5.
59. Patmonoaji, A.; Tahta, M.A.; Tuasikal, J.A.; She, Y.; Hu, Y.; Suekane, T. Dissolution mass transfer of trapped gases in porous media: A correlation of Sherwood, Reynolds, and Schmidt numbers. *Int. J. Heat Mass Transf.* **2023**, *205*, 123860. [CrossRef]

60. Donaldson, J.H.; Istok, J.D.; Humphrey, M.D.; O'Reilly, K.T.; Hawelka, C.A.; Mohr, D.H. Development and Testing of a Kinetic Model for Oxygen Transport in Porous Media in the Presence of Trapped Gas. *Groundwater* **1997**, *35*, 270–279. [CrossRef]
61. Patmonoaji, A.; Hu, Y.; Nasir, M.; Zhang, C.; Suekane, T. Effects of Dissolution Fingering on Mass Transfer Rate in Three-Dimensional Porous Media. *Water Resour. Res.* **2021**, *57*, e2020WR029353. [CrossRef]

**Disclaimer/Publisher's Note:** The statements, opinions and data contained in all publications are solely those of the individual author(s) and contributor(s) and not of MDPI and/or the editor(s). MDPI and/or the editor(s) disclaim responsibility for any injury to people or property resulting from any ideas, methods, instructions or products referred to in the content.

Review

# Carbon Capture and Storage in Depleted Oil and Gas Reservoirs: The Viewpoint of Wellbore Injectivity

Reyhaneh Ghorbani Heidarabad and Kyuchul Shin *

School of Chemical Engineering and Applied Chemistry, Kyungpook National University, 80 Daehak-ro, Buk-gu, Daegu 41566, Republic of Korea; reyhanghorbani@knu.ac.kr
* Correspondence: kyuchul.shin@knu.ac.kr; Tel.: +82-53-950-5587

**Abstract:** Recently, there has been a growing interest in utilizing depleted gas and oil reservoirs for carbon capture and storage. This interest arises from the fact that numerous reservoirs have either been depleted or necessitate enhanced oil and gas recovery (EOR/EGR). The sequestration of $CO_2$ in subsurface repositories emerges as a highly effective approach for achieving carbon neutrality. This process serves a dual purpose by facilitating EOR/EGR, thereby aiding in the retrieval of residual oil and gas, and concurrently ensuring the secure and permanent storage of $CO_2$ without the risk of leakage. Injectivity is defined as the fluid's ability to be introduced into the reservoir without causing rock fracturing. This research aimed to fill the gap in carbon capture and storage (CCS) literature by examining the limited consideration of injectivity, specifically in depleted underground reservoirs. It reviewed critical factors that impact the injectivity of $CO_2$ and also some field case data in such reservoirs.

**Keywords:** CCUS; injectivity; depleted oil and gas reservoirs; $CO_2$ injectivity

## 1. Introduction

Global warming is one of the main concerns of human beings currently and the global average temperature and sea level will reach 3.5 °C and 95 cm, respectively, by 2100 if the anthropogenic greenhouse gas (GHG) emissions continue to increase [1–7]. $CO_2$ accounts for ~64–76% of the total global GHG emissions and is one of the pollutants that endanger public health and welfare [6–11].

Several strategies have been introduced to reduce the carbon footprint, including shifting the energy mix to less carbon-intensive sources, reducing energy consumption, replacing fossil fuels with fuels that have shorter carbon chains, improving energy efficiency, and long-term carbon capture and sequestration in underground structures [5,6,12,13]. Among these, carbon capture and storage (CCS) has garnered attention because it decelerates the rate of the increase in atmospheric $CO_2$ concentrations [5,7,12,14–23]. CCS can contribute to ~12% of the global decarbonization target by 2050 and stabilize atmospheric GHG concentrations to ~450 ppm $CO_2$ eq by 2100 [5,7,12,14–23].

Deep saline aquifers, depleted hydrocarbon reservoirs, and hydrocarbon reservoirs of enhanced oil recovery (EOR) and enhanced gas recovery (EGR) projects, unmineable coal seams of enhanced coal bed methane (ECBM) projects, and salt domes/mined caverns are primary mediums of $CO_2$ sequestration [3,6–8,12,13,18,22,24–34]. $CO_2$ is trapped in these storage mediums via (i) structural trapping, where an impermeable caprock stops the $CO_2$ plume; (ii) capillary trapping, where $CO_2$ remains in pore spaces as residual $CO_2$ gas saturation; (iii) solubility trapping, where $CO_2$ dissolution produces dense $CO_2$-saturated brine; and (iv) mineral trapping caused by the reaction of minerals with $CO_2$-saturated brine [8,27,35–37]. Table 1 shows storage media based on their capacity, cost, integrity, and technical feasibility.

Citation: Heidarabad, R.G.; Shin, K. Carbon Capture and Storage in Depleted Oil and Gas Reservoirs: The Viewpoint of Wellbore Injectivity. Energies 2024, 17, 1201. https://doi.org/10.3390/en17051201

Academic Editors: Reza Rezaee and Dameng Liu

Received: 31 January 2024
Revised: 16 February 2024
Accepted: 29 February 2024
Published: 2 March 2024

**Copyright:** © 2024 by the authors. Licensee MDPI, Basel, Switzerland. This article is an open access article distributed under the terms and conditions of the Creative Commons Attribution (CC BY) license (https://creativecommons.org/licenses/by/4.0/).

Table 1. Evaluating geologically suitable storage reservoirs [Reproduced with permission [27] Copyright 2015 Elsevier].

| Storage Option | Relative Capacity | Relative Cost | Storage Integrity | Technical Feasibility |
|---|---|---|---|---|
| Active Oil Well (EOR) | Small | Very Low | Good | High |
| Coal Beds | Unknown | Low | Unknown | Unknown |
| Depleted oil/gas wells | Moderate | Low | Good | High |
| Deep Aquifers | Large | Unknown | Unknown | Unknown |
| Mined caverns/salt domes | Large | Very High | Good | High |

The majority of studies on carbon capture, and storage (CCS) focused on assessing the storage potential of deep saline aquifers; however, oil and gas reservoirs, despite their relatively smaller storage potential, are ideal for CCS owing to their high capacity, containment, reservoir structure, and surface facilities that can be adapted for $CO_2$ storage operations [3,4,27,38]. However, it should be noted that oil is considered as a hydrophobic fluid and has no harmful effect on pipeline walls, but $CO_2$ is a moisture content that in contact with water causes sweet corrosion in pipelines [39–44]. The facilities already existing are designed for hydrocarbons so it makes the $CO_2$ corrosion a significant problem in oil and gas production and transportation facilities that the cost of remediation can be higher than replacing the facilities [41,42,45–47].

Oil and gas reservoirs have also trapped hydrocarbons under caprock sealing for millions of years at high pressures, ensuring rock integrity for long-term $CO_2$ geosequestration with less environmental impact [48–50]. $CO_2$ sequestration in these reservoirs can provide an economic incentive, which is additional revenue from oil and gas recovered from $CO_2$ injection with EOR/EGR technology [7,51–55]. Technologies for CCS can increase the oil recovery through EOR/EGR up to 40%, whereas a recovery factor of 1–18% can be obtained after $CO_2$ injection in gas reservoirs [6,7,28,48,56]. A study using a Silurian core drilled from Door District in northeast Wisconsin, United States revealed an oil recovery of 34% during $CO_2$ flooding [57]. Studies conducted on gas reservoirs and $CO_2$–EGR field pilots have revealed that common depleted gas reservoirs have $CO_2$ storage capacities of 390–750 gigatons [51,58].

The feasibility of CCS operation mainly depends on adequate storage capacity and threshold well injectivity, which ensure that the desired amount of $CO_2$ is injected at acceptable rates through a minimum number of wells [3,5,29,59,60]. In other words, even with enormous storage capacities and high-quality overlying seals, CCS operation might not be financially viable without obtaining a minimum level of $CO_2$ injectivity into a formation [3,5,29,59,60]. Injectivity refers to the fluid's ability to be injected into a geological formation, i.e., the rate at which it can be injected into a storage medium without causing caprock fractures [7,12]. An adequate $CO_2$ injectivity is a prerequisite for CCS projects and considerably influences their economics.

$CO_2$ injectivity is strongly affected by the interactions between injected $CO_2$ and rock minerals or fluids in storage sites [5,7,9,12,28]. Although $CO_2$ storage mechanisms in oil and gas reservoirs have been more extensively studied, the implementation of CCS at a field scale is met with some limitations [12,27]. Moreover, although the storage and factors affecting oil/gas recovery have been explored, factors influencing wellbore injectivity in depleted hydrocarbon reservoirs remain understudied [12,27]. Herein, wellbore injectivity was investigated along with the main factors affecting the pressure build-up in CCS projects in active/depleted oil and gas reservoirs.

## 2. Factors Affecting the Injectivity of $CO_2$

Well injectivity issues are detrimental to CCS projects because large volumes of $CO_2$ must be stored for long periods in geological time scales [61]. $CO_2$ injectivity is mainly influenced by innate reservoir properties, residual gas/water saturation, residual oil/condensate saturation, injected fluid properties, mineral dissolution, salt precipitation, asphaltene precipitation, thermodynamic phase behavior of $CO_2$ in the wellbore, clay swelling, injection rate, and wettability alteration [7,9,12,27,60,62–65]. These factors and their influences on the wellbore injectivity of $CO_2$ geosequestration are discussed in subsequent sections.

## 2.1. Innate Reservoir Properties

$CO_2$ injection affects petrophysical characteristics, necessitating in-depth investigation [49]. Therefore, the innate reservoir properties that influence $CO_2$ injectivity were investigated herein: permeability, porosity, rock strength, composition of fluids, and heterogeneity level of the storage medium [12,18,66,67].

### 2.1.1. Permeability

Permeability is the ability of a porous medium to facilitate fluid flow and is measured in $m^2$ (in the metric system) or Darcy (D) or milli-Darcy (mD in the oilfield system) [68]. The permeability of a formation is closely linked to injectivity [7]. However, Effective wellbore permeability is also a critical parameter for estimating wellbore leakage potential, which significantly influences the $CO_2$ leak rate [69]. Given the close tie between well permeability and $CO_2$ leakage, quantitatively assessing well permeability uncertainty is crucial for evaluating $CO_2$ leakage risk [27,69]. Geochemical reactions as well as temperature can cause changes in permeability at both microscopic and mesoscale [9,70].

For effective $CO_2$ storage, low permeability is essential, but >100 mD is needed for good injectivity near the well [67]. However, the permeability of a medium should be low for permanent $CO_2$ storage assurance [67]. $CO_2$ flow in tight, low-permeability rocks is controlled by reservoir heterogeneity and permeability, which demand significant capillary pressure for $CO_2$ to penetrate the pores [48,71]. Positive total skin factor and wellbore permeability decline can result from partial penetration and formation damage [72].

Relative permeability significantly controls the injectivity, and pressure build-up in the reservoir [61,64,73]. $CO_2$ migration through caprocks involves a two-phase flow with capillary effects; however, measuring relative permeability curves is challenging due to low caprock permeability [62].

### 2.1.2. Porosity

Petroleum reservoirs are porous rocks containing hydrocarbons and connate water [68]. Porosity quantifies the pore volume relative to the total volume [68]. Naturally fractured reservoirs (NFRs) have matrices and fracture zones [68]. Matrices have higher porosity, whereas natural fractures (NF) have higher permeability [68]. The alteration of rock permeability and porosity is a case-specific phenomenon and is influenced by the composition of injection rate of fluid, rock mineral type, pore geometry, and thermodynamics [61].

### 2.1.3. Pressure

During CCS processes, a significant pressure difference between discharge and target pressure is created, which causes high storage density in depleted reservoirs [38]. Fluid injection into reservoirs is a complicated process, which is a thermoelastic, poroelastic, and chemoelastic coupled problem, and is accompanied by the state change of in situ stresses in reservoirs [19,27,62,74]. Considering the effect of injection on the storage site, changes in pore pressure have a direct impact on rocks' poroelastic properties [27].

High reservoir pressure in $CO_2$ injection can lead to mechanical stress and deformation, impacting both the reservoir and caprock sealing, potentially reducing injectivity and emphasizing the importance of preventing reservoir pressure from surpassing caprock fracture pressure for maintaining $CO_2$ containment and evaluating seal integrity to prevent fracture initiation [22,67]. The amount of oil produced rises and the time it takes for gas to break through reduces with increasing $CO_2$ injection pressure [55].

According to a study on how reservoir depletion affects stress, as pore pressure drops during reservoir depletion, effective horizontal in situ stress rises by 50 to 80% [74]. $CO_2$ enhances oil recovery in low-permeability reservoirs due to low viscosity and miscibility [49]. The injectivity is also controlled by the integrity of well bores, the failure of which causes rapid escape of injected $CO_2$. In turn, the integrity of well bores also depends on the injecting pressure and the upper limit for injecting pressure.

### 2.1.4. Well Configuration

The reliability analysis of storage sites emphasizes seal capacity, geometry, and integrity as crucial factors [66,67]. The sealing effectiveness of faults, influenced by pore throat size and mineral properties like water-wettability, particularly benefits from minerals such as mica, muscovite, and phlogopite, enhancing the ability to contain $CO_2$ plumes [67]. By increasing the pressure during the injection, the normal stress on a fault surface decreases and can lead to mechanical breakdown (reactivation) [27]. Moreover, $CO_2$ leakage is a significant challenge in CCS, occurring through various pathways like casing-cement interfaces, cement-rock contacts, and degraded or fractured materials. Wellbore integrity is critical for preventing these leaks [75].

### 2.1.5. Heterogeneity Level of the Storage Medium

It is unknown how the storage reservoir's injectivity is affected by compartmentalization and geological heterogeneities [9,66]. Geological heterogeneities are often classified into two main categories: the presence of alternating layers with varied mechanical properties, pore pressure, and/or lithology and permeability, and the presence of faults and compartmentalization within the specified storage reservoir [9,66]. Reservoir compartmentalization is influenced by fault structures and deposition history, impacting permeability between deposition units [9]. Existing faults and fractures before $CO_2$ injection may also enhance and/or postpone fluid migration rates [76]. This is the reason why it is included as a vital factor in fully understanding the locations, geometries, and permeabilities of the reservoirs [76].

The multiphase flow of $CO_2$-brine can be expressed in terms of permeability variation influenced by heterogeneity [77]. Storage capacity depends on heterogeneities and porosity; however, injectivity relies on petrophysical properties such as permeability [77]. Heterogeneities can cause unexpected outcomes in simulating injection processes and $CO_2$ plume behavior in storage reservoirs, especially in depleted oil or gas fields [9]. Remediation options include acid injection to create high-permeability pathways and surfactant formulations to alter wettability and counteract $CO_2$ trapping [9].

## 2.2. Capillary Trapping

Capillary trapping is a physical phenomenon in which $CO_2$ is trapped as residual gas saturation ($SgrCO_2$) in pore spaces due to capillary force [27]. Capillary trapping, a mechanism impacting injectivity, leads to residual $CO_2$ saturation in rock pores, influencing storage capacity [78]. The minimum saturation level at which gas can start to flow is called residual gas saturation [79]. In other words, injected $CO_2$ trapped in rock pores surrounded by water forms residual $CO_2$ saturation during capillary trapping [78]. When there is no mobility threshold above the saturation level, this parameter is equal to the residual gas saturation [79]. Residual gas in depleted reservoirs can increase or decrease storage capacity, and decrease brine mobility, density, and viscosity of gas mixtures when dissolved in supercritical $CO_2$ [67]. Based on a flooding experiment on four core samples (one composite and three Berea samples drilled from the Waarre C Formation in the $CO_2$CRC's Naylor Filed) it was concluded that early on in the $CO_2$ injection process, residual natural gas in the depleted reservoir lowered the $CO_2$ injectivity [80].

## 2.3. Residual Oil/Condensate Saturation

After primary and secondary oil recovery, residual oil saturation in reservoirs often ranges from 50 to 60% of the original oil-in-place (OOIP) [56,81]. Higher oil saturation linearly decreases storage capacity, but at 40% oil saturation, storage capacity does not vary very much [73].

## 2.4. Fluid Properties

Carbon dioxide transitions between gas, liquid, and solid states based on temperature and pressure variations [7]. Low levels of residual gas, water, and condensate in the reservoir are required for efficient $CO_2$ injectivity [48]. Injecting supercritical or liquid $CO_2$ in low-pressure reservoirs may cause evaporation in the tubing or wellbore [4,48]. Some properties are presented in Table 2 and the following section to understand more about the effect of fluid properties on wellbore injectivity.

### 2.4.1. Viscosity and Density of Injected $CO_2$

Density and viscosity are crucial properties in the storage process, receiving significant attention for their impact on storage capacity and enhanced oil recovery rate [82]. The low-density gas phase reduces hydrostatic pressure, impacting flow stability and potentially causing cavitation during phase changes [7]. In the sites with a depth higher than 800 m, the pressures and temperatures reach above the critical points of 7.38 MPa and 31.1 °C, respectively, and $CO_2$ can be injected as a supercritical fluid [12,22]. Therefore, sequestering $CO_2$ in shallow reservoirs (<2600 ft. (800 m)) is discouraged because of nonsupercritical conditions, although such complexities can be avoided with an understanding of changing multiphase behavior [7]. Theoretically, $CO_2$ temperature at the injector well bottom correlates positively with $CO_2$ injectivity [18]. This is due to density and viscosity decrease with rising bottom hole temperature, enhancing $CO_2$ mobility and injectivity [18]. The density of the supercritical $CO_2$ is more like liquid, but the viscosity is like gas [71]. Furthermore, residual $CH_4$ in reservoirs alters supercritical $CO_2$ density and viscosity in pore space [48]. In a study conducted by Nicot, et al. [83] on the impact of viscosity on the geologic storage capacity in shallow depth, results showed that a decrease in viscosity of the $CO_2$ mixtures leads to approximately the same proportion loss in the storage capacity [83].

Above 31.1 °C and critical pressure, $CO_2$ is supercritical with a mass density of $\sim$0.3–0.8 g/cm$^3$ (less dense than coexisting brine), which is a crucial behavior for storage considerations [7,84]. $CO_2$'s higher density promotes the stability of displacement fronts, and its supercritical state enhances efficient subsurface storage [85]. $CO_2$ density influences hydrocarbon extraction; heavier hydrocarbons are extracted at higher densities [71]. In addition, the efficiency of $CO_2$ storage increases at higher $CO_2$ densities, enhancing safety by reducing the buoyancy force [12]. $CO_2$ density at reservoir condition (the temperature between 293 K and the pressure between 25 bar and 700 bar) can be estimated by [67]:

$$\rho = \alpha + \beta T + \gamma T^2 + \theta T^3 \tag{1}$$

$$\alpha = (A_1 + B_1 P + C_1 P^2 + D_1 P^3) \tag{2}$$

$$\beta = (A_2 + B_2 P + C_2 P^2 + D_2 P^3) \tag{3}$$

$$\gamma = (A_3 + B_3 P + C_3 P^2 + D_3 P^3) \tag{4}$$

$$\theta = (A_4 + B_4 P + C_4 P^2 + D_4 P^3) \tag{5}$$

In the mentioned equations, $\alpha$, $\beta$, and $\gamma$ are the temperature coefficients. P is the pressure in bar scale and T is the temperature in Kelvin. $\rho$ is the density with the scale of kg/m$^3$, and $\theta$ is the contact angle representing the medium wettability [67].

### 2.4.2. Injected $CO_2$ Purity

$CO_2$ can be captured from fossil fuel power plants but it comprises a variation of impurities such as $N_2$, $NO_x$, Ar, $O_2$, and $SO_2$ in different concentrations [86]. Injecting this $CO_2$ may affect the amount of storage in a geological medium [27]. According to

Wang, et al. [87], who performed a study on the effect of $H_2S$ and $SO_2$ on $CO_2$ injectivity, coinjecting impurities with $CO_2$ for storage is cost-effective, but impurities negatively affect transport, injection, and storage [87]. Acid impurities such as $SO_x$ and $NO_x$ react with rocks, impacting injectivity and storage integrity; hazardous impurities pose environmental risks in the case of a $CO_2$ leakage [87].

The coinjection of these impurities may also impact well injectivity and wellbore integrity, thereby reducing porosity, cap rock integrity, $CO_2$, and water containment altering formation salinity near the well and mineral composition in the reservoir [9,83,86]. It also affects static capacity by altering the density and viscosity of the $CO_2$ mixture [83]. Lower density impacts $CO_2$ capacity due to impurity space and generally lower impurity density [83]. Impurity type influences thermal front location, delineating the radial zone with significant induced reservoir cooling [88]. They have a more pronounced impact on plume shape at shallow depths [67]. Separating impurities before injection is crucial for maintaining storage capacity, and removing $CO_2$ moisture is necessary to prevent corrosion and hydration-related costs [67].

Impurity removal is expensive; therefore, their coinjection will considerably reduce the cost of $CO_2$ capture [86]. The cost of a CCS project depends on $CO_2$ separation, with the impurity level affecting storage capacity [59]. The higher the level of impurity, the lower the storage capacity for $CO_2$ and the lower the $CO_2$ injected [59]. In addition, injecting a pure $CO_2$ stream that is free of impurities and water, prevents corrosion and formation damage caused by insoluble iron precipitates, thereby preserving injectivity in porous media [89].

2.4.3. Injectant Temperature

Injectant temperature directly influences total horizontal stress, with lower temperatures reducing near-wellbore stress significantly [19]. Repetitive thermal loading can cause the failure of well barrier material, particularly in $CO_2$ injection wells that experience temperature variations from injected fluid and rock [21]. The temperature fluctuations can range from 15 °C to 25 °C and 6 °C to 7 °C for onshore and offshore transport, respectively [21]. Thermal stress-induced wellbore damage is influenced by factors such as injection and formation temperature, formation stress state, and the thermal/mechanical properties of well barrier materials [21].

Table 2. The impacts of fluid properties on wellbores injectivity.

| Reference | Study Remarks |
| --- | --- |
| Jin, Pekot, Smith, Salako, Peterson, Bosshart, Hamling, Mibeck, Hurley and Beddoe [37] | $CO_2$ saturated Mead-Strawn stock-tank oil at 135° F showed that the density of oil increases when $CO_2$ is dissolved in the oil [90]. The gas storage rate in the Bell Creek oil field is linked to the injection rate, decreasing as the injection stabilizes. |
| Kazemzadeh, et al. [91] | The minimum miscible pressure (MMP) is the ideal pressure for cost-effective injection in oil recovery. |
| Barrufet, Bacquet and Falcone [59] | The duration of a project on a gas condensate fluid from the Cusiana field located in the northeast of Bogota, Colombia, in the Lianos basin is determined by injectivity, injection rates, and the number of wells; injection rates do not affect the eventual storage capacity. |
| Tawiah, Duer, Bryant, Larter, O'Brien and Dong [18] | Injection rates in reservoir rocks near the wellbore are influenced by injection pressures, fluid saturation, and fluid mobility. |

2.5. Mineral Dissolution/Precipitation

Mineral dissolution/precipitation in $CO_2$ (supercritical) and water–rock interactions enhance $CO_2$ trapping and alter mineral surface wettability, which is crucial for residual trapping [16]. CO$_2$ injection induces mineral dissolution and precipitation determined via the compositions of the original formation water and rock samples [49,92]. Increasing temperature and decreasing fluid pressure led to reduced carbonate solubility and CO$_2$ degassing [93]. An increased concentration of Ca or Mg ions from the dissolution of rock can lead to the rapid mineralization of CO$_2$ [36]. CO$_2$ dissolution in brine for storage triggers mineral reactions, which transform reservoir mineralogy and influence petrophysical properties such as porosity and permeability [94,95]. Mineral dissolution (chemical mineral dissolution of pore-filling cements such as carbonate and anhydrite [94]) improves formation permeability [7,93,94,96], porosity [93,94,96] and the proportions of pore-exposed grain-rimming clay coatings [94]. Dissolution near the injection site may increase CO$_2$ storage capacity within the medium, increasing the amount of localized CO$_2$ storage [97], altering the transport of CO$_2$ [16], and inducing surface cracking, which can increase the reactive surface area [70]. Uniform mineral dissolution was observed during the simulation of pure quartz sandstone, with slow surface reaction during CO$_2$ injection into sandstone [98]. The porosity and permeability of the reservoir are increased when rock minerals dissolve, whereas they are decreased when carbonate or sulfate compounds precipitate [49]. Sokama-Neuyam and Ursin's [99] study presents evidence that mineral dissolution negatively impacts CO$_2$ injectivity, reducing the efficiency of CO$_2$ injection. Injectivity impairment experienced a reduction of 9% when brine salinity was halved. The experimented sample belonged to Kocurek Industries, Caldwell, TX, USA, and the impairment of injectivity was measured via the pressure drop measurements [99].

Hydrated well cement, composed primarily of C-S-H and Portandite, is a reactive component in the near-well zone [95]. However, dissolved minerals that aggregate into fine particles in rock formations can enhance chemical reactions, causing pore throat blockage, reducing permeability, and impacting CO$_2$ mobility near plume boundaries [7,16,94,95]. Moreover, the permeability and porosity increase because of calcite [7], anhydrite [7,94], and carbonate dissolution; it also provides improved fluid pathways [94]. Pore throat sealing minimally affects total rock porosity but significantly deteriorates its permeability [94].

The main reactions that occur during the dissolution of CO$_2$ in water are as follows [98]:

$$CO_2 + H_2O \leftrightarrow H_2CO_3 \tag{6}$$

$$H_2CO_3 \leftrightarrow HCO_3^- + H^+ \tag{7}$$

$$HCO_3^- \leftrightarrow CO_3^{2-} + H^+ \tag{8}$$

CO$_2$ dissolves into formation water and generates H$^+$, HCO$_3^-$, and CO$_3^{2-}$ ions, which then react with specific ions in the formation water and rocks [49]. The concentrations of CO$_3^{2-}$ and HCO$_3^-$, are increased by a larger volume fraction of injected water and an increased Ca$^{2+}$ content in the formation water [49].

*2.6. Salt Precipitation*

Salt precipitation due to water evaporation has been recognized both experimentally and numerically and proven to lead to the abandonment of wells [61,95]. Formation water that is rich in ions reacts with injected CO$_2$, vaporizing and precipitating salts in reservoirs [7]. Salt precipitation also occurs in gas condensate reservoirs [61] and is a longstanding issue in the gas and petroleum industry [100]. CO$_2$ partially dissolves in brine and vaporizes saline water [63]. Evaporated water increases brine salt content, potentially leading to halite deposition if the solubility limit exceeds (~26.5% by weight) [63]. If the salt concentration exceeds the solubility limit under reservoir conditions, minerals precipitate [63,101]. The precipitated salt fills the porous space and clogs the pore network of the formation, modifying the flow channels [7,63,101,102]. It can reduce porosity and permeability, change capillary forces, and cause injectivity loss around the wellbore [5,7,24,27,61,63,95,100–103] completely blocking the injection [101] and reducing oil/gas productivity [63]. The decrease in permeability reduces the overall porosity and affects the pore space geometry as well as the precipitate distribution within the pore space [101]. The impact of precipitation on permeability varies based on reservoir chemistry and pore structure [103].

Concentrated halite precipitation results from sufficient brine mobility caused by a capillary pressure gradient, which impacts injectivity [101]. Simulations demonstrated that even when dry $CO_2$ is introduced into a depleted gas reservoir that contains medium-salinity brine, halite can still precipitate [101]. In addition, the combined effect of salt precipitation and fine migration can decrease the permeability three-fold compared to salt precipitation alone [102].

Salt precipitation is high around the injection well because the fluxes, concentrations, and saturation gradients of injected fluids are highest [5]. Even minimal salt deposition near the injection zone can cause significant $CO_2$ injectivity impairment [5]. However, this effect varies based on initial liquid saturation [103]. With sufficiently mobile brine, the continuously recharged precipitation front that is driven by capillary pressure considerably reduces the formation permeability [103].

For reducing salt precipitation, the most common remediating injectivity method is to inject chemicals [9]. In addition, salt precipitation in production wells can be reduced by diluting produced water with low-salinity water downhole and in production systems [5].

### 2.6.1. Effects of $CO_2$ Flow Rate

A critical $scCO_2$ injection flow rate influences particle migration in porous media, affecting salt precipitation and permeability [102]. Evaporation rate, directly linked to injection rate, influences brine concentration; increased gas volumes enhance halite precipitation [63]. High $CO_2$ injection rates may detach formation fines, causing pore clogging and reduced injectivity [5]. Optimal brine salinity and injection rate mitigate salt precipitation efficiently [5]. Because of the availability of NaCl, solid saturation rises with initial brine saturation, except at very low rates (0.1 kg/s) [101]. Moreover, high injection rates create a higher pressure gradient, suppressing capillary backflow and reducing the possibility of intensive salt accumulation [100]. Injectivity significantly decreases to a low of 40% point at a $CO_2$ injection rate of >5 mL/min possibly due to the uneven mineral distribution of a Berea sandstone core with a diameter of 3.81 cm and length of 7.62 cm ± 0.05 [102].

### 2.6.2. Effects of Brine Salinity

Brine salinity intricately affects $CO_2$ injectivity and reservoir properties [63]. Initial brine salinity is crucial for determining the residual salinity and salt precipitation; it also influences porosity and reduces permeability [63]. Higher brine salinity increases both salt precipitation and permeability reduction [100]. Injectivity of the Berea sandstone cores improves with reduced brine salinity but declines below 21.102 g/L [102]. Significantly reducing brine salinity by almost half results in only a marginal 9% injectivity decline, highlighting nonlinear mineral precipitation and injectivity dependency [5]. Low brine salinity, such as LSW, induces colloidal particle detachment, causing pore bridging and reduced $CO_2$ injectivity [5]. Extremely low salinity may cause simulation deviations, causing chemical interactions and impairing injectivity [5]. The positive correlation observed between brine salinity and injectivity impairment emphasizes the importance of brine salinity in assessing the effectiveness of $CO_2$ injection [102].

### 2.6.3. Effects of Pore Size

Initial permeability considerably influences subsequent permeability losses during $CO_2$ injection; lower initial permeability causes more significant reductions in permeability [63]. Higher permeability rocks have larger, less susceptible pore throats; thus, the permeability decreases only slightly [63]. Moreover, the pore channel size plays a crucial role in the dissolution of precipitates [5]. The impact of precipitates on injectivity is expected to be more pronounced in low-permeability rocks due to the potential plugging of narrow pore channels, even with the same precipitation rate [5].

2.6.4. Effects of Particle Size

While the correlation between particle size and pore throat size is a crucial factor, there is a theoretical understanding that the concentration of the suspension may significantly influence plugging mechanisms [102]. This is attributed to the shortened distance between suspended particles as the concentration increases, thereby intensifying the multiparticle blocking of invaded pores [102]. Injected dry $CO_2$ induces persistent water evaporation and halite precipitation around the well [101]. Despite potential halite effects, a high $CO_2$ injection rate can mitigate injectivity issues depending on reservoir factors such as the formation's characteristics, initial reservoir thermodynamic conditions, initial brine saturation, and salinity [101].

2.6.5. Effects of Water Saturation

Irreducible water saturation (Swi) determines the maximum relative permeability of $CO_2$; higher Swi results in lower maximum permeability [63]. A decrease in water saturation triggers brine capillary backflow, sustaining evaporation and precipitation [63].

2.6.6. Effects of Temperature

Elevated temperatures decrease water saturation and increase halite and salt precipitation in high-temperature gas reservoirs [63]. Rising temperature increases $CO_2$ phase water solubility, causing rapid saturation and salt precipitation [100]. Temperature only slightly impacts salt precipitation compared with factors such as injection rate and capillary pressure [100]. Numerical simulations were conducted to assess the dry-out process during $CO_2$ injection; however, the impact of halite precipitation on field operations could not be accurately predicted due to undefined relations between changes in porosity and permeability [103].

2.7. Asphaltene Precipitation

Crude oil is a complex mixture containing different hydrocarbons, primarily saturates, aromatics, resins, and asphaltenes (SARA) [56,104,105]. The most polar components are asphaltenes because they contain heteroatoms like oxygen, sulfur, or nitrogen, while saturated and aromatic compounds are nonpolar [55,106]. The stability of asphaltene depends on aromatic-to-saturate and resin-to-asphaltene ratios in crude oil [105]. Resins are an excellent bridging agent for all components of crude oil because they contain both polar and nonpolar sites [55,105,107].

Asphaltenes, the heaviest and most polar crude oil fractions, have a molecular weight of 100 to 10,000 g/mol [55,56,81,104–111]. They are composed mainly of carbon and hydrogen, with condensed aromatic rings surrounded by insoluble aliphatic chains [55,56,81,104–111]. While insoluble in light hydrocarbon solvents, asphaltenes are soluble in aromatics such as toluene and xylene [55,56,81,104–111]. They are generally incompatible with light petroleum fractions leading to undesirable effects in many stages of the petroleum industry such as pipeline routes, oil and gas production, and EOR/EGR [56]. The nature of asphaltenes is not clearly understood due to their complex nature and because various parameters affect their precipitation [56,104]. Thus, a universal model to predict and simulate their precipitation has not been developed yet [56,104,109]; their exact chemical structure also remains unknown [109].

Crude oil components are in equilibrium under reservoir conditions, but precipitation can occur due to pressure or temperature changes during oil field operations, such as high flow rates or gas introduction [55,104,107]. Asphaltene precipitation can be problematic and has caused major issues with flow assurance for the oil industry in recent decades [106,111]. For instance, when $CO_2$ is injected into the oil reservoir formation, the equilibrium state of the asphaltene–oil colloidal system in porous media is altered by variations in thermodynamic conditions, such as pressure (e.g., during primary depletion of highly under-saturated reservoirs [109]), temperature, asphaltene concentration, $CO_2$ content, oil composition, flow condition, and chemical additives; this results in a liquid-like solid precipitation with high viscosity [52,54,81,106,107,109]. Asphaltene precipitation can occur from reservoirs containing even very small amounts of asphaltene [107]. This will be far more severe when the reservoir pressure drops below the bubble point if the reservoir pressure is initially above the bubble point [55,104]. The molar volume of oil increases as the pressure decreases, subsequently decreasing the asphaltene solubility [105]. The maximum amount of asphaltene is precipitated at the bubble point [104,105] both with/without $CO_2$ injection [104] because of high amounts of dissolved gas by volume in oil [104,105]. In addition, precipitation is more at higher $CO_2$ injection rates [54].

The interaction coefficient between $CO_2$ and asphaltene is considerably higher than other components (natural gas and nitrogen [110]) with asphaltene [56,110]. Elevated $CO_2$ concentration increases the bubble point, precipitation, and interaction coefficient with asphaltene due to shared polarity [56,104].

Asphaltenes precipitated during $CO_2$ flooding, can either remain suspended or deposit onto surfaces, particularly high-specific-area clay minerals [28,52,54–56,81,104,105, 107,109,110]. This deposition can lead to reservoir and wellbore pore plugging, reduced porosity, permeability, and $CO_2$ injectivity, and altered wettability from water-wet to oil-wet [28,52,54–56,81,104,105,107,109,110]. Formation damage and a negative impact on production efficiency may result from pressure drop increase and changes in pore surface wettability [28,52,54–56,81,104,105,107,109,110]. Precipitation also threatens the capacity of surface facilities by plugging tubular and flow lines and clogging production separators, such as damage to pumps [56,104,106]. Despite all these disadvantages, asphaltene precipitation yields oil with less asphaltene content [81,110], making it lighter and less viscous than crude oil [56].

The degree of asphaltene precipitation during $CO_2$ injection is influenced by factors such as injection method, pressure, and miscibility [55]. It also depends on the asphaltene content in crude oil, correlated with injected $CO_2$, reservoir conditions, pressure, and temperature [28,112].

2.7.1. Effects of Permeability

Asphaltene deposition reduces permeability [112] and improves oil displacement by water due to increased oil relative permeability [108]. In particular, permeability affects the stability of asphaltenes and has a significant impact on deposition [111]. Deposited asphaltenes are more stable and migrate significantly less in more permeable reservoirs [111]. With increasing rock permeability, the impact of asphaltene deposition on the decreasing core permeability and oil recovery reduces [113]. Oil reservoirs with lower permeability exhibit more severe formation damages caused by simultaneous sulfur and asphaltene deposition than those with higher permeability [112,113]. Permeability reduction fluctuates between 40% and 90%, influenced by fluid composition, porous medium pore structure, and proximity to the core inlet or outlet [111]. Maximum asphaltene precipitation was reported at the core inlet, with a linear relation between permeability-reduction factor and asphaltene amount [107].

2.7.2. Effects of Pore Size Distribution

The characteristics of crude oil, the pore-plugging mechanism, and the deposition process are all impacted by the size distribution of precipitated asphaltene [56]. The analysis of pore size distribution can be utilized to determine how much asphaltene deposition reduces permeability [112]. Pore size distribution must be determined to identify fluid transport properties of porous media [108]. With increasing pore sizes (pore diameter > 8 μm [111]), the weight percent of asphaltene decreased and the oil recovery rate increased [55,111]. Small pores (<8.0 μm [111] or 9.0 μm [112] in radius) showed higher sensitivity to more asphaltene deposition, indicating their significant contribution to reducing permeability [111,112]. Asphaltene precipitation in unconventional reservoirs may be more severe than in conventional reservoirs due to considerable differences in pore sizes [55].

### 2.7.3. Effects of Temperature

Temperature is recognized as a relevant parameter for asphaltene stability [106]. When the temperature is raised during the $CO_2$-EOR process, asphaltene from crude oil can become unstable since $CO_2$ is usually in a supercritical state [54]. In multiphase flows, low heat transfer increases asphaltene deposition depending on the concentration and velocity of reactants [28]. The temperature of the reservoir considerably influences the kinetics of asphaltene precipitation, controlling the precipitation onset time and the size of deposits [54].

Oil recovery increases with increasing temperature; however, $CO_2$ breakthrough time decreases due to low oil viscosity [55]. Higher temperature also leads to increased asphaltene weight percent in bypassed oil due to resin instability [55]. Moreover, the asphaltene precipitation onset point increases with $CO_2$ injection at higher temperatures, which also increases the solubility parameter difference value [114]. In contrast, lower temperatures with liquid $CO_2$ decrease asphaltene stability due to its nonpolar behavior [106]. At higher temperatures, when $CO_2$ becomes supercritical, polar–polar interactions are enhanced and asphaltene precipitation is reduced due to steric repulsion and increased side-chain motion [106].

### 2.7.4. Effects of $CO_2$ Concentration

The amount of asphaltene precipitation depends on the concentration of injected $CO_2$ gas and rapidly increases when the $CO_2$ concentration exceeds its critical value [107]. The formation of a gas phase can cause the asphaltene phase volume to reduce at very high $CO_2$ concentrations [109]. Maximum amounts of asphaltene precipitates were obtained at saturation pressure, which gradually increased with an increase in the mole of $CO_2$ gas [107]. Higher $CO_2$ concentration increases asphaltene precipitation in the single-phase region and bubble point [115]. At higher concentrations of injected gas and reservoir pressures above and below the bubble point, asphaltene precipitation increases [104].

### 2.7.5. Effects of Porosity

The porosity and asphaltene accumulation along the core are correlated [111]. Porosity decreases in response to any increase in asphaltene deposits in a particular core region, and vice versa [111]. When the injection pressure was roughly at the MMP, there was an increase in oil recovery from the smaller pores [112]. $CO_2$ penetrated the small pores during miscible flooding, improving the oil recovery [112]. This highlights how crucial the MMP is for improving oil recovery from small pores [112].

### 2.7.6. Effects of Pressure

Asphaltene precipitation increases with injection pressure [55,112,114]. By increasing the injection pressure, more $CO_2$ molar percentage is required to achieve asphaltene precipitation [114]. For samples of crude oil, there was a small increase in the mean asphaltene particle size as the pressure was raised [116]. In contrast, Lei, Pingping, Ying, Jigen, Shi and Aifang [52] reported that at a constant injection pressure, the asphaltene precipitate amount first increases and decreases as the injected $CO_2$ amount increases; the asphaltene precipitate amount reaches its maximum when the gas phase occurs in the $CO_2$–oil system [52].

### 2.7.7. Effects of Viscosity

The gas phase, the liquid phase rich in hydrocarbons, and the liquid phase rich in asphaltene have different viscosities depending on the phase composition, temperature, and pressure [109]. The dissolution of $CO_2$ in crude oil and reduction in its asphaltene content considerably reduce oil viscosity, thereby decreasing the weight percent of asphaltene [109]. This is one of the main reasons for lower amounts of oil production at higher viscosity [55]. The weight percentage of asphaltene showed a positive correlation with the rise in viscosity for both the oil extracted and the oil that bypassed the system [55].

2.7.8. Effects of Flow Rate

By increasing the $CO_2$ flow rate, pressure decreases considerably, indicating asphaltene deposition and permeability reduction [28,113,117]. Thus, by reducing the flow rate, the formation damage can be considerably reduced when producing crude oils with medium and high contents of sulfur and asphaltene [113].

2.8. Fine Mobilization

Mobile particles with an equivalent diameter of <40 μm are known as fines [9]. They are assumed to be initially located on the surface of quartz grains [118]. The nonswelling and swelling clays that detach from the pore-grain interface can migrate [9,119]. Fines mobilize through chemical or mechanical interaction with pore fluid, including clay swelling and fluid flow-induced mobilization [9]. During $CO_2$ injection, fine particles are lifted into the reservoir and possibly plug the pore channels depending on the petrophysical characteristics of the rock, particle sizes, hydrodynamic conditions, solid concentrations, reservoir properties, and ionic strength and/or pH of the carrying fluid [7,9,60,89,94,95,120,121]. Born repulsion, electrical double-layer repulsion, and London–van der Waals attraction contribute to the potential energy that characterizes fine detachment [122]. The short-range potential known as the Born repulsive potential is created by the electron clouds' overlap [122]. Fines can originate in the injected fluid due to its contamination by contact with casing cement, drilling fluid filter cake residue on the wellbore wall, or from the formation itself [9,60,94]. The result of experiments on the Berea core sample with permeability (60–100 mD) and porosity (19–20%) showed that fines mobilization harms injectivity more than drying or high salt concentration [60].

The critical salt concentration (CSC) is defined as a specific salt concentration at which the fine particles may be released [122]. It can impair formation permeability (which some laboratory observations suggest may be permanent [119]) by blocking/bridging the pore throats [7,60,94,118,120,121] and productivity [7], but negligible porosity change [121]. Significant injectivity reduction could be caused by minor particles in the pore fluid or wellbore fluid in the immediate injection area [9,60,94].

The modeling of fine migration and effects of mineral dissolution can be expressed by [123]:

$$\frac{k}{k_0} = \frac{1}{[1 + \beta_s \sigma_s - \beta(\varphi - \varphi_i)]} \tag{9}$$

where $\beta$ and $\beta_s$ are the dissolution coefficient and the formation damage coefficient, respectively, and $\sigma_s$ is the volumetric concentration of retained fines [123]. Their values for the same brine salinity can be achieved by Wang, Bedrikovetsky, Yin, Othman, Zeinijahromi and Le-Hussain [123]. $k$ and $k_0$ are permeability and initial permeability, respectively. $\varphi$ is defined as porosity and $\varphi_i$ is the initial porosity [123].

Well impairment can lead to many challenges, including the possibility of an uncontrolled rise in cap rock pore pressure above the permissible fracturing pressure, the start and spread of fractures, and preferred flow pathways for $CO_2$ leaks [120]. Injectable wells may need to be abandoned in severe cases [120]. Clogging caused by fine release, migration, and capture is commonly considered an irreversible process [120]. However, some remedies can be suggested such as high injection velocities for deposition away from the wellbore due to radial flow because of the radially divergent nature of the injection flow [9]. In addition, greater permeability impairment is caused by core plugs with higher clay contents, most likely as a result of larger clay particles and smaller pore throats that make blocking and bridging easier [121]. Injecting at higher permeability intervals may further reduce the possibility of injectivity loss [121]. Clay fines can only detach when the pore space's salinity falls below a particular critical salinity, which can be prevented by raising the injected water's salinity above the CSC [119].

2.8.1. Effects of Permeability

In typically highly permeable sandstones, fine migration has been studied as the main cause of permeability decrease [9]. The injectivity will rapidly decrease due to a decrease in permeability resulting from increasing pore pressure [120]. Until the permeability is restored through additional physical or chemical treatments, the fluid can no longer be injected into the damaged formation at the required high flow rates [120]. A decrease in permeability implies that higher pressure will be required to inject the fluid into the formation, causing changes in injectivity [7]. The low-permeability reservoir rocks showed a more noticeable decrease in permeability [121]. Therefore, low-permeability reservoirs show lower injectivity loss [7,121]. Even within a short timeframe, changes in reservoir permeability affect injectivity, productivity, and $CO_2$ flow dynamics, impacting saturation distribution [120].

2.8.2. Particle Concentration

Particle concentration considerably affects $CO_2$ injectivity [7]. The concentration of fine particles in the fluid has a direct impact on the deposition rate [28]. The interaction between the particle and the pore throat increases as the particle concentration rises [7]. $CO_2$ injectivity impairment increased with an increase in the concentration of fine particles [7,60].

2.8.3. Injection Rate

Injectivity loss increased with increasing $CO_2$ injection rate [7]. Turbulence from a higher flow rate makes the particle stack tighter and enables their even redistribution in the pore network [7]. In contrast, Sokama-Neuyam, Ginting, Timilsina and Ursin [60] concluded that the $CO_2$ injection rate did considerably influence the reduction in $CO_2$ injectivity induced by fine entrapment [60]. However, at high injection velocities, the strong fluxes may dislodge the particles bridging the pore channels, opening some of the clogged channels [60]. The fluid's density, the square of its velocity, viscosity, and compressibility, as well as the size and form of the particles, all affect how much force is needed to raise them [60]. Hydrodynamic force acting on the particles is increased by supercritical $CO_2$'s gas-like viscosity and liquid-like density [60]. Typical $CO_2$ injection rates under storage conditions are approximately 1 Mt/pa [60].

2.8.4. Particle Size

Particle size and pore constriction, or more pertinently, the size of the fine particle to pore constriction, is crucial for determining the entrapment or piping mechanism occurring in the pore throat [7,60]. The larger the size of the particle, the higher the injectivity loss [7].

2.9. Clay Swelling

Many targeted geological storage sites are sealed by shale or mudstone rich in clay minerals such as sandstone gas reservoirs, which are heterogeneous and mainly composed of clay minerals and silica with small amounts of carbonates [35,58,124,125]. Clays are interlayer aluminosilicate minerals that have different structures [58]. They are crucial for the geological storage of $CO_2$ [126]. $CO_2$ is stored for a long term by overlying caprocks that act as low-permeability barriers to upward fluid flow [17,124–127].

Clay minerals are generally <1 mm in diameter and are therefore known as micro to nanocrystalline materials [128]. Clay minerals' layered structure and atom arrangement give them a platy morphology [128]. Illite, chlorite, smectite, and kaolinite are the main types of clay found in typical sandstone rocks [58]. A property that sets swelling clay minerals apart from other clays and micas is their easy interchangeability with the surrounding environment [125]. Of the different common rock-forming minerals, calcite is the most reactive because of its high solubility and kinetics [24]. These rocks are frequently composed of smectites, or expanding clays like montmorillonites, which are frequently found in faults that laterally seal possible storage reservoirs at depths and temperatures of up to 3.5 km and 100 °C, respectively [17,124]. At greater depths, smectites generally begin to transform into nonswelling clay minerals [124]. Smectites, a mixed layer of illite and smectite, and a mixed layer of chlorite and smectite are most sensitive to water and are hydrophilic [129]. These minerals, when present in the pore network, can drastically reduce the intrinsic permeability [129]. When the clay is exposed to solutions containing cations, the present cations may be exchanged with other cations [129]. The highest amount of total cations in clays that can be exchanged with a solution of a particular pH is known as the cation exchange capacity of clays [17,129]. The different capabilities of $CO_2$ adsorption and selectivity of mixed clays are caused by a variety of elements in clay structures, including cation exchange capacity (which is low in nonswelling forms such as kaolinite), charge on the clay surface, and interlayer distance [58]. Smectite clays are often used as samples due to their high swelling capacity [129]. The quantity and thermodynamic properties of water in the system have a significant impact on the intercalation and retention of $CO_2$ in smectite interlayers [130]. Illite and kaolinite, not known for interlayer expansion under any experimental conditions, can adsorb significant amounts of $CO_2$ [17]. If the surface area of the clays contacting the solution is large, the activities of divalent cations may be reduced either via sorption onto clay surfaces or via cation exchange [128]. Clays also increase the adsorption capacity of sandstone rocks because of their high surface area [58].

Clay swelling is a major cause of damage formation in hydrocarbon reservoirs and can substantially reduce nanoparticle mobility in porous media [127]. It can cause substantial changes in the reservoir structure, blocking the swelling pores, and causing dynamical behavior of the intercalated fluid molecules concerning the bulk fluid phase [17,125,130]. Another way to clog pore throats and reduce injection rates is by the mobilization of detrital or diagenetic clays [67]. The term "swelling clay" is derived from the ability of clay particles to increase their molar volume, thereby shrinking or contracting the interlayer pores based on the migration of polar molecules, such as water or organic molecules, into and out of the interlayers [9,124,125,127,130]. This migration depends on the clay mineral type and its hydration state [9,124,125,127,130]. The impact of clay swelling is most severe when incompatible fluids (e.g., oil) come in contact with swelling clays, leading to reduced formation permeability [127]. The same phenomenon is observed in the presence of $CO_2$ in brine solution or in contact with resident brine [9,131]. $CO_2$ can laterally displace the fluid and brine from the rock to assist in the intercalation of other chemical species in the clay, particularly during the early stages of injection [35,125,130].

The long-term efficacy of impermeable cap rocks in sealing the reservoir against the leakage of injected $CO_2$ needs further evaluation [125]. To ensure a safe $CO_2$ injection into reservoir formations, caprock failure and subsidence must be prevented, and the overlaying caprock must operate effectively [132].

2.9.1. Effects of Pressure

Pressure and temperature changes do not impact the clay nanoscale structure, suggesting stable seal quality during $CO_2$ injection [84]. At lower pressures, the $CO_2$ adsorption selectivity is enhanced, whereas at higher pressures, the adsorption amount increases [53]. Nonswelling clays can exhibit swelling under high pressure and temperature with $CO_2$ and water coexistence [133]. Supercritical $CO_2$ can alter capillary entry pressure in swelling clays, affecting injectivity, and may reduce long-term $CO_2$ storage capacity due to mineral dissolution and precipitation [9,131].

2.9.2. Effects on Strain

Because of the strain that the intercalated molecules may produce, the pore can swell or shrink [133]. Strain changes induced by $CO_2$ injection are considerably larger than those from pore-pressure changes alone [132,133]. $CO_2$ injection initially increases the strain rapidly and then stabilizes [35]. Strain changes vary spatially due to sedimentary structure, with local low-porosity layers acting as barriers [35]. Swelling strain rapidly increases with $CO_2$ pressure and decreases at higher pressures [17]. Linear strain is the length change relative to the initial length, and volumetric strain is the volume change relative to the initial volume [134]. Swelling strain can be predicted when micropore volume is known [133].

Material strain varies with the consolidation state [132]. Consolidated conditions exhibit higher strains than overconsolidated conditions due to microstructural differences [132]. Larger strain changes are observed due to $CO_2$ adsorption, particularly in kaolinite-rich rock [35]. Moreover, swelling strain can be substantial, leading to differential matrix swelling during brine displacement [35]. $CO_2$-induced swelling stress in Na-SWy-1 montmorillonite decreases as the effective stress and burial depth increase [124].

2.9.3. Effects of Temperature

At higher temperatures, $CO_2$ adsorption on sandstone rocks decreases [58]. Bandera sandstone showed the least reduction due to favorable $CO_2$ adsorption conditions on carbonate minerals [58]. By increasing the treatment temperature, the aromatic and aliphatic hydrocarbons in clay-rich shales are reduced [131]. Hot $CO_2$ injection doubles natural gas production and improves $CO_2$ sequestering in depleted gas reservoirs [58].

2.10. Hydrate Formation

The expansion of $CO_2$ is linked with the Joule-Thomson phenomenon, potentially causing the formation of dry ice or hydrates, consequently diminishing the injectivity of $CO_2$ [4]. Hydrates are formed due to the interaction of water with $CO_2$ and gaseous hydrocarbons such as methane during injection into depleted reservoirs [89]. Such formations depend on specific conditions such as injection rate, pressure, and lower temperatures [89].

Figure 1 compares the impact of injectivity loss with cases involved with salt deposition, mineral dissolution, fine migration, hydrates, and also without injectivity disruption [89].

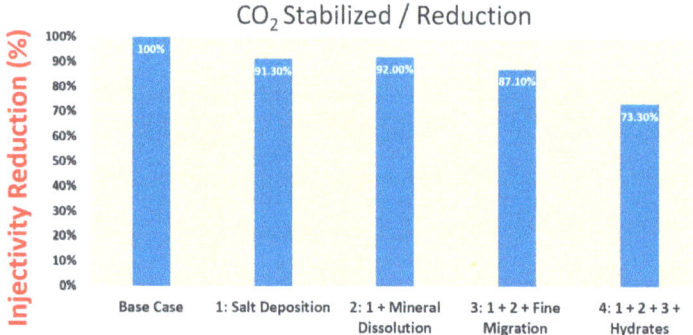

**Figure 1.** Comparison between the impact of injectivity reduction in hydrates with some other effects on injectivity [Reproduced with permission [89] Copyright 2023 Elsevier].

In Figure 1 case 1 was modeled using the following reaction in CMG simulators [89]:

$$Halite \leftrightarrow Cl^- + Na^+ \tag{10}$$

The value for case 2 was obtained from modeling the below reactions based on the reservoir rock mineralogy of a sample composed of 1.1% Calcite and 0.4% Dolomite [89].

$$Calcite + H^+ \leftrightarrow Ca^{2+} + HCO_3^- \tag{11}$$

$$Dolomite + H^+ \leftrightarrow Ca^{2+} + Mg^{2+} + 2HCO_3^- \tag{12}$$

Case 3 was calculated from reaction (11) and in the case of hydrate, it was modeled in CMG-STARS assuming that the permeability of the block was zero from the initial hydrate formation time [89]. Injectivity decreased to 73% from the base case in specific conditions [89].

## 3. $CO_2$ Injectivity in Field Cases

This field case study aimed to improve a general understanding of $CO_2$ injectivity into depleted hydrocarbon reservoirs by analyzing practical examples. The results are expected to provide concrete insights and useful recommendations to those who lead the responsibility of influencing environmental and energy management. However, the rapidly evolving confidential matters in CCS political settings may have resulted in uncertainties, leading to the potential for outdated information.

### 3.1. Niagaran Pinnacle Reef Oil Field

In the depleted pinnacle reef fields in Michigan, USA, CCS operations are performed by the Midwest Regional Carbon Sequestration Partnership (MRCSP) [135–138]. In the Michigan Basin, there are several hundred pinnacle reef structures, including this field [136]. It is a pinnacle reef that is late Silurian in age and has undergone significant primary and secondary recovery phases [136]. The shallow shelf carbonate depositional system that covered the Lower Peninsula of Michigan, northern Indiana, northeastern Illinois, eastern Wisconsin, northwest Ohio, and the Bruce Peninsula of Ontario gave rise to the reefs along the Niagaran Pinnacle Reef Trend oil fields [135].

MRCSP studies an extensive reef trend with over 880 closely spaced, highly compartmentalized, and laterally discontinuous reefs (Figure 2) [138]. The northern reef trend is divided into gas, oil, and water zones [138].

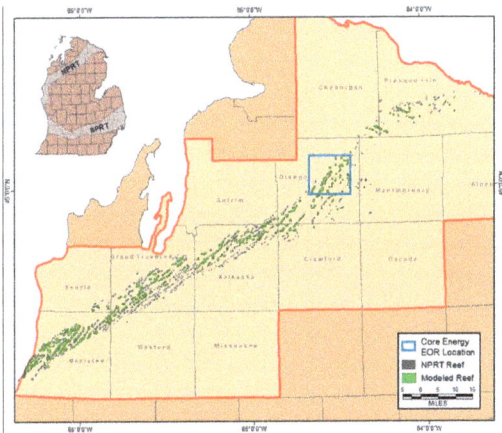

**Figure 2.** Northern pinnacle reef trend modeled; spatial extent is highlighted in green [Reproduced [138] Copyright 2020 Elsevier].

### 3.2. Netherlands Fields

The Netherlands leads the European Union in natural gas production with over 190 exploited gas fields. Fields in mature or depleted phases in the Netherlands are used for $CO_2$ storage [139]. Induced seismicity affects 15% with ML w3.6. $CO_2$ storage in depleted gas fields is favored due to proven seal quality and there is no record of seal integrity failure due to fault reactivation in seismically active Netherlands [139]. The most suitable sites to store $CO_2$ are depleted gas fields (79%), followed by aquifers (19%) [139].

The number of wells, injection rates, duration of the project, and reservoir thickness and depth all affect drilling costs [23]. The costs increase linearly with the length of drilling up to a 3 km depth [23]. After 3 km, costs increase exponentially with depth [23]. The injectivity index assesses the ability of wells to inject fluids into a porous, permeable formation in petroleum reservoir engineering [140]. The injectivity index is the ratio of injection rate to pressure difference [140]. Because data are not available for every site, average injection rates based on the stratigraphic unit were utilized instead of site-specific injection rates (Table 3) [23].

Table 3. The average $CO_2$ injection rate of hydrocarbon fields per stratigraphic unit in Mt/y per well [Reproduced with permission [23] Copyright 2010 Elsevier].

|  | Formation | Hydrocarbon Field |
| --- | --- | --- |
| Lower Cretaceous Group | Vlieland Sandstone Fm | 1 |
| Lower Germanic Trias Group | Lower Buntsandstein Fm | 0.4 |
| Niedersachsen Group | Friese front Fm (sandstone members) | 0.4 |
| Upper Rotliegend Group | Zechstein Fm (carbonate members) | 0.2 |
|  | Slochteren Fm (sandstone members) | 1 |
| Limburg Group |  | 0.4 |

### 3.3. Malaysia

Malaysia explores CCS to cut greenhouse gas emissions and environmental impact [141]. Malaysia's largest gas field was drilled with about 20 wells and they are largely depleted [142]. The structure of the field is defined by a mild anticline, exhibiting a typical north–south orientation [142]. The average porosity of the reservoir exceeds 26%, and its permeability is favorable at an average of 1000 mD for $CO_2$ storage [142].

Fractures may initiate during injection if pressure surpasses the caprock fracture threshold [33]. To address the challenges with injectivity, the optimal injection rate must be determined before field-scale injection [50]. $CO_2$ injection within safe limits prevents formation damage from fine migration in heterogeneous formations around the wellbore [50]. By maintaining injection below the critical limit, considering rock permeability, wellbore formation damage can be prevented [50].

### 3.4. Goldeneye

Goldeneye, a depleted gas field and platform in the northern North Sea, UK, was operated by Shell (2004–2011) about 100 km northeast of Moray Firth, Scotland [143,144]. This site is ideal for $CO_2$ storage because of factors such as young facilities, a dedicated pipeline, excellent Darcy sandstone quality, and good containment [145]. This storage site is located at 58°0′10.8″ N, 0°22′48″ W, and has a 120 m water depth [144]. The seafloor, besides pipelines and offshore platforms, is mostly flat with occasional pockmarks and abandoned wells [144]. The current infrastructure facilitates cost-effective appraisal and expansion into connected saline aquifer systems near or overlying the Goldeneye field [145].

Depleted field storage projects may stop injection below preproduction pressures to ensure containment and address uncertainties in capacity [146]. In cases such as Goldeneye, the "pressure-sink" effect acts as a risk-mitigating factor, causing water to flow back into the pressure anomaly [146]. Demonstrating long-term stability for site closure in the EU, particularly with potential pressure increases, can be challenging [146].

### 3.5. Cranfield

The depleted deep clastic Cranfield field in southwestern Mississippi, USA, presents potential for CCS operations [34,64,146,147]. From its discovery in 1943 until 1966, oil was produced there [34,64,146,147]. In 2008, the EOR $CO_2$ injection project began [34]. At a depth of 3300 m, the Detailed Area Study (DAS) pilot project aims to reach the Cretaceous Lower Tuscaloosa Formation [34]. The $CO_2$ injection pilot, which commenced on 1 December 2009, began at ~175 kg/min; the value was increased to ~300 kg/min after 20 days and ~500 kg/min after 156 days [34]. Over 3 1/4 years, approximately 3 million metric tons of $CO_2$ have been injected into the Cranfield field [148]. The $CO_2$ injection zone in the Lower Tuscaloosa Formation forms a four-way anticline with a diameter of 6.4 km (4 mi), shaped by an inactive salt dome [147].

Injected $CO_2$ flow rates are directly proportional to its relative permeability, which is a crucial parameter [34]. Low gas relative permeabilities lead to slower $CO_2$ plume spreading and increased pressure build-up [34]. Moreover, salt precipitation from water evaporation may impact injectivity, thereby increasing the $CO_2$ flow and reducing the porosity [149]. Pressure build-up is yet to be quantitatively determined [149].

## 4. Conclusions

The assessment of injectivity is vital for the secure sequestration of $CO_2$ in targeted depleted reservoirs. This study addressed various factors affecting injectivity and presents field case data. It was concluded that hydrophysical, chemical, and geo-mechanical processes significantly influence injectivity. The reviewed papers in the literature highlight that asphaltene precipitation, mineral dissolution, and fine mobilization can disturb injectivity by causing core plugging. Additionally, properties like purity, density, and viscosity of injected $CO_2$ play a crucial role in ensuring effective injection and storage.

A comprehensive understanding of diverse trapping mechanisms and the adept modeling of CCS in depleted reservoirs is mandatory for achieving successful $CO_2$ injection. Therefore, it is advisable to pursue further research in this domain.

**Author Contributions:** R.G.H.: Conceptualization, Writing—original draft. K.S.: Writing—review & editing, Supervision. All authors have read and agreed to the published version of the manuscript.

**Funding:** This work was supported by NRF Korea [grant numbers 2021R1F1A1047221].

**Data Availability Statement:** Not applicable.

**Acknowledgments:** We acknowledge the support by NRF Korea (2021R1F1A1047221) funded by the Ministry of Science and ICT of Korea.

**Conflicts of Interest:** The authors declare that they have no known competing financial interests.

## References

1. Iddphonce, R.; Wang, J.; Zhao, L. Review of $CO_2$ injection techniques for enhanced shale gas recovery: Prospect and challenges. *J. Nat. Gas Sci. Eng.* **2020**, *77*, 103240. [CrossRef]
2. Yan, H.; Zhang, J.; Rahman, S.S.; Zhou, N.; Suo, Y. Predicting permeability changes with injecting $CO_2$ in coal seams during $CO_2$ geological sequestration: A comparative study among six SVM-based hybrid models. *Sci. Total Environ.* **2020**, *705*, 135941. [CrossRef]
3. Park, Y.-C.; Kim, S.; Lee, J.H.; Shinn, Y.J. Effect of reducing irreducible water saturation in a near-well region on $CO_2$ injectivity and storage capacity. *Int. J. Greenh. Gas Control* **2019**, *86*, 134–145. [CrossRef]
4. Hoteit, H.; Fahs, M.; Soltanian, M.R. Assessment of $CO_2$ injectivity during sequestration in depleted gas reservoirs. *Geosciences* **2019**, *9*, 199. [CrossRef]
5. Sokama-Neuyam, Y.A.; Ursin, J.R. Experimental and theoretical investigations of $CO_2$ injectivity. *AGH Drill. Oil Gas* **2016**, *33*, 245–258. [CrossRef]
6. Abba, M.K. Enhanced gas recovery by $CO_2$ injection: Influence of monovalent and divalent brines and their concentrations on $CO_2$ dispersion in porous media. *J. Nat. Gas Sci. Eng.* **2020**, *84*, 103643. [CrossRef]
7. Ginting, P. *Effect of Colloidal Transport on $CO_2$ Injectivity*; University of Stavanger: Stavanger, Norway, 2016.
8. Wang, X.; Alvarado, V.; Swoboda-Colberg, N.; Kaszuba, J.P. Reactivity of dolomite in water-saturated supercritical carbon dioxide: Significance for carbon capture and storage and for enhanced oil and gas recovery. *Energy Convers. Manag.* **2013**, *65*, 564–573. [CrossRef]
9. Torsæter, M.; Cerasi, P. Geological and geomechanical factors impacting loss of near-well permeability during $CO_2$ injection. *Int. J. Greenh. Gas Control* **2018**, *76*, 193–199. [CrossRef]
10. Peysson, Y.; André, L.; Azaroual, M. Well injectivity during $CO_2$ storage operations in deep saline aquifers—Part 1: Experimental investigation of drying effects, salt precipitation and capillary forces. *Int. J. Greenh. Gas Control* **2014**, *22*, 291–300. [CrossRef]
11. Abba, M.K.; Al-Otaibi, A.; Abbas, A.J.; Nasr, G.G.; Burby, M. Influence of permeability and injection orientation variations on dispersion coefficient during enhanced gas recovery by $CO_2$ injection. *Energies* **2019**, *12*, 2328. [CrossRef]
12. Raza, A.; Gholami, R.; Rezaee, R.; Rasouli, V.; Rabiei, M. Significant aspects of carbon capture and storage—A review. *Petroleum* **2019**, *5*, 335–340. [CrossRef]

13. Al-Hasami, A.; Ren, S.; Tohidi, B. $CO_2$ Injection for Enhanced Gas Recovery and Geo-Storage: Reservoir Simulation and Economics. In Proceedings of the SPE Europec/EAGE Annual Conference, Madrid, Spain, 13–16 June 2005.
14. Kasahara, J.; Tsuruga, K. An Innovative Method for the 4D Monitor of Storage in CCS(Carbon Dioxide Capture and Storage) and Oil and Gas Reservoirs and Aqufers. In Proceedings of the 2nd Joint BCSR-JCCP Environmental Symposium, Bahrain, 8–10 February 2010.
15. Valle, L.; Rodríguez, R.; Grima, C.; Martínez, C. Effects of supercritical $CO_2$ injection on sandstone wettability and capillary trapping. *Int. J. Greenh. Gas Control* **2018**, *78*, 341–348. [CrossRef]
16. Min, Y.; Kim, D.; Jun, Y.-S. Effects of Na+ and K+ exchange in interlayers on biotite dissolution under high-temperature and high-$CO_2$-pressure conditions. *Environ. Sci. Technol.* **2018**, *52*, 13638–13646. [CrossRef]
17. De Jong, S.; Spiers, C.; Busch, A. Development of swelling strain in smectite clays through exposure to carbon dioxide. *Int. J. Greenh. Gas Control* **2014**, *24*, 149–161. [CrossRef]
18. Tawiah, P.; Duer, J.; Bryant, S.L.; Larter, S.; O'Brien, S.; Dong, M. $CO_2$ injectivity behaviour under non-isothermal conditions—Field observations and assessments from the Quest CCS operation. *Int. J. Greenh. Gas Control* **2020**, *92*, 102843. [CrossRef]
19. Jung, H.; Espinoza, D.N.; Hosseini, S.A. Wellbore injectivity response to step-rate $CO_2$ injection: Coupled thermo-poro-elastic analysis in a vertically heterogeneous formation. *Int. J. Greenh. Gas Control* **2020**, *102*, 103156. [CrossRef]
20. Ren, B.; Ren, S.; Zhang, L.; Chen, G.; Zhang, H. Monitoring on $CO_2$ migration in a tight oil reservoir during CCS-EOR in Jilin Oilfield China. *Energy* **2016**, *98*, 108–121. [CrossRef]
21. Roy, P.; Morris, J.P.; Walsh, S.D.; Iyer, J.; Carroll, S. Effect of thermal stress on wellbore integrity during $CO_2$ injection. *Int. J. Greenh. Gas Control* **2018**, *77*, 14–26. [CrossRef]
22. Zhao, Y.; Yu, Q. $CO_2$ breakthrough pressure and permeability for unsaturated low-permeability sandstone of the Ordos Basin. *J. Hydrol.* **2017**, *550*, 331–342. [CrossRef]
23. Ramírez, A.; Hagedoorn, S.; Kramers, L.; Wildenborg, T.; Hendriks, C. Screening $CO_2$ storage options in the Netherlands. *Int. J. Greenh. Gas Control* **2010**, *4*, 367–380. [CrossRef]
24. Rohmer, J.; Pluymakers, A.; Renard, F. Mechano-chemical interactions in sedimentary rocks in the context of $CO_2$ storage: Weak acid, weak effects? *Earth Sci. Rev.* **2016**, *157*, 86–110. [CrossRef]
25. Mackay, E.J. 3—Modelling the Injectivity, Migration and Trapping of $CO_2$ in Carbon Capture and Storage (CCS). In *Geological Storage of Carbon Dioxide ($CO_2$)*; Gluyas, J., Mathias, S., Eds.; Woodhead Publishing: Sawston, UK, 2013; pp. 45–70.e.
26. Al-Khdheeawi, E.A.; Vialle, S.; Barifcani, A.; Sarmadivaleh, M.; Iglauer, S. Influence of injection well configuration and rock wettability on $CO_2$ plume behaviour and $CO_2$ trapping capacity in heterogeneous reservoirs. *J. Nat. Gas Sci. Eng.* **2017**, *43*, 190–206. [CrossRef]
27. Raza, A.; Rezaee, R.; Gholami, R.; Rasouli, V.; Bing, C.H.; Nagarajan, R.; Hamid, M.A. Injectivity and quantification of capillary trapping for $CO_2$ storage: A review of influencing parameters. *J. Nat. Gas Sci. Eng.* **2015**, *26*, 510–517. [CrossRef]
28. Khurshid, I.; Choe, J. Analysis of asphaltene deposition, carbonate precipitation, and their cementation in depleted reservoirs during $CO_2$ injection. *Greenh. Gases Sci. Technol.* **2015**, *5*, 657–667. [CrossRef]
29. Kalra, S.; Wu, X. $CO_2$ Injection for Enhanced Gas Recovery. In Proceedings of the SPE Western North American and Rocky Mountain Joint Meeting, Denver, CO, USA, 17–18 April 2014.
30. Gao, S.; Wang, Y.; Jia, L.; Wang, H.; Yuan, J.; Wang, X. $CO_2$–$H_2O$–coal interaction of $CO_2$ storage in coal beds. *Int. J. Min. Sci. Technol.* **2013**, *23*, 525–529. [CrossRef]
31. Santibanez-Borda, E.; Govindan, R.; Elahi, N.; Korre, A.; Durucan, S. Maximising the dynamic $CO_2$ storage capacity through the optimisation of $CO_2$ injection and brine production rates. *Int. J. Greenh. Gas Control* **2019**, *80*, 76–95. [CrossRef]
32. Borda, E.S.; Govindan, R.; Elahi, N.; Korre, A.; Durucan, S. The Development of a Dynamic $CO_2$ Injection Strategy for the Depleted Forties and Nelson Oilfields Using Regression-based Multi-objective Programming. *Energy Procedia* **2017**, *114*, 3335–3342. [CrossRef]
33. Raza, A.; Gholami, R.; Rezaee, R.; Bing, C.H.; Nagarajan, R.; Hamid, M.A. Well selection in depleted oil and gas fields for a safe $CO_2$ storage practice: A case study from Malaysia. *Petroleum* **2017**, *3*, 167–177. [CrossRef]
34. Soltanian, M.R.; Amooie, M.A.; Cole, D.R.; Graham, D.E.; Hosseini, S.A.; Hovorka, S.; Pfiffner, S.M.; Phelps, T.J.; Moortgat, J. Simulating the Cranfield geological carbon sequestration project with high-resolution static models and an accurate equation of state. *Int. J. Greenh. Gas Control* **2016**, *54*, 282–296. [CrossRef]
35. Zhang, Y.; Xue, Z.; Park, H.; Shi, J.Q.; Kiyama, T.; Lei, X.; Sun, Y.; Liang, Y. Tracking $CO_2$ plumes in clay-rich rock by distributed fiber optic strain sensing (DFOSS): A laboratory demonstration. *Water Resour. Res.* **2019**, *55*, 856–867. [CrossRef]
36. Jeon, P.R.; Lee, C.-H. Effect of surfactants on $CO_2$ solubility and reaction in $CO_2$-brine-clay mineral systems during $CO_2$-enhanced fossil fuel recovery. *Chem. Eng. J.* **2020**, *382*, 123014. [CrossRef]
37. Jin, L.; Pekot, L.J.; Smith, S.A.; Salako, O.; Peterson, K.J.; Bosshart, N.W.; Hamling, J.A.; Mibeck, B.A.; Hurley, J.P.; Beddoe, C.J. Effects of gas relative permeability hysteresis and solubility on associated $CO_2$ storage performance. *Int. J. Greenh. Gas Control* **2018**, *75*, 140–150. [CrossRef]
38. Valbuena, E.; Barrufet, M. A generalized partial molar volume algorithm provides fast estimates of $CO_2$ storage capacity in depleted oil and gas reservoirs. *Fluid Phase Equilibria* **2013**, *359*, 45–53. [CrossRef]
39. Chauhan, D.S.; Quraishi, M.; Qurashi, A. Recent trends in environmentally sustainable Sweet corrosion inhibitors. *J. Mol. Liq.* **2021**, *326*, 115117. [CrossRef]

40. Usman, B.J.; Ali, S.A. Carbon dioxide corrosion inhibitors: A review. *Arab. J. Sci. Eng.* **2018**, *43*, 1–22. [CrossRef]
41. Perez, T.E. Corrosion in the oil and gas industry: An increasing challenge for materials. *Jom* **2013**, *65*, 1033–1042. [CrossRef]
42. Nejad, A.M. A review of contributing parameters in corrosion of oil and gas wells. *Anti Corros. Methods Mater.* **2018**, *65*, 73–78. [CrossRef]
43. Florez, J.J.A.; Ferrari, J.V. Fluid flow effects on $CO_2$ corrosion: A review of applications of rotating cage methodology. *Anti-Corros. Methods Mater.* **2019**, *66*, 507–519. [CrossRef]
44. Yang, H.-M. Role of organic and eco-friendly inhibitors on the corrosion mitigation of steel in acidic environments—A state-of-art review. *Molecules* **2021**, *26*, 3473. [CrossRef]
45. Liu, Q.; Mao, L.; Zhou, S. Effects of chloride content on $CO_2$ corrosion of carbon steel in simulated oil and gas well environments. *Corros. Sci.* **2014**, *84*, 165–171. [CrossRef]
46. Askari, M.; Aliofkhazraei, M.; Jafari, R.; Hamghalam, P.; Hajizadeh, A. Downhole corrosion inhibitors for oil and gas production—A review. *Appl. Surf. Sci. Adv.* **2021**, *6*, 100128. [CrossRef]
47. Chamkalani, A.; Nareh'ei, M.A.; Chamkalani, R.; Zargari, M.H.; Dehestani-Ardakani, M.R.; Farzam, M. Soft computing method for prediction of $CO_2$ corrosion in flow lines based on neural network approach. *Chem. Eng. Commun.* **2013**, *200*, 731–747. [CrossRef]
48. Hamza, A.; Hussein, I.A.; Al-Marri, M.J.; Mahmoud, M.; Shawabkeh, R.; Aparicio, S. $CO_2$ enhanced gas recovery and sequestration in depleted gas reservoirs: A review. *J. Pet. Sci. Eng.* **2021**, *196*, 107685. [CrossRef]
49. Cui, G.; Yang, L.; Fang, J.; Qiu, Z.; Wang, Y.; Ren, S. Geochemical reactions and their influence on petrophysical properties of ultra-low permeability oil reservoirs during water and $CO_2$ flooding. *J. Pet. Sci. Eng.* **2021**, *203*, 108672. [CrossRef]
50. Mat Razali, N.Z.; Mustapha, K.A.; Kashim, M.Z.; Misnan, M.S.; Md Shah, S.S.; Abu Bakar, Z.A. Critical rate analysis for $CO_2$ injection in depleted gas field, Sarawak Basin, offshore East Malaysia. *Carbon Manag.* **2022**, *13*, 294–309. [CrossRef]
51. Mahmoud, M.; Hussein, I.; Carchini, G.; Shawabkeh, R.; Eliebid, M.; Al-Marri, M.J. Effect of rock mineralogy on Hot-$CO_2$ injection for enhanced gas recovery. *J. Nat. Gas Sci. Eng.* **2019**, *72*, 103030. [CrossRef]
52. Lei, H.; Pingping, S.; Ying, J.; Jigen, Y.; Shi, L.; Aifang, B. Prediction of asphaltene precipitation during $CO_2$ injection. *Pet. Explor. Dev.* **2010**, *37*, 349–353. [CrossRef]
53. Zhou, W.; Wang, H.; Yan, Y.; Liu, X. Adsorption mechanism of $CO_2/CH_4$ in kaolinite clay: Insight from molecular simulation. *Energy Fuels* **2019**, *33*, 6542–6551. [CrossRef]
54. Li, X.; Chi, P.; Guo, X.; Sun, Q. $CO_2$-induced asphaltene deposition and wettability alteration on a pore interior surface. *Fuel* **2019**, *254*, 115595. [CrossRef]
55. Fakher, S.; Imqam, A. Asphaltene precipitation and deposition during $CO_2$ injection in nano shale pore structure and its impact on oil recovery. *Fuel* **2019**, *237*, 1029–1039. [CrossRef]
56. Zanganeh, P.; Dashti, H.; Ayatollahi, S. Visual investigation and modeling of asphaltene precipitation and deposition during $CO_2$ miscible injection into oil reservoirs. *Fuel* **2015**, *160*, 132–139. [CrossRef]
57. Baban, A.; Hosseini, M.; Keshavarz, A.; Ali, M.; Hoteit, H.; Amin, R.; Iglauer, S. Robust NMR Examination of the Three-Phase Flow Dynamics of Carbon Geosequestration Combined with Enhanced Oil Recovery in Carbonate Formations. *Energy Fuels* **2024**, *38*, 2167–2176. [CrossRef]
58. Hamza, A.; Hussein, I.A.; Al-Marri, M.J.; Mahmoud, M.; Shawabkeh, R. Impact of clays on $CO_2$ adsorption and enhanced gas recovery in sandstone reservoirs. *Int. J. Greenh. Gas Control* **2021**, *106*, 103286. [CrossRef]
59. Barrufet, M.A.; Bacquet, A.; Falcone, G. Analysis of the storage capacity for $CO_2$ sequestration of a depleted gas condensate reservoir and a saline aquifer. *J. Can. Pet. Technol.* **2010**, *49*, 23–31. [CrossRef]
60. Sokama-Neuyam, Y.A.; Ginting, P.U.R.; Timilsina, B.; Ursin, J.R. The impact of fines mobilization on $CO_2$ injectivity: An experimental study. *Int. J. Greenh. Gas Control* **2017**, *65*, 195–202. [CrossRef]
61. Jeddizahed, J.; Rostami, B. Experimental investigation of injectivity alteration due to salt precipitation during $CO_2$ sequestration in saline aquifers. *Adv. Water Resour.* **2016**, *96*, 23–33. [CrossRef]
62. Li, Z.; Dong, M.; Li, S.; Huang, S. $CO_2$ sequestration in depleted oil and gas reservoirs—Caprock characterization and storage capacity. *Energy Convers. Manag.* **2006**, *47*, 1372–1382. [CrossRef]
63. Tang, Y.; Yang, R.; Kang, X. Modeling the effect of water vaporization and salt precipitation on reservoir properties due to carbon dioxide sequestration in a depleted gas reservoir. *Petroleum* **2018**, *4*, 385–397. [CrossRef]
64. Hosseini, S.A.; Lashgari, H.; Choi, J.W.; Nicot, J.-P.; Lu, J.; Hovorka, S.D. Static and dynamic reservoir modeling for geological $CO_2$ sequestration at Cranfield, Mississippi, USA. *Int. J. Greenh. Gas Control* **2013**, *18*, 449–462. [CrossRef]
65. Goudarzi, A.; Meckel, T.A.; Hosseini, S.A.; Treviño, R.H. Statistical analysis of historic hydrocarbon production data from Gulf of Mexico oil and gas fields and application to dynamic capacity assessment in $CO_2$ storage. *Int. J. Greenh. Gas Control* **2019**, *80*, 96–102. [CrossRef]
66. Ambrose, W.; Lakshminarasimhan, S.; Holtz, M.; Núñez-López, V.; Hovorka, S.D.; Duncan, I. Geologic factors controlling $CO_2$ storage capacity and permanence: Case studies based on experience with heterogeneity in oil and gas reservoirs applied to $CO_2$ storage. *Environ. Geol.* **2008**, *54*, 1619. [CrossRef]
67. Raza, A.; Rezaee, R.; Gholami, R.; Bing, C.H.; Nagarajan, R.; Hamid, M.A. A screening criterion for selection of suitable $CO_2$ storage sites. *J. Nat. Gas Sci. Eng.* **2016**, *28*, 317–327. [CrossRef]

68. Shirbazo, A.; Taghavinejad, A.; Bagheri, S. CO$_2$ Capture and Storage Performance Simulation in Depleted Shale Gas Reservoirs as Sustainable Carbon Resources. *J. Constr. Mater.* **2021**. [CrossRef]
69. Gan, M.; Nguyen, M.C.; Zhang, L.; Wei, N.; Li, J.; Lei, H.; Wang, Y.; Li, X.; Stauffer, P.H. Impact of reservoir parameters and wellbore permeability uncertainties on CO$_2$ and brine leakage potential at the Shenhua CO$_2$ Storage Site, China. *Int. J. Greenh. Gas Control* **2021**, *111*, 103443. [CrossRef]
70. Akono, A.T.; Druhan, J.L.; Dávila, G.; Tsotsis, T.; Jessen, K.; Fuchs, S.; Crandall, D.; Shi, Z.; Dalton, L.; Tkach, M.K. A review of geochemical-mechanical impacts in geological carbon storage reservoirs. *Greenh. Gases Sci. Technol.* **2019**, *9*, 474–504. [CrossRef]
71. Yu, W.; Lashgari, H.R.; Wu, K.; Sepehrnoori, K. CO$_2$ injection for enhanced oil recovery in Bakken tight oil reservoirs. *Fuel* **2015**, *159*, 354–363. [CrossRef]
72. Kelley, M.; Abbaszadeh, M.; Mishra, S.; Mawalkar, S.; Place, M.; Gupta, N.; Pardini, R. Reservoir characterization from pressure monitoring during CO$_2$ injection into a depleted pinnacle reef-MRCSP commercial-scale CCS demonstration project. *Energy Procedia* **2014**, *63*, 4937–4964. [CrossRef]
73. Islam, M.R.; Chakma, A. Storage and utilization of CO$_2$ in petroleum reservoirs—A simulation study. *Energy Convers. Manag.* **1993**, *34*, 1205–1212. [CrossRef]
74. Streit, J.E.; Hillis, R.R. Building Geomechanical Models for the Safe Underground Storage of Carbon Dioxide in Porous Rock. In Proceedings of the Greenhouse Gas Control Technologies—6th International Conference, Kyoto, Japan, 1–4 October 2002; pp. 495–500.
75. Recasens, M.; Garcia, S.; Mackay, E.; Delgado, J.; Maroto-Valer, M.M. Experimental study of wellbore integrity for CO$_2$ geological storage. *Energy Procedia* **2017**, *114*, 5249–5255. [CrossRef]
76. Kaldi, J.; Daniel, R.; Tenthorey, E.; Michael, K.; Schacht, U.; Nicol, A.; Underschultz, J.; Backe, G. Containment of CO$_2$ in CCS: Role of Caprocks and Faults. *Energy Procedia* **2013**, *37*, 5403–5410. [CrossRef]
77. Raza, A.; Gholami, R.; Rezaee, R.; Rasouli, V.; Bhatti, A.A.; Bing, C.H. Suitability of depleted gas reservoirs for geological CO$_2$ storage: A simulation study. *Greenh. Gases Sci. Technol.* **2018**, *8*, 876–897. [CrossRef]
78. Karine, S.; Gérald, H.; Joël, B.; Vincent, T. SCA2003-14: Residual Gas Saturation of Sample Originally at Residual Water Saturation in Heterogeneous Sandstone Reservoirs. Available online: https://www.jgmaas.com/SCA/2003/SCA2003-14.pdf (accessed on 21 September 2003).
79. Raza, A.; Gholami, R.; Rezaee, R.; Bing, C.H.; Nagarajan, R.; Hamid, M.A. CO$_2$ storage in depleted gas reservoirs: A study on the effect of residual gas saturation. *Petroleum* **2018**, *4*, 95–107. [CrossRef]
80. Saeedi, A.; Rezaee, R. Effect of residual natural gas saturation on multiphase flow behaviour during CO$_2$ geo-sequestration in depleted natural gas reservoirs. *J. Pet. Sci. Eng.* **2012**, *82*, 17–26. [CrossRef]
81. Cao, M.; Gu, Y. Oil recovery mechanisms and asphaltene precipitation phenomenon in immiscible and miscible CO$_2$ flooding processes. *Fuel* **2013**, *109*, 157–166. [CrossRef]
82. Tan, Y.; Nookuea, W.; Li, H.; Thorin, E.; Yan, J. Property impacts on Carbon Capture and Storage (CCS) processes: A review. *Energy Convers. Manag.* **2016**, *118*, 204–222. [CrossRef]
83. Nicot, J.-P.; Solano, S.; Lu, J.; Mickler, P.; Romanak, K.; Yang, C.; Zhang, X. Potential Subsurface Impacts of CO$_2$ Stream Impurities on Geologic Carbon Storage. *Energy Procedia* **2013**, *37*, 4552–4559. [CrossRef]
84. Rother, G.; Ilton, E.S.; Wallacher, D.; Hauβ, T.; Schaef, H.T.; Qafoku, O.; Rosso, K.M.; Felmy, A.R.; Krukowski, E.G.; Stack, A.G. CO$_2$ sorption to subsingle hydration layer montmorillonite clay studied by excess sorption and neutron diffraction measurements. *Environ. Sci. Technol.* **2013**, *47*, 205–211. [CrossRef] [PubMed]
85. Oldenburg, C.M.; Benson, S.M. CO$_2$ Injection for Enhanced Gas Production and Carbon Sequestration. In Proceedings of the SPE International Petroleum Conference and Exhibition in Mexico, Villahermosa, Mexico, 10–12 February 2002.
86. Vu, H.P.; Black, J.R.; Haese, R.R. The geochemical effects of O$_2$ and SO$_2$ as CO$_2$ impurities on fluid-rock reactions in a CO$_2$ storage reservoir. *Int. J. Greenh. Gas Control* **2018**, *68*, 86–98. [CrossRef]
87. Wang, J.; Ryan, D.; Anthony, E.J.; Wildgust, N.; Aiken, T. Effects of impurities on CO$_2$ transport, injection and storage. *Energy Procedia* **2011**, *4*, 3071–3078. [CrossRef]
88. Ziabakhsh-Ganji, Z.; Kooi, H. Sensitivity of Joule-Thomson cooling to impure CO$_2$ injection in depleted gas reservoirs. *Appl. Energy* **2014**, *113*, 434–451. [CrossRef]
89. Machado, M.V.B.; Delshad, M.; Sepehrnoori, K. Injectivity assessment for CCS field-scale projects with considerations of salt deposition, mineral dissolution, fines migration, hydrate formation, and non-Darcy flow. *Fuel* **2023**, *353*, 129148. [CrossRef]
90. Holm, L.; Josendal, V. Mechanisms of oil displacement by carbon dioxide. *J. Pet. Technol.* **1974**, *26*, 1427–1438. [CrossRef]
91. Kazemzadeh, Y.; Parsaei, R.; Riazi, M. Experimental study of asphaltene precipitation prediction during gas injection to oil reservoirs by interfacial tension measurement. *Colloids Surf. A Physicochem. Eng. Asp.* **2015**, *466*, 138–146. [CrossRef]
92. Gaus, I.; Audigane, P.; André, L.; Lions, J.; Jacquemet, N.; Durst, P.; Czernichowski-Lauriol, I.; Azaroual, M. Geochemical and solute transport modelling for CO$_2$ storage, what to expect from it? *Int. J. Greenh. Gas Control* **2008**, *2*, 605–625. [CrossRef]
93. Bacci, G.; Korre, A.; Durucan, S. An experimental and numerical investigation into the impact of dissolution/precipitation mechanisms on CO$_2$ injectivity in the wellbore and far field regions. *Int. J. Greenh. Gas Control* **2011**, *5*, 579–588. [CrossRef]
94. Pudlo, D.; Henkel, S.; Reitenbach, V.; Albrecht, D.; Enzmann, F.; Heister, K.; Pronk, G.; Ganzer, L.; Gaupp, R. The chemical dissolution and physical migration of minerals induced during CO$_2$ laboratory experiments: Their relevance for reservoir quality. *Environ. Earth Sci.* **2015**, *73*, 7029–7042. [CrossRef]

95. Belgodere, C.; Sterpenich, J.; Pironon, J.; Randi, A.; Birat, J.-P. Experimental Study of $CO_2$ Injection in the Triassic Sandstones of Lorraine (Eastern France)–Investigation of Injection Well Injectivity Impairment by Mineral Precipitations. In Proceedings of the Le Studium Conference. Geochemical Reactivity in $CO_2$ Geological Storage Sites, Orleans, France, 25–26 February 2014; pp. 25–26.
96. Pudlo, D.; Henkel, S.; Enzmann, F.; Heister, K.; Werner, L.; Ganzer, L.; Reitenbach, V.; Albrecht, D.; Gaupp, R. The Relevance of Mineral Mobilization and Dissolution on the Reservoir Quality of Sandstones in $CO_2$ Storage Sites. *Energy Procedia* **2014**, *59*, 390–396. [CrossRef]
97. Wellman, T.P.; Grigg, R.B.; McPherson, B.J.; Svec, R.K.; Lichtner, P.C. Evaluation of $CO_2$-Brine-Reservoir Rock Interaction with Laboratory Flow Tests and Reactive Transport Modeling. In Proceedings of the International Symposium on Oilfield Chemistry, Houston, TX, USA, 5–7 February 2003.
98. Al-Yaseri, A.; Zhang, Y.; Ghasemiziarani, M.; Sarmadivaleh, M.; Lebedev, M.; Roshan, H.; Iglauer, S. Permeability Evolution in Sandstone Due to $CO_2$ Injection. *Energy Fuels* **2017**, *31*, 12390–12398. [CrossRef]
99. Sokama-Neuyam, Y.A.; Ursin, J.R. $CO_2$ Well Injectivity: Effect of Viscous Forces on Precipitated Minerals. In Proceedings of the International Petroleum Technology Conference, Doha, Qatar, 6–9 December 2015.
100. Miri, R.; Hellevang, H. Salt precipitation during $CO_2$ storage—A review. *Int. J. Greenh. Gas Control* **2016**, *51*, 136–147. [CrossRef]
101. Giorgis, T.; Carpita, M.; Battistelli, A. 2D modeling of salt precipitation during the injection of dry $CO_2$ in a depleted gas reservoir. *Energy Convers. Manag.* **2007**, *48*, 1816–1826. [CrossRef]
102. Yusof, M.A.M.; Mohamed, M.A.; Akhir, N.A.M.; Ibrahim, M.A.; Mardhatillah, M.K. Combined Impact of Salt Precipitation and Fines Migration on $CO_2$ Injectivity Impairment. *Int. J. Greenh. Gas Control* **2021**, *110*, 103422. [CrossRef]
103. Bacci, G.; Korre, A.; Durucan, S. Experimental investigation into salt precipitation during $CO_2$ injection in saline aquifers. *Energy Procedia* **2011**, *4*, 4450–4456. [CrossRef]
104. Ashoori, S.; Balavi, A. An investigation of asphaltene precipitation during natural production and the $CO_2$ injection process. *Pet. Sci. Technol.* **2014**, *32*, 1283–1290. [CrossRef]
105. Hemmati-Sarapardeh, A.; Ahmadi, M.; Ameli, F.; Dabir, B.; Mohammadi, A.H.; Husein, M.M. Modeling asphaltene precipitation during natural depletion of reservoirs and evaluating screening criteria for stability of crude oils. *J. Pet. Sci. Eng.* **2019**, *181*, 106127. [CrossRef]
106. Cruz, A.A.; Amaral, M.; Santos, D.; Palma, A.; Franceschi, E.; Borges, G.R.; Coutinho, J.A.P.; Palácio, J.; Dariva, C. $CO_2$ influence on asphaltene precipitation. *J. Supercrit. Fluids* **2019**, *143*, 24–31. [CrossRef]
107. Kalantari-Dahaghi, A.; Gholami, V.; Moghadasi, J.; Abdi, R. Formation damage through asphaltene precipitation resulting from $CO_2$ gas injection in Iranian carbonate reservoirs. *SPE Prod. Oper.* **2008**, *23*, 210–214. [CrossRef]
108. Shedid, S.A. Influences of asphaltene precipitation on capillary pressure and pore size distribution of carbonate reservoirs. *Pet. Sci. Technol.* **2001**, *19*, 503–519. [CrossRef]
109. Nasrabadi, H.; Moortgat, J.; Firoozabadi, A. New three-phase multicomponent compositional model for asphaltene precipitation during $CO_2$ injection using CPA-EOS. *Energy Fuels* **2016**, *30*, 3306–3319. [CrossRef]
110. Mohammed, S.; Gadikota, G. The influence of $CO_2$ on the structure of confined asphaltenes in calcite nanopores. *Fuel* **2019**, *236*, 769–777. [CrossRef]
111. Papadimitriou, N.; Romanos, G.; Charalambopoulou, G.C.; Kainourgiakis, M.; Katsaros, F.; Stubos, A. Experimental investigation of asphaltene deposition mechanism during oil flow in core samples. *J. Pet. Sci. Eng.* **2007**, *57*, 281–293. [CrossRef]
112. Wang, C.; Li, T.; Gao, H.; Zhao, J.; Gao, Y. Quantitative study on the blockage degree of pores due to asphaltene precipitation in low-permeability reservoirs with NMR technique. *J. Pet. Sci. Eng.* **2018**, *163*, 703–711. [CrossRef]
113. Shedid, S.A.; Zekri, A.Y. Formation damage caused by simultaneous sulfur and asphaltene deposition. *SPE Prod. Oper.* **2006**, *21*, 58–64. [CrossRef]
114. Saeedi Dehaghani, A.; Shadman, M.; Ahmadi, S.; Assaf, M. Modeling the onset point of asphaltene precipitation using the solubility parameter in $CO_2$ injection. *Energy Sources Part A Recovery Util. Environ. Eff.* **2017**, *45*, 1–7.
115. Srivastava, R.; Huang, S.; Dong, M. Asphaltene deposition during $CO_2$ flooding. *SPE Prod. Facil.* **1999**, *14*, 235–245. [CrossRef]
116. Nielsen, B.B.; Svrcek, W.Y.; Mehrotra, A.K. Effects of Temperature and Pressure on Asphaltene Particle Size Distributions in Crude Oils Diluted with n-Pentane. *Ind. Eng. Chem. Res.* **1994**, *33*, 1324–1330. [CrossRef]
117. Jafari Behbahani, T.; Ghotbi, C.; Taghikhani, V.; Shahrabadi, A. Investigation on Asphaltene Deposition Mechanisms during $CO_2$ Flooding Processes in Porous Media: A Novel Experimental Study and a Modified Model Based on Multilayer Theory for Asphaltene Adsorption. *Energy Fuels* **2012**, *26*, 5080–5091. [CrossRef]
118. Rosenbrand, E.; Kjøller, C.; Riis, J.F.; Kets, F.; Fabricius, I.L. Different effects of temperature and salinity on permeability reduction by fines migration in Berea sandstone. *Geothermics* **2015**, *53*, 225–235. [CrossRef]
119. Cihan, A.; Petrusak, R.; Bhuvankar, P.; Alumbaugh, D.; Trautz, R.; Birkholzer, J.T. Permeability Decline by Clay Fines Migration around a Low-Salinity Fluid Injection Well. *Groundwater* **2022**, *60*, 87–98. [CrossRef] [PubMed]
120. Sbai, M.A.; Azaroual, M. Numerical modeling of formation damage by two-phase particulate transport processes during $CO_2$ injection in deep heterogeneous porous media. *Adv. Water Resour.* **2011**, *34*, 62–82. [CrossRef]
121. Xie, Q.; Saeedi, A.; Delle Piane, C.; Esteban, L.; Brady, P.V. Fines migration during $CO_2$ injection: Experimental results interpreted using surface forces. *Int. J. Greenh. Gas Control* **2017**, *65*, 32–39. [CrossRef]

122. Takahashi, S.; Kovscek, A.R. Wettability estimation of low-permeability, siliceous shale using surface forces. *J. Pet. Sci. Eng.* **2010**, *75*, 33–43. [CrossRef]
123. Wang, Y.; Bedrikovetsky, P.; Yin, H.; Othman, F.; Zeinijahromi, A.; Le-Hussain, F. Analytical model for fines migration due to mineral dissolution during $CO_2$ injection. *J. Nat. Gas Sci. Eng.* **2022**, *100*, 104472. [CrossRef]
124. Zhang, M.; de Jong, S.; Spiers, C.J.; Busch, A.; Wentinck, H.M. Swelling stress development in confined smectite clays through exposure to $CO_2$. *Int. J. Greenh. Gas Control* **2018**, *74*, 49–61. [CrossRef]
125. Giesting, P.; Guggenheim, S.; Van Groos, A.F.K.; Busch, A. Interaction of carbon dioxide with Na-exchanged montmorillonite at pressures to 640 bars: Implications for $CO_2$ sequestration. *Int. J. Greenh. Gas Control* **2012**, *8*, 73–81. [CrossRef]
126. Narayanan Nair, A.K.; Cui, R.; Sun, S. Overview of the Adsorption and Transport Properties of Water, Ions, Carbon Dioxide, and Methane in Swelling Clays. *ACS Earth Space Chem.* **2021**, *5*, 2599–2611. [CrossRef]
127. Pham, H.; Nguyen, Q.P. Effect of silica nanoparticles on clay swelling and aqueous stability of nanoparticle dispersions. *J. Nanoparticle Res.* **2014**, *16*, 1–11. [CrossRef]
128. Bibi, I.; Icenhower, J.; Niazi, N.K.; Naz, T.; Shahid, M.; Bashir, S. Clay minerals: Structure, chemistry, and significance in contaminated environments and geological $CO_2$ sequestration. *Environ. Mater. Waste* **2016**, 543–567. [CrossRef]
129. Anderson, R.; Ratcliffe, I.; Greenwell, H.; Williams, P.; Cliffe, S.; Coveney, P. Clay swelling—A challenge in the oilfield. *Earth-Sci. Rev.* **2010**, *98*, 201–216. [CrossRef]
130. Loganathan, N.; Bowers, G.M.; Yazaydin, A.O.; Schaef, H.T.; Loring, J.S.; Kalinichev, A.G.; Kirkpatrick, R.J. Clay swelling in dry supercritical carbon dioxide: Effects of interlayer cations on the structure, dynamics, and energetics of $CO_2$ intercalation probed by XRD, NMR, and GCMD simulations. *J. Phys. Chem. C* **2018**, *122*, 4391–4402. [CrossRef]
131. Fatah, A.; Mahmud, H.B.; Bennour, Z.; Hossain, M.; Gholami, R. Effect of supercritical $CO_2$ treatment on physical properties and functional groups of shales. *Fuel* **2021**, *303*, 121310. [CrossRef]
132. Favero, V.; Laloui, L. Impact of $CO_2$ injection on the hydro-mechanical behaviour of a clay-rich caprock. *Int. J. Greenh. Gas Control* **2018**, *71*, 133–141. [CrossRef]
133. Pang, J.; Liang, Y.; Masuda, Y.; Matsuoka, T.; Zhang, Y.; Xue, Z. Swelling phenomena of the nonswelling Clay induced by $CO_2$ and Water cooperative adsorption in janus-surface micropores. *Environ. Sci. Technol.* **2020**, *54*, 5767–5773. [CrossRef] [PubMed]
134. Mukherjee, M.; Misra, S. A review of experimental research on Enhanced Coal Bed Methane (ECBM) recovery via $CO_2$ sequestration. *Earth Sci. Rev.* **2018**, *179*, 392–410. [CrossRef]
135. Miller, J.; Sullivan, C.; Larsen, G.; Kelley, M.; Rike, W.; Gerst, J.; Gupta, N.; Paul, D.; Pardini, R.; Modroo, A. Alternative conceptual geologic models for $CO_2$ injection in a Niagaran pinnacle reef oil field, Northern Michigan, USA. *Energy Procedia* **2014**, *63*, 3685–3701. [CrossRef]
136. Ganesh, P.R.; Mishra, S.; Mawalkar, S.; Gupta, N.; Pardini, R. Assessment of $CO_2$ injectivity and storage capacity in a depleted pinnacle reef oil field in northern Michigan. *Energy Procedia* **2014**, *63*, 2969–2976. [CrossRef]
137. Barnes, D.; Harrison, B.; Grammer, G.M.; Asmus, J. $CO_2$/EOR and geological carbon storage resource potential in the Niagara Pinnacle Reef Trend, lower Michigan, USA. *Energy Procedia* **2013**, *37*, 6786–6799. [CrossRef]
138. Mishra, S.; Haagsma, A.; Valluri, M.; Gupta, N. Assessment of $CO_2$-enhanced oil recovery and associated geologic storage potential in the Michigan Northern Pinnacle Reef Trend. *Greenh. Gases Sci. Technol.* **2020**, *10*, 32–49. [CrossRef]
139. Orlic, B. Geomechanical effects of $CO_2$ storage in depleted gas reservoirs in the Netherlands: Inferences from feasibility studies and comparison with aquifer storage. *J. Rock Mech. Geotech. Eng.* **2016**, *8*, 846–859. [CrossRef]
140. Gupta, N.; Kelley, M.; Place, M.; Conner, A.; Mawalkar, S.; Mishra, S.; Sminchak, J. *Integrated Monitoring Volume: A Summary of Monitoring Studies Conducted in Niagaran Carbonate Pinnacle Reefs During Enhanced Oil Recovery with $CO_2$*; Battelle Memorial Inst.: Columbus, OH, USA, 2020.
141. Sazali, Y.; Sazali, W.; Ibrahim, J.; Dindi, M.; Graham, G.; Gödeke, S. Investigation of high temperature, high pressure, scaling and dissolution effects for carbon capture and storage at a high $CO_2$ content carbonate gas field offshore Malaysia. *J. Pet. Sci. Eng.* **2019**, *174*, 599–606. [CrossRef]
142. Brownsort, P.; Scott, V.; Sim, G.; Haszeldine, S. Carbon Dioxide Transport Plans for Carbon Capture and Storage in the North Sea Region. Available online: https://era.ed.ac.uk/bitstream/handle/1842/16481/wp-2015-02.pdf?sequence=1&isAllowed=y (accessed on 20 July 2015).
143. Dale, A.W.; Sommer, S.; Lichtschlag, A.; Koopmans, D.; Haeckel, M.; Kossel, E.; Deusner, C.; Linke, P.; Scholten, J.; Wallmann, K. Defining a biogeochemical baseline for sediments at Carbon Capture and Storage (CCS) sites: An example from the North Sea (Goldeneye). *Int. J. Greenh. Gas Control* **2021**, *106*, 103265. [CrossRef]
144. Esposito, M.; Martinez-Cabanas, M.; Connelly, D.P.; Jasinski, D.; Linke, P.; Schmidt, M.; Achterberg, E.P. Water column baseline assessment for offshore Carbon Dioxide Capture and Storage (CCS) sites: Analysis of field data from the Goldeneye storage complex area. *Int. J. Greenh. Gas Control* **2021**, *109*, 103344. [CrossRef]
145. Tucker, O.; Holley, M.; Metcalfe, R.; Hurst, S. Containment risk management for $CO_2$ storage in a depleted gas field, UK North Sea. *Energy Procedia* **2013**, *37*, 4804–4817. [CrossRef]
146. Hannis, S.; Lu, J.; Chadwick, A.; Hovorka, S.; Kirk, K.; Romanak, K.; Pearce, J. $CO_2$ storage in depleted or depleting oil and gas fields: What can we learn from existing projects? *Energy Procedia* **2017**, *114*, 5680–5690. [CrossRef]
147. Núñez-López, V.; Gil-Egui, R.; Hosseini, S.A. Environmental and operational performance of $CO_2$-EOR as a CCUS technology: A Cranfield example with dynamic LCA considerations. *Energies* **2019**, *12*, 448. [CrossRef]

148. Choi, J.-W.; Nicot, J.-P.; Hosseini, S.A.; Clift, S.J.; Hovorka, S.D. $CO_2$ recycling accounting and EOR operation scheduling to assist in storage capacity assessment at a US gulf coast depleted reservoir. *Int. J. Greenh. Gas Control* **2013**, *18*, 474–484. [CrossRef]
149. Kim, S.; Hosseini, S.A. Above-zone pressure monitoring and geomechanical analyses for a field-scale $CO_2$ injection project in Cranfield, MS. *Greenh. Gases Sci. Technol.* **2014**, *4*, 81–98. [CrossRef]

**Disclaimer/Publisher's Note:** The statements, opinions and data contained in all publications are solely those of the individual author(s) and contributor(s) and not of MDPI and/or the editor(s). MDPI and/or the editor(s) disclaim responsibility for any injury to people or property resulting from any ideas, methods, instructions or products referred to in the content.

MDPI  
St. Alban-Anlage 66  
4052 Basel  
Switzerland  
www.mdpi.com

*Energies* Editorial Office  
E-mail: energies@mdpi.com  
www.mdpi.com/journal/energies

Disclaimer/Publisher's Note: The statements, opinions and data contained in all publications are solely those of the individual author(s) and contributor(s) and not of MDPI and/or the editor(s). MDPI and/or the editor(s) disclaim responsibility for any injury to people or property resulting from any ideas, methods, instructions or products referred to in the content.

www.ingramcontent.com/pod-product-compliance
Lightning Source LLC
LaVergne TN
LVHW070358100526
838202LV00014B/1344